MW00592920

Proceedings of the
XXI International Conference on
Atomic Physics

PUSHING THE FRONTIERS OF ATOMIC PHYSICS

Proceedings of the
XXI International Conference on
Atomic Physics

PUSHING THE FRONTIERS OF ATOMIC PHYSICS

Storrs, Connecticut, USA   27 July – 1 August 2008

editors

Robin Côté, Phillip L Gould,
Michael Rozman & Winthrop W Smith

University of Connecticut, USA

World Scientific

NEW JERSEY · LONDON · SINGAPORE · BEIJING · SHANGHAI · HONG KONG · TAIPEI · CHENNAI

Published by

World Scientific Publishing Co. Pte. Ltd.
5 Toh Tuck Link, Singapore 596224
USA office: 27 Warren Street, Suite 401-402, Hackensack, NJ 07601
UK office: 57 Shelton Street, Covent Garden, London WC2H 9HE

British Library Cataloguing-in-Publication Data
A catalogue record for this book is available from the British Library.

PUSHING THE FRONTIERS OF ATOMIC PHYSICS
Proceedings of the XXI International Conference on Atomic Physics

ISBN-13 978-981-4271-99-8
ISBN-10 981-4271-99-3

Printed in Singapore by World Scientific Printers

PREFACE

The 21st International Conference on Atomic Physics (ICAP 2008) was held July 27 — August 1, 2008 at the University of Connecticut, Storrs, CT, USA. Approximately 600 participants from 36 countries attended. This conference was part of an ongoing series of conferences devoted to fundamental studies of atoms, broadly defined. A Web site with the Conference Program, Abstracts, Proceedings and other archival information can be found at: http://www.phys.uconn.edu/icap2008/.

The ICAP papers encompass forefront research on basic atomic, molecular and optical (AMO) physics, emphasizing atoms and their interactions with each other and with external fields, including various kinds of laser fields. These meetings grew out of the molecular beams conferences of the Rabi group. The first ICAP was held at NYU in 1968. Later conferences have been held in all even-numbered years, alternating between North America and other locations, including Europe and recently Brazil, with the next conference planned for Cairns, Australia in 2010. The Web site for the Cairns meeting is: http://www.swin.edu.au/icap2010/. The growth of ICAP in recent years reflects the health and vitality of the AMO field and the continuing emergence of exciting and often surprising new developments and connections with other areas of physics.

Historically, topics have included quantum electrodynamics, tests of fundamental symmetries, precision measurements (including atomic clocks and fundamental constants), laser spectroscopy, ultracold atoms and molecules, Bose-Einstein condensates, degenerate Fermi gases, optical lattices, quantum computing/quantum information with atoms and ions, coherent control, and ultrafast and intense field interactions. As per tradition, all 50 invited talks were plenary, with approximately 400 contributed papers presented at one of three afternoon poster sessions. The program included lectures by Nobel Laureates Phillips, Cornell, Glauber, Chu and Ketterle, as well as two in memoriam talks, commemorating the scientific lives and broad impact on atomic physics of Willis Lamb and Herbert Walther. Two "hot topics" sessions of five invited papers each were held, emphasizing the most recent research and including the 2008 winner of the IUPAP Young Scientist Prize in this area of physics. The conference was preceded by a one-week Summer School for new AMO researchers, organized by the Harvard-MIT Center for Ultracold Atoms in Cambridge, MA. The ICAP Co-Chairs were Robin Côté, Phillip Gould and Winthrop Smith of the Physics Department at the University of Connecticut. The University and particularly its Conference Services group provided excellent institutional and logistical support and facilities.

The conference was sponsored by IUPAP, NSF, NIST, ARO and the Department of Physics, the College of Liberal Arts and Sciences, and the Research Foundation of the University of Connecticut, as well as by several industrial companies. The organizers acknowledge with thanks the generous support of all these organizations and companies.

The Local Organizing Committee worked tirelessly to make the conference a success and the Program Committee did a great job in selecting the invited speakers. We also thank the members of the International Advisory Committee for their advice on conference organization, publicity and promotion of ICAP 2008.

Robin Côté
Phillip L. Gould
Michael G. Rozman
Winthrop W. Smith

ICAP 2008 COMMITTEES

Co-Chairs

Winthrop Smith, Phillip Gould, Robin Côté

Program Committee

Alain Aspect	France
Joachim Burgdoerfer	Austria
Paul Corkum	Canada
Nir Davidson	Israel
David DeMille	USA
Thomas Gallagher	USA
Peter Hannaford	Australia
Wonho Jhe	Korea
Paul Julienne	USA
Hidetoshi Katori	Japan
Kate Kirby	USA
Maciej Lewenstein	Spain
Margaret Murnane	USA
Eugene Polzik	Denmark
Gerhard Rempe	Germany
Christophe Salomon	France
Sandro Stringari	Italy
John Thomas	USA
Vladan Vuletic	USA
Ian Walmsley	England
Susanne Yelin	USA
Peter Zoller	Austria

Local Organizing Committee

David DeMille	Yale University
Edward Eyler	University of Connecticut
George Gibson	University of Connecticut
Lutz Hüwel	Wesleyan University
Juha Javanainen	University of Connecticut
Vasili Kharchenko	University of Connecticut
Harvey Michels	University of Connecticut
David Perry	University of Connecticut
Michael Rozman	University of Connecticut
William Stwalley	University of Connecticut
Susanne Yelin	University of Connecticut

International Advisory Committee

Ennio Arimondo	Italy
Vanderlei Bagnato	Brazil
Victor Balykin	Russia
Rainer Blatt	Austria
Claude Cohen-Tannoudji	France
Gordon Drake	Canada
Norval Fortson	USA
Theodor Hänsch	Germany
Serge Haroche	France
Ed Hinds	UK
Massimo Inguscio	Italy
Wolfgang Ketterle	USA
Daniel Kleppner	USA
Indrek Martinson	Sweden
William Phillips	USA
Lev Pitaevskii	Russia
David Pritchard	USA
Fujio Shimizu	Japan
Jook Walraven	The Netherlands
David Wineland	USA
Tsutomu Yabuzaki	Japan

INVITED SPEAKERS

Paul Berman	Michigan
François Biraben	ENS, Paris
Rainer Blatt	Innsbruck
Immanuel Bloch	Mainz
Philippe Bouyer	Institut d'Optique, Palaiseau
Antoine Browaeys	Orsay
Cheng Chin	Chicago
Steven Chu	Lawrence Berkeley
Isaac Chuang	MIT
Eric Cornell	JILA
Louis DiMauro	Ohio State
Michael Drewsen	Aarhus
Peter Drummond	Queensland
Nirit Dudovich	Weizmann Institute
Stephan Dürr	Munich
Gerald Gabrielse	Harvard
Kurt Gibble	Penn State
Roy Glauber	Harvard
Rudolf Grimm	Innsbruck
Serge Haroche	ENS, Paris
Jack Harris	Yale
Randall Hulet	Rice
Massimo Inguscio	LENS, Florence
Deborah Jin	JILA
Mark Kasevich	Stanford
Tobias Kippenberg	Munich
Alex Kuzmich	Georgia Tech
Wolfgang Ketterle	MIT

Misha Lukin	Harvard
Pierre Meystre	Arizona
Giovanni Modugno	LENS, Florence
Markus Oberthaler	Heidelberg
Belen Paredes	Mainz
Tilman Pfau	Stuttgart
William Phillips	NIST, Gaithersburg
Pierre Pillet	Orsay
Trey Porto	NIST, Gaithersburg
Till Rosenband	NIST, Boulder
Dan Stamper-Kurn	Berkeley
Yong-Il Shin	MIT
Robert Schoelkopf	Yale
William Stwalley	Connecticut
John Thomas	Duke
Thomas Udem	Munich
Vladan Vuletic	MIT
David Weiss	Penn State
Yoshihisa Yamamoto	Stanford
Jun Ye	JILA
Linda Young	Argonne
Peter Zoller	Innsbruck

CONTENTS

In Memoriam

Nobel Laureate Session

Precision Measurements

Quantum Information and Quantum Optics

Quantum Degenerate Systems

Optical Lattices and Cold Molecules

Ultrafast Phenomena

HERBERT WALTHER, SCIENTIST EXTRAORDINAIRE
(January 19, 1935 — July 22, 2006)

PIERRE MEYSTRE

College of Optical Sciences, Department of Physics and B2 Institute
The University of Arizona, Tucson, AZ 85721, USA

People who have a transforming influence on science come in many variations: there are of course the extraordinary researchers whose discoveries help open and define a new field of investigations; the great teachers who can inspire generations of students; and the visionary administrators who provide the financial and infrastructure support needed to carry out our work. Many physicists excel at one of these tasks, significantly less at two of them, and only very few at all three. Herbert Walther was one of these rare few.

As a researcher, he produced advances and insights of the highest value. Amongst his many achievements, he helped define the field of cavity quantum electrodynamics, a field that is in turn pivotal to the emerging science of quantum information; as a teacher, he not only distinguished himself in the classroom, but trained and mentored an extraordinary palette of graduate students who went on to pursue highly distinguished academic careers — Nobel Prize winner Wolfgang Ketterle and Max-Planck directors Gerd Leuchs and Gerhard Rempe immediately come to mind — or alternatively became influential industry leaders, and here I am thinking for instance of Rainer Schlicher and Andreas Dorsel. Last, but not least, starting from an empty leased building and a couple of "theory containers", he built a world-leading quantum optics research institute, producing Nobel-quality research and a mecca for both very senior scientists and budding young researchers from all over the world. By attracting to Garching the likes of Marlan Scully and Ted Haensch and hosting a palette of scientific world stars, the absolute who-is-who of our field, he made the Max-Planck

Institute for Quantum Optics an undisputed center of gravity for AMO science, a must-stop destination.

Herbert Walther's scientific career took him from his PhD studies at the University of Heidelberg to Hannover, where he obtained his Habilitation in 1968. After extended foreign stays at JILA in Boulder and at the Laboratoire Aimé Cotton in Orsay, he became a Professor at the University of Bonn, followed soon thereafter by a move to Cologne, where he stayed until 1975. He then became a Professor at the University of Munich, becoming at the same time a founding Director of the Projektgruppe fuer Laserforschung with Siegbert Witkowski and Karl Kompa. This research group then morphed into the permanent Max-Planck Institute for Quantum Optics in 1981.

The record indicates that during his Munich years, Herbert Walther graduated 94 PhD students, and had 10 of his collaborators receive their Habilitation. He is the author or coauthor of over 600 publications, covering topics from applications of narrow linewidth dye lasers to spectroscopy, a field where he was a true pioneer, to multiphoton processes; from the study of Rydberg atoms to cavity QED and micromasers; from trapped ion research to molecular spectroscopy; and from surface physics to more

applied topics such as the development of techniques to monitor the ozone layer in the atmosphere.

Herbert Walther's impact on science policy is just as remarkable: he served on a number of national and international committees, where he profoundly influenced the development and support of science in Germany, Europe and worldwide. These activities notably included a stint on the Science Council of the German Federal Republic as well as membership on the Executive Council of the European Science Foundation. Most importantly perhaps, in his role as Vice-President of the Max-Planck Society he was charged with the difficult and enormously challenging task of initiating the reorganization of science in the former East Germany following the reunification of Germany. This was without a doubt a task that would have fully consumed lesser mortals, but that he accomplished tirelessly with grace, fairness and good taste, while at the same time continuing all of his other activities. I remember once asking him how he could possibly carry out essentially three full-time jobs, while still finding time to be an extraordinary host and mentor, always giving the impression of having all the time in the world for his visitors. His answer was that "yes, one can turn any job into a full-time job, but there is no reason why this should be the case". He also explained to me one of his important rules, which was to touch any piece of paper that landed on his desk only once. Still, I remain immensely impressed by the fact that whenever I would talk to him, he gave me the impression that I was the most important person in the world and that he had unlimited time to talk and listen to me. To this day, I still don't know how he did it.

Herbert Walther had such a profound impact on so many aspects of science that it is not surprising that a long list of prestigious rewards, awards, and honors were bestowed upon him. They include the Max Born Prize, the Charles Townes Award, the King Faisal Prize in Physics, the Michelson Medal, the Humboldt Medal, the Stern-Gerlach Medal, the Verdienstkreuz 1. Klasse des Verdienstordens der Bundesrepublik Deutschland, the Willis E. Lamb Medal for Laser Physics, the Quantum Electronics Prize of the European Physical Society, the Alfred Krupp Prize for Science, the Order of Merit of the State of Bavaria, and the Frederic Ives Medal/Jarus W. Quinn Endowment. He was also a Member or Honorary Member of Academia Sinica, the Bavarian Academy of Sciences, the Akademie der Naturforscher Leopoldina, the Roland Eötvös Physical Society of Hungary, the American Academy of Arts and Sciences, the Heidelberg Academy of Sciences, the Romanian Academy, the Nordrhein-Westflische Akademie der Wissenschaften, Academia Europaea, the Russian Academy of Sciences, the Hungarian Academy of Sciences, the Convent for Technical Sciences of the German Academies, the German Physical Society, and the Belarusian Physical Society.

But as impressive as they may be, these lists don't even come close to expressing the impact that Herbert Walther has had on AMO science and most importantly, on generations of students and colleagues worldwide. When confronted with a difficult situation or a thorny problem, I still find myself asking "how would Herbert deal with that?" And this, for me, is the biggest compliment I can make to this unforgettable mentor. I still miss him terribly, and I know that there are very many of us who feel that way.

Last but not least: As the saying goes, behind every great man there is a great woman. In Herbert Walther's case, everybody who knew him also knows how very true that was. It is very difficult indeed to imagine that he could have achieved even a fraction of what he did without the support and love of his wife Margot. Not to mention her wonderful hospitality and her great dinners, which many of us have enjoyed so much!

Acknowledgements

I am thankful to Dr. T. W. Haensch for providing me with a number of pictures of Herbert Walther, including those shown here.

WILLIS E. LAMB
July 12, 1913 — May 15, 2008

PAUL BERMAN

Physics Department, University of Michigan, 450 Church Street
Ann Arbor, Michigan 48109-1040, USA

The atomic and optical physics community lost one of its pioneers with the death of Willis E. Lamb, Jr. on May 15, 2008. Lamb was born on July 12, 1913, received the BS degree in Chemistry at Berkeley in 1934, and obtained his PhD under the tutelage of J. Robert Oppenheimer at Berkeley in 1938. He served on the faculties of Columbia University, Stanford University, Oxford University, Yale University, and the University of Arizona. Lamb received the Nobel prize in 1955 for his work on the fine structure of hydrogen and was awarded the President's National Medal for Science in 2000.

The paper by Lamb and R. Retherford entitled *Fine Structure of the Hydrogen Atom by a Microwave Method* that appeared in *The Physical Review* in 1947 ushered in the field of quantum electrodynamics. Lamb and Retherford used microwave spectroscopy to measure a splitting "of about 1000 MHz" between the $2S_{1/2}$ and $2P_{1/2}$ levels of hydrogen, levels that were predicted to be degenerate on the basis of the Dirac theory. This paper was followed by a series of six papers in which both the theory and accuracy of the measurements were refined, and by a paper with N. Kroll that contained the first relativistic calculation of the level splitting. This work provided the first step into what was to become an incredible journey involving high precision tests of quantum electrodynamics.

Somewhat ironically, the "Lamb shift" paper is not Lamb's most cited work; his most cited work by far is his paper on the *Theory of an Optical Maser*, published in 1964. This paper contains a detailed theory of gas laser operation, predicting features such as a dip in the output power at

line center, frequency pulling, and phase locking of different modes. This work was followed by a quantum theory of laser operation (with M. Scully) and papers on the Zeeman laser, ring laser, and laser pulse propagation.

Early in his career (1939), Lamb calculated the neutron capture cross section by nucleons bound in a lattice and showed that the recoil of the nucleons could be suppressed, a precursor of the Mossbauer effect. Based on this work and a paper by Dicke in 1953, the interaction of light with atoms confined to distances less than a wavelength is referred to as the Lamb-Dicke limit.

In his later years, Lamb concentrated his efforts on fundamental problems in non-relativistic quantum mechanics, a field which was his true passion. He continued an "anti-photon" campaign, stating that there were just a precious few who qualify for a "license" to use the word "photon" in its proper sense. He also worked on the theory of measurement and the "classical" underpinnings of the Schrödinger equation.

Anyone who has read Lamb's papers is impressed by his clarity of presentation. He always stressed physical understanding over mathematical formalism. He displayed a profound mastery of scientific literature and had a deep respect for the historical development of physical theories. Lamb tried to instill in his students the ethic that integrity was an essential component of any publication. His advice in writing papers was, "if it's worth putting in, it's worth explaining." Willis Lamb leaves with us a legacy that will be appreciated for decades to come.

WHEN IS A QUANTUM GAS A QUANTUM LIQUID?

J. M. PINO, R. J. WILD, S. B. PAPP, S. RONEN, D. S. JIN, and E. A. CORNELL*

NIST/JILA, University of Colorado,
Boulder, CO, 80309, USA
* E-mail: cornell@jila.colorado.edu

We report on measurements of the excitation spectrum of a strongly interacting Bose-Einstein condensate (BEC). A magnetic-field Feshbach resonance is used to tune atom-atom interactions in the condensate and to reach a regime where quantum depletion and beyond mean-field corrections to the condensate chemical potential are significant. We use two-photon Bragg spectroscopy to probe the condensate excitation spectrum; our results demonstrate the onset of beyond mean-field effects in a gaseous BEC.

Keywords: Bragg Spectroscopy; Beyond mean-field; Feshbach resonance.

The concept of an interacting but dilute Bose gas was originally developed fifty years ago as a theoretically tractable surrogate for superfluid liquid helium. For some years after the eventual experimental realization of dilute-gas Bose-Einstein condensates (BEC), experiments were performed mainly in the extreme dilute limit, in which atom-atom correlations were of negligible significance. Such correlations again assumed a central role, however, in the 2002 experiments on a Mott state for bosons in an optical lattice[1] and on atom-molecule coherence near a Feshbach resonance.[2] Atom-atom correlations are also central to the current hot-topic field of resonant fermionic condensates.[3,4]

In this paper we describe an experimental study of elementary excitations in a system which harkens back to a previous century, in that, like liquid helium, it is a strongly interacting, bulk, bosonic superfluid. Unlike liquid helium, our gas of Bose-condensed ^{85}Rb has the modern virtue of Feshbach-tunable interactions well-described by a scattering length a that is much larger than the reach of the actual interatomic potential. Our tool for characterizing the sample is Bragg spectroscopy.[5,6]

Our experiments are performed using a ^{85}Rb BEC near a Feshbach resonance at 155 G.[7,8] A gas of ^{85}Rb atoms in the $|F = 2, m_F = -2\rangle$ state is first sympathetically cooled with ^{87}Rb in a magnetic trap and then evaporated directly to ultralow temperature in an optical dipole trap.[9] We create a single-species ^{85}Rb condensate,[10] with 40,000 atoms and a condensate fraction greater than 85%, in a weakly confining optical dipole trap at a magnetic field above the Feshbach resonance where the scattering length is 150 a_0. Curvature of the magnetic field enhances confinement along the axial direction of the optical trap. Following evaporative cooling, the optical dipole trap is recompressed and the final trap has a measured radial (axial) trap frequency of $2\pi \times 134$ Hz ($2\pi \times 2.9$ Hz), yielding a condensate mean density of 2.1×10^{13} cm^{-3}.

Bragg spectroscopy via stimulated two-photon transitions provides a direct probe of the condensate excitation spectrum. Two counter-propagating, near-resonant laser beams are aligned along the long axis of the condensate. The momentum imparted to a condensate excitation is given by $\hbar k = 2\hbar k_L$ where $k_L = \frac{2\pi}{780\,nm}$ is the wave vector of a beam. The excitation energy is scanned by adjusting the frequency difference of the two laser beams. The average of the two frequencies is red detuned from atomic resonance by 4.2 GHz. The intensity and pulse duration of the Bragg beams are chosen so that the fraction of the condensate excited is less than 10%.

Just before performing the Bragg spectroscopy, we transiently enhance the condensate density by means of large amplitude radial and axial breathing modes, which we excite by modulating the magnetic field and thus the Feshbach-modified scattering length. The rates of the ramps are limited so the $\dot{a}/a$ never exceeds $0.06\hbar/(ma^2)$. The scattering length is derived from measurements of the magnetic field and a previous measurement of the ^{85}Rb Feshbach resonance.[11] Synchronized with the inner turning point of the radial oscillation, we ramp the scattering length to the value for a given measurement and then pulse on the Bragg beams. During the pulse, the cloud's inward motion is checked and it begins to breathe outward. We model the resulting time-dependent condensate density using a variational solution to the Gross-Pitaevskii equation,[12] which predicts that the density of the cloud does not change by more than 30% during the Bragg pulse. We can meet this goal only by using progressively shorter Bragg pulses for higher values of desired a. The time- and space-averaged density during the pulse is approximately 7.6×10^{13} cm^{-3}, but this depends weakly on the final value of a.

After the Bragg pulse, we ramp a to 917 a_0 in order to ensure that the momentum of the excitations is spread via collisions[13,14] to the entire condensate sample. We then infer the total momentum, and thus excitation fraction, from the amplitude of the resulting axial slosh, measured via an absorption image taken of the cloud at a time near its axial turning point.

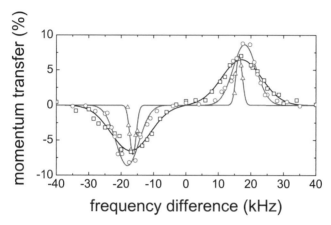

Fig. 1. Typical Bragg spectra at a scattering length of 100 a_0 (blue triangles), 585 a_0 (red circles), and 890 a_0 (black squares). The excitation fraction is determined from the measured momentum transferred to the BEC and plotted as a function of the frequency difference between the two Bragg beams. Lines are fits of the data as described in the text. Mean-field theory predicts a continuous increase in the line shift with increasing a, however by 890 a_0 our data display a *decreasing* shift with stronger interactions.

Figure 1 shows measured Bragg spectra for three values of a. We fit each Bragg spectrum to an antisymmetric function assuming a Gaussian peak and extract a center frequency and an RMS width. The Bragg line shift is the difference between the fitted center and the ideal gas result $\frac{1}{2\pi}\frac{\hbar k^2}{2m} = 15.423$ kHz. In Fig. 2(a) we plot our measured line shifts as a function of the scattering length a. For $a \lesssim 300a_0$ (where the predicted LHY[25] correction is already a 10% effect), the measured line shift ($\bullet$ in Fig. 2(a) agrees with the simple mean-field result. However, as the scattering length is increased further, the resonance line shift deviates significantly from the mean-field prediction. The measured line shift reaches a maximum near $a = 500a_0$ and then *decreases* as the scattering length is increased further.

At large a we find that our measured line shift exhibits a systematic dependence on the temperature of the sample.[16] Non-condensed ^{85}Rb atoms also respond to the Bragg pulse, and this causes an observable effect in the

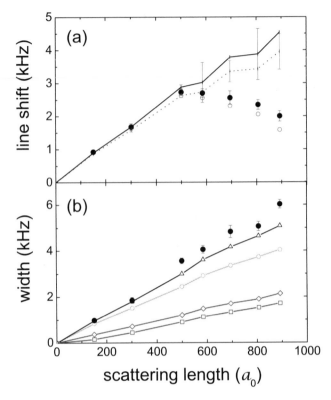

Fig. 2.   (a) Bragg line shift and (b) width as a function of scattering length. In (a) the hollow circles are our observations. The solid circles are data corrected for a fitting systematic associated with the broad thermal atom background, and the error bars represent fit uncertainties. The solid black theory line corresponds to the simple mean-field shift $\propto na$, and the blue dotted line gives the full Bogoliubov[15] theory, which includes phonon interactions. The theory lines are calculated for the trapped gas using a local density approximation for each of the corresponding data points. The mean BEC density ranges from 6.3×10^{13} cm^{-3} to 7.6×10^{13} cm^{-3}. Error bars on the theory lines reflect uncertainty in these densities. Some of the error bars have been omitted for clarity. In (b) the solid black circles are the rms width of a gaussian fit to the Bragg spectra. Black triangles are from a fit to a convolution of various contributions to the width calculated under the conditions of our measurements. The remaining symbols characterize constituent contributions to the convolution including the Lorentzian FWHM width due to collisions (blue squares), and the RMS width of a Gaussian fitted to the contributions due to the inhomogeneous density (red diamonds) and the pulse duration (green circles). The largest contribution to the width comes from the pulse duration; because the jump to large a initiates rapid expansion of the BEC, ever shorter pulses are used to obtain the spectra at larger a.

measured line shift when the spectral width of the condensate response becomes comparable to that of the non-condensed atoms (for $a > 500a_0$). We vary the temperature of the gas to characterize this effect and we apply a small empirical correction to our data to represent the expected line shift at zero temperature (● in Fig. 2(a)).

Figure 2(b) shows the measured width of the Bragg peak as a function of a. Several effects contribute to the total width of the Bragg resonance including the finite duration of the pulse, the inhomogeneous density of the trapped condensate,[5] collisions between the excitations and the condensate,[13] and Doppler broadening.[5] In our case, Doppler broadening is negligible since the axial size of the condensate is relatively large. To understand the total width we convolve the various calculated lineshapes of the remaining three effects. We fit a Gaussian to the convolution (to match the Gaussian fit to our data) and the RMS width from this fit is shown in Fig. 2(b) (black △). Fig. 2(b) also shows the expected contributions to the width from each of the three effects. For the lineshape due to collisions we expect a Lorentzian with a full width at half maximum $\delta\nu = \frac{1}{2\pi} \frac{n\sigma k}{m}$, where σ is the elastic cross section for collisions between the excitations and low momentum atoms. In calculating σ we include the suppression in the phonon regime predicted by Beliaev.[13,17] The measured Bragg width exceeds the predicted width in the strongly interacting regime. However, many of the theoretical difficulties in describing the line shift apply also to predicting the width that arises from inhomogeneous density and excitation lifetime.

A key future goal of our work is to experimentally explore different timescales for the establishment of local many-body quasi-equilibrium, and for longer-term evolution. The extreme aspect ratio of our sample hastens the loss of density that occurs during the expansion caused by the ramp to high a. At present, we are required to use short duration Bragg pulses, which limits our spectral resolution, and we are prevented from tracking the time evolution of line shifts. An ongoing redesign to a more spherical geometry will help. In addition, the Bragg beams are being reconfigured to allow access to the low-k, pure-phonon regime for which $1/(k\xi) \gg 1$.

A more complete account of this work has appeared in S. B. Papp, J. M. Pino, R. J. Wild, S. Ronen, C. E. Wieman, D. S. Jin, and E. A. Cornell, Phys. Rev. Lett. **101**, 135301 (2008). We gratefully acknowledge useful conversations with J. Bohn, M. Holland, R. Ballagh, S. Stringari and the JILA ultracold atom collaboration. This work is supported by NSF and ONR.

References

1. M. Greiner *et al.*, *Nature* **415**, 39 (2002).
2. E. A. Donley *et al.*, *Nature* **417**, 529 (2002).
3. C. A. Regal, M. Greiner and D. S. Jin, *Phys. Rev. Lett.* **92**, 040403 (2004).
4. M. W. Zwierlein *et al.*, *Phys. Rev. Lett.* **92**, 120403 (2004).
5. J. Stenger *et al.*, *Phys. Rev. Lett.* **82**, 4569 (1999).
6. D. M. Stamper-Kurn *et al.*, *Phys. Rev. Lett.* **83**, 2876 (1999).
7. S. L. Cornish *et al.*, *Phys. Rev. Lett.* **85**, 1795 (2000).
8. J. L. Roberts *et al.*, *Phys. Rev. Lett.* **85**, 728 (2000).
9. S. B. Papp and C. E. Wieman, *Phys. Rev. Lett.* **97**, 180404 (2006).
10. The number of trapped ^{87}Rb atoms remaining is $< 10\%$ of the ^{85}Rb number.
11. N. R. Claussen *et al.*, *Phys. Rev. A* **67**, 060701(R) (2003).
12. V. M. Pérez-García *et al.*, *Phys. Rev. A* **56**, 1424 (1997).
13. N. Katz *et al.*, *Phys. Rev. Lett.* **89**, 220401 (2002).
14. A. P. Chikkatur *et al.*, *Phys. Rev. Lett.* **85**, 483 (2000).
15. N. Bogoliubov, *J. Phys. (Moscow)* **11**, 23 (1947).
16. A. Brunello *et al.*, *Phys. Rev. A* **64**, 063614 (2001).
17. S. T. Beliaev, *Sov. Phys. JETP* **34**, 299 (1958).
18. L. Pitaevskii and S. Stringari, *Phys. Rev. Lett.* **81**, 4541 (1998).
19. A. Altmeyer *et al.*, *Phys. Rev. Lett.* **98**, 040401 (2007).
20. J. Kinast, A. Turlapov and J. E. Thomas, *Phys. Rev. A* **70**, 051401(R) (2004).
21. F. Dalfovo *et al.*, *Rev. Mod. Phys.* **71**, 463 (1999).
22. J. Steinhauer *et al.*, *Phys. Rev. Lett.* **88**, 120407 (2002).
23. R. Ozeri *et al.*, *Rev. Mod. Phys.* **77**, 187 (2005).
24. T. D. Lee and C. N. Yang, *Phys. Rev.* **105**, 1119 (1957).
25. T. D. Lee, K. Huang and C. N. Yang, *Phys. Rev.* **106**, 1135 (1957).
26. P. Noziéres and D. Pines, *The Theory of Quantum Liquids* Superfluid Bose Liquids Vol. II (Addison-Wesley, Redwood City, CA, 1990).
27. A. Griffin, *Excitations in a Bose-Condensed Liquid* (Cambridge University Press, Cambridge, England, 1993).
28. J. Steinhauer *et al.*, *Phys. Rev. A* **72**, 023608 (2005).
29. E. Braaten, H. W. Hammer and T. Mehen, *Phys. Rev. Lett.* **88**, 040401 (2002).
30. F. Mohling and A. Sirlin, *Phys. Rev.* **118**, 370 (1960).
31. S. Ronen, arXiv:0809.1448.

COOPERATIVE EMISSION OF LIGHT QUANTA: A THEORY OF COHERENT RADIATION DAMPING*

R. J. GLAUBER

Harvard University, Cambridge, MA 02138
Email: glauber@physics.harvard.edu

A quantum emitted by any of a collection of identical atoms may be absorbed and re-emitted by other atoms many times before it eventually emerges. The radiation process is thus best described as collective or cooperative in nature. The atomic excitations are shown to attenuate as linear combinations of certain characteristic decay modes that lend a complex structure to the spectrum radiated. Instead of a single line, it becomes a closely-spaced multiplet of lines, the elements of which have a variety of lifetimes, line-shifts and line-widths. We calculate these quantities, first with an abstract two-state model for the atoms and then with an isotropic four-state model that accommodates the full polarization dependence of the radiation.

Keywords: Cooperative emission; Coherent radiation damping.

1. Introduction

We consider a collection of identical atoms interacting with the radiation field. If one of the atoms is raised to an excited state and radiates a single quantum in returning to its ground state, the other unexcited atoms nearby will have a resonantly large probability of absorbing the quantum and then reradiating it, a cycle which may be repeated in many ways indefinitely many times. Radiation by the atomic system is thus a collective process. The single quantum is radiated not by any one atom but cooperatively by the full collection. We shall show that this process is best described by introducing certain collective excitation modes for the atomic system, which lead to exponential decay with a range of different lifetimes. These modes are found then to radiate with different line-shifts as well as different

---

*A paper based on the first five sections of the present work was presented at the Conference "Photons, Atoms and Qubits," PAQO7, at the Royal Society, London, September 4, 2007.

line-widths. An arbitrary initial excitation of the atomic system decays in general as a linear combination of these characteristic modes.

The atoms, we assume occupy fixed positions $\mathbf{r}_j$, $(j = 1, \ldots n)$, and we take them, as an initial simplifying assumption, to have just two quantum states, a ground state $|g_j\rangle$ and an excited state $|e_j\rangle$. We represent the operators that bring about transitions between those states for the j-th atom as $\sigma_j^{\pm}$, so that

$$\sigma_j^+ |g_j\rangle = |e_j\rangle, \quad \sigma_j^+ |e_j\rangle = 0$$
$$\sigma_j^- |e_j\rangle = |g_j\rangle, \quad \sigma_j^- |g_j\rangle = 0. \tag{1.1}$$

We label an oscillation mode of the electromagnetic field with an index k, which will later be taken to specify both the propagation vector and the polarization of a plane wave. A single quantum state for the k-th mode will be $|k\rangle$ and will have energy $\hbar\omega_k$. The annihilation and creation operators for this quantum are a_k and $a_k^\dagger$, which obey the familiar commutation relations

$$a_k a_{k'}^\dagger - a_{k'}^\dagger a_k = \delta_{kk'}. \tag{1.2}$$

If we take the excitation energy of the states $|e_j\rangle$ to be $\hbar\omega_0$ then we may write the Hamiltonian for the system of atoms plus field, before their interaction is taken into account, as

$$H_0 = \hbar\omega_0 \sum_j \sigma_j^+ \sigma_j^- + \sum_k \hbar\omega_k a_k^\dagger a_k. \tag{1.3}$$

The ground state of this system, which has no excitation or quanta present, has energy zero. That energy changes, however, when the atom-field interactions are taken into account, and we shall later add a constant to H_0 that readjusts the ground state energy to zero.

We need, for the present, only to note that the atom-field interaction permits an atom in either state to go to the other state while either emitting or absorbing a quantum. The interaction Hamiltonian can then be written as

$$H_1 = \hbar \sum_{jk} (\sigma_j^+ + \sigma_j^-)(\lambda_{jk} a_k + \lambda_{jk}^* a_k^\dagger), \tag{1.4}$$

in which the coupling coefficients λ_{jk} and λ_{jk}^* are atomic matrix elements for the radiative transition between the two atomic states.

States of the atom-field system that evolve from a single atomic excitation will tend not to have many excited atoms or many quanta present. The simplest way of labeling the states then will be to indicate only the atoms or modes containing excitations, leaving unmentioned all that remain in their unexcited states. Thus $|e_j\rangle$ will be a state with only the j-th atom excited, while $|k\rangle$ will be a state with one quantum in the k-th mode and no other excitations. We shall also encounter the states $|e_i e_j k\rangle$ with two atoms excited, $j \neq i$, and one quantum present, and the states $|e_i kk'\rangle$ with one excited atom and two quanta present, which may occupy the same or different modes.

To approximate the time-dependent Schrödinger state of the system, we introduce a set of time-dependent coefficients to expand it in terms of the succession of time-independent basis states we have introduced, as follows:

$$|t\rangle = \sum_j \beta_j(t)|e_j\rangle + \sum_k \alpha_k(t)|k\rangle +$$

$$\sum_{i<j,k} \zeta_{ijk}(t)|e_i e_j k\rangle + \frac{1}{2} \sum_{jkk'} \eta_{jkk'}(t)|e_{jkk'}\rangle. \tag{1.5}$$

In the last term $\eta_{jkk'}$ is defined to be symmetric in the indices k and k'. Further terms with more excitations will not be necessary. The time dependence of the coefficients $\beta_j, \cdots \eta_{ikk'}$ must be determined from the overall Schrödinger equation

$$i\hbar \frac{d}{dt}|t\rangle = (H_0 + H_1)|t\rangle. \tag{1.6}$$

We can project out of this equation the four equations obeyed by the time derivatives $\dot\beta_i \ldots \dot\eta_{jkk'}$. The first, for example, is

$$i\hbar\dot\beta_j = \langle e_j|H_0 + H_1|t\rangle \tag{1.7}$$

and the other three are constructed analogously. The four coupled differential equations are thus

$$i\dot\beta_i = \omega_0 \beta_i + \sum_k \lambda_{ik}\alpha_k + \sum_{j\neq i,k} \lambda_{jk}\zeta_{ijk} \tag{1.8}$$

$$i\dot\alpha_k = \omega_k \alpha_k + \sum_j \lambda_{jk}^* \beta_j + \sum_{jk'} \lambda_{jk'}\eta_{jk'k} \tag{1.9}$$

$$i\dot\zeta_{ijk} = (2\omega_0 + \omega_k)\zeta_{ijk} + \lambda_{ik}^*\beta_j + \lambda_{jk}^*\beta_i \qquad (i \neq 1) \tag{1.10}$$

$$i\dot\eta_{jkk'} = (\omega_0 + \omega_k + \omega_{k'})\eta_{jkk'} + \lambda_{jk}^*\alpha_{k'} + \lambda_{jk'}^*\alpha_k. \tag{1.11}$$

In constructing these equations we have dropped the terms that link the amplitudes ζ_{ijk} and $\eta_{jkk'}$, since they would introduce higher order corrections than we need. The amplitudes ζ_{ijk} and $\eta_{jkk'}$ themselves however must not be neglected. To drop them would be to fall back on what has been called the "rotating wave" approximation which, in effect, omits those terms of the interaction in Eq. (1.4) that cannot immediately conserve energy. The importance of including such transitions in the calculation of interatomic forces has been indicated by M. Stephen[1] and extended to other radiative interactions by several authors.[2,3]

2. The Coupling of the Excitation Amplitudes

We assume that our system starts out with no quanta present at $t = 0$, so that $\alpha_k(0)$, $\zeta_{ijk}(0) = 0$, and $\eta_{jkk'}(0) = 0$. Then the time integrals of Eqs. (1.9)–(1.11) can be written as

$$\alpha_k(t) = -i \int_0^t \left\{ \sum_j \lambda_{jk}^* \beta_j(t') + \sum_{jk'} \lambda_{jk'} \eta_{jk'k}(t') \right\} e^{-i\omega_k(t-t')} dt' \quad (2.1)$$

$$\zeta_{ijk}(t) = -i \int_0^t \left\{ \lambda_{ik}^* \beta_j(t') + \lambda_{jk}^* \beta_i(t') \right\} e^{-i(2\omega_0+\omega_k)(t-t')} dt' \quad (2.2)$$

$$\eta_{jkk'}(t) = -i \int_0^t \left\{ \lambda_{jk}^* \alpha_{k'}(t') + \lambda_{jk'}^* \alpha_k(t') \right\} e^{-i(\omega_0+\omega_k+\omega_{k'})(t-t')} dt'. \quad (2.3)$$

We can use the first two of these equations to find an equation relating only the excitation amplitudes β_j. First however, let us take away the rapid oscillation of these amplitudes by defining

$$\beta_j' = \beta_j e^{i\omega_0 t}. \quad (2.4)$$

Then by substituting the expressions in Eqs. (2.1) and (2.2) into Eq. (1.8) we find

$$\dot{\beta}_i' = - \int_0^t \sum_{jk} \lambda_{ik} \lambda_{jk}^* \beta_j'(t') e^{i(\omega_0-\omega_k)(t-t')} dt'$$

$$- \int_0^t \sum_{j \neq i, k} \left\{ |\lambda_{jk}|^2 \beta_i'(t') + \lambda_{jk} \lambda_{ik}^* \beta_j'(t') \right\} e^{-i(\omega_0+\omega_k)(t-t')} dt'. \quad (2.5)$$

The coupling constants λ_{ik}, as we have noted, are matrix elements of the combined transitions of an atom and the field. In a single electron transition, for example, in the i-th atom we would have

$$\lambda_{ik}^* = -\frac{e}{mc} \langle k | \mathbf{p} \cdot \mathbf{A}(\mathbf{r}) | e_i \rangle, \quad (2.6)$$

where $\mathbf{p}$ is the electron momentum and $\mathbf{A}(\mathbf{r})$ the vector potential at its position $\mathbf{r}$. The matrix element thus contains the integral

$$\int \psi_g^* \mathbf{p} e^{i\mathbf{k}\cdot\mathbf{r}} \psi_e \, d\mathbf{r}, \quad (2.7)$$

where ψ_e and ψ_g are the excited state and ground state wave functions, respectively. If, for each atom, these wave functions are spread over a volume with radius of magnitude R, then for $kR \gg 1$ the phase factor $\exp(i\mathbf{k} \cdot \mathbf{p})$ will oscillate many times within each atom. In the limit $kR \to \infty$ then we expect the integral (2.7) to go rapidly to zero. The summations in Eq. (2.5), which are carried out over all modes of field excitation, will then converge to finite values.

It will be helpful at this point to recall briefly several elements of the radiation damping calculation for a single atom.[4] If only the i-th atom is present, Eq. (2.5) takes the simpler form

$$\dot{\beta}'_i = - \int_0^t \sum_k |\lambda_{ik}|^2 e^{i(\omega_0 - \omega_k)(t-t')} \beta'_i(t') dt'. \tag{2.8}$$

The mode summation in its expression is ultimately an integration over the values of $\mathbf{k}$. With the factor

$$e^{-i\omega_k(t-k')} = e^{-ikc(t-t')}$$

in its integrand, it behaves very much like a Fourier integral over k. Since the matrix elements λ_{ik} tend to vanish for $kR \gg 1$, the mode summation tends to vanish for $c(t-t') \gg 1/R$. In that case nearly all of the contributions to the t' integration in Eq. (2.8) come from the brief time interval $(t-t') \sim R/c$ near the upper limit of integration. The interval R/c, the passage time of light through the atom is, of course shorter than the oscillation period $1/\omega_0$, and much shorter than the damping time in which we may expect the amplitude $\beta'_i(t')$ to vary appreciably. It becomes accurate then simply to evaluate the function β'_i in the integrand of Eq. (2.8) at the upper limit $t' = t$, and factor it out of the integral.

There are two other familiar steps in the treatment of a single atom. For times $t \gg R/c$ the lower limit of integration in Eq. (2.8) can be displaced from 0 to $-\infty$ without materially altering the integral. Furthermore the integrals of the individual terms of the mode summation can be defined by adding to ω_0 a positive imaginary infinitesimal $i\epsilon$ and taking the limit as $\epsilon \to 0$. These steps lead us to the differential equation

$$\dot{\beta}'_i = - \lim_{\epsilon \to 0} \sum_k \frac{|\lambda_{ik}|^2}{\epsilon - i(\omega_0 - \omega_k)} \cdot \beta'_i. \tag{2.9}$$

The limiting form of this relation can be written as

$$\dot{\beta}'_i = -(\gamma + i\delta\omega_0)\beta'_i, \tag{2.10}$$

where

$$\gamma = \pi \sum_k |\lambda_{ik}|^2 \delta(\omega_k - \omega_0) \tag{2.11}$$

is the damping constant, and the radiative frequency shift $\delta\omega_0$ is given by the sum

$$\delta\omega_0 = P \sum_k \frac{|\lambda_{ik}|^2}{\omega_0 - \omega_k}, \tag{2.12}$$

in which the symbol P indicates the principal value of the sum. The results we have reached in Eqs. (2.10)–(2.12) are the familiar ones for the radiation by a single atom. If $\beta_i'(0) = 1$, then

$$\beta_i(t) = e^{-[\gamma + i(\omega_0 + \delta\omega_0)]t}, \tag{2.13}$$

and the damping constant γ is both the half-width of the spectrum line radiated and one-half the transition rate as calculated in perturbation theory.

Our problem now is to generalize this analysis to deal with the full set of n atoms, which have different positions $\mathbf{r}_j$, and may in general have differently oriented electric dipole moment vectors $\boldsymbol{\mu}_i$. It seems simplest to deal with these generalizations in two stages. In the first stage we shall assume that our two-state atoms have dipole moment matrix-element vectors $\boldsymbol{\mu}_i$ that are all the same $\boldsymbol{\mu}_j = \boldsymbol{\mu}$, $j = 1 \cdots n$. With this assumption it is clear from the structure of the matrix element (2.6) that we can write

$$\lambda_{jk} = \lambda_k e^{i\mathbf{k}\cdot\mathbf{r}_j}, \tag{2.14}$$

where the coupling constants λ_k are the same for all atoms $j = 1 \cdots n$. In Section 6 we shall extend our model to let all the electric dipole vectors vary freely. For the present however, it suffices to consider the summations that occur in Eq. (2.5), by using the expression (2.14) for the coupling constants. We have then, for example,

$$\sum_{j\neq i,k} \lambda_{ik}\lambda_{jk}^* e^{i(\omega_0 - \omega_k)(t-t')} = \sum_{j\neq i,k} |\lambda_k|^2 e^{i[\mathbf{k}\cdot(\mathbf{r}_j - \mathbf{r}_j) + (\omega_0 - \omega_k)(t-t')]}, \tag{2.15}$$

in which the phase factors $\exp[i\mathbf{k}\cdot(\mathbf{r}_i - \mathbf{r}_j)]$ result from the time delays for the passage of light from one atom to another. If the distances between the atoms are small compared to the distance c/γ that light travels in a decay period, we may again factor the amplitudes $\beta_j'(t')$ out of the integrands in Eq. (2.5), evaluating them at $t' = t$.

The assumption that the interatomic propagation times are small compared to the decay time does impose a limit on the size of the atomic systems we can discuss, but it is quite a generous one. The distances $r_{ij} = |\mathbf{r}_i - \mathbf{r}_j|$ can be as large as 10^4 or 10^6 times the resonant wavelength.

As a further simplification, motivated as before, we can let the lower limits of the t' integrations go to $-\infty$, so that Eq. (2.5) becomes

$$\dot{\beta}'_i = - (\gamma + i\delta\omega_0)\beta'_i$$

$$- i \int_{-\infty}^{t} \sum_{j \neq i, k} \left\{ \lambda_{ik}\lambda^*_{jk} e^{i(\omega_0 - \omega_k)(t-t')} + \lambda^*_{ik}\lambda_{jk} e^{-i(\omega_0 + \omega_k)(t-t')} \right\} dt' \beta'_j(t)$$

$$- i \int_{0}^{\infty} \sum_{j \neq i, k} |\lambda_{jk}|^2 e^{-i(\omega_0 + \omega_k)(t-t')} dt' \beta'_i(t). \tag{2.16}$$

The integrations over t' can then be carried out as before by letting $\omega_0 \to \omega_0 + i\epsilon$ in the first of these three t'-integrations to keep the singularity in the mode summation well-defined. The result is

$$\dot{\beta}'_i = - (\gamma + i\delta\omega_0)\beta'_i - \sum_{j \neq i, k} \left\{ \frac{\lambda_{ik}\lambda^*_{jk}}{\epsilon - i(\omega_0 - \omega_k)} + \frac{\lambda_{ik}\lambda^*_{jk}}{i(\omega_0 + \omega_k)} \right\} \beta'_j$$

$$+ (n-1)i \sum_{k} \frac{|\lambda_k|^2}{\omega_0 + \omega_k} \beta'_i. \tag{2.17}$$

We can immediately recognize the last of these terms. The energy level shift of the ground state of an atom due to its interaction with the radiation field is just

$$\hbar\delta\omega_g = -\hbar \sum_{k} \frac{|\lambda_k|^2}{\omega_0 + \omega_k}, \tag{2.18}$$

according to second-order perturbation theory. The last term of Eq. (2.17) thus represents the frequency shift of β'_i due to the coupling to the field of the $n-1$ atoms that remain unexcited.

It is convenient at this point to return to the discussion of $\beta_i = \beta'_i e^{-i\omega_0 t}$ and to introduce the notation

$$k_0 = \frac{\omega_0}{c}, \quad k = |\mathbf{k}| = \frac{\omega_k}{c}, \quad \epsilon' = \frac{\epsilon}{c}, \tag{2.19}$$

so that we can rewrite Eq. (2.17) in the form

$$\dot{\beta}_i = -[\gamma + i(\omega_0 + \delta\omega_0) + i(n-1)\delta\omega_g]\beta_i - \gamma \sum_{j \neq i} S_{ij}\beta_j \tag{2.20}$$

in which we have defined S_{ij} for $i \neq j$ as

$$S_{ij} = -\frac{i}{\gamma} \sum_k \left\{ \frac{\lambda_{ik}\lambda_{jk}^*}{k - k_0 - i\epsilon'} + \frac{\lambda_{ik}^*\lambda_{jk}}{k + k_0} \right\}. \tag{2.21}$$

3. The Radiation Amplitudes and Energy Renormalization

It is helpful, before solving more explicitly for the coupled excitation amplitudes, to show how the field amplitudes $\alpha_k(t)$ may be derived from them. We return, for this purpose, to Eq. (1.9) for $\dot{\alpha}_k(t)$ and substitute in it the expression (2.3) for the amplitudes $n_{jkk'}(t)$. By then carrying out in it the same sequence of approximations we made in deriving the equation (2.17) that couples the amplitudes β_i', we find

$$i\dot{\alpha}_k = (\omega_k + n\delta\omega_g)\alpha_k + \sum_j \lambda_{jik}^*\beta_j - \frac{1}{\omega_0 + \omega_k} \sum_{j,k'} \lambda_{jk}^*\lambda_{jk'}\alpha_{k'}. \tag{3.1}$$

The term $n\delta\omega_g$ that is added to ω_k in this equation has essentially the same origin as the term $(n-1)\delta\omega_g$ added to ω_0 in Eq. (2.20). We have begun by taking the state of zero energy for our system to be the one in which no photons are present and all n atoms are in their "bare" ground states. Those are the ground states in the absence of interaction with the field. When the atoms are coupled to the field their ground state energy becomes $n\hbar\delta\omega_g$, and it is this state to which an excited atom decays.

We can reset the final state energy of the n atoms to zero by subtracting from the Hamiltonian H_0, given by Eq. (1.3), the "self-energy" $n\hbar\delta\omega_g$. That subtraction multiplies all the amplitudes α_k and β_i, in effect, by the same phase factor $\exp(in\delta\omega_g t)$ which then subtracts away the $n\delta\omega_g$ terms in Eqs. (2.20) and (3.1).

A further simplification is that the last term of Eq. (3.1) leads to higher order corrections than we need for the discussion of single photon radiation, so we may drop it for the present, (although it does play a role in the scattering of pre-existing photons.) With that term omitted, and the ground state energies suitably adjusted, we may integrate Eq. (3.1), noting $\alpha_k(0) = 0$, to find that

$$\alpha_k(t) = -ie^{-i\omega_k t} \int_0^t \sum_j \lambda_{jk}^*\beta_j(t')e^{i\omega_k t'} dt'. \tag{3.2}$$

With the $n\delta\omega_g$ term subtracted away Eq. (2.20) takes the form

$$\dot{\beta}_i = -[\gamma + i(\omega_0 + \delta\omega_0 - \delta\omega_g)]\beta_i - \gamma \sum_{j \neq i} S_{ij}\beta_j. \tag{3.3}$$

If we write the renormalized resonant frequency as

$$\omega_0' = \omega_0 + \delta\omega_0 - \delta\omega_g \tag{3.4}$$

and redefine β_i' as

$$\beta_i' = \beta_i e^{i\omega_0't}, \tag{3.5}$$

then the equation for the coupled excitation amplitudes becomes

$$\dot{\beta}_i' = -\gamma\beta_i' - \gamma\sum_{j\neq i} S_{ij}\beta_i'. \tag{3.6}$$

4. Solution for the Excitation Amplitudes

Only the off-diagonal matrix elements S_{ij} for $i \neq j$ have been defined by Eq. (2.21). If we now define the diagonal elements as

$$S_{ii} = 1, \qquad i = 1\cdots n \tag{4.1}$$

then the equation for the coupled amplitudes assumes the compact form

$$\dot{\beta}_i' = -\gamma\sum_j S_{ij}\beta_j'. \tag{4.2}$$

If we think of the set of n amplitudes $\beta_1' \cdots \beta_n'$ as the components of an n-dimensional complex vector B, we may write the set of coupled equations in the still more compact form

$$\dot{B} = -\gamma SB. \tag{4.3}$$

The matrix S in these equations is complex-valued in general and in fact symmetric, $S_{ij} = S_{ji}$. The symmetry property follows from Eq. (2.14) and the fact that for each propagation vector $\mathbf{k}$ in the mode sum given by Eq. (2.21) there is also a propagation vector $-\mathbf{k}$.

The total excitation probability of the atomic system is the scalar product

$$B^* \cdot B = \sum_i |\beta_i|^2, \tag{4.4}$$

and its time derivative is

$$\frac{d}{dt}(B^* \cdot B) = -\gamma B^* \cdot (S + S^*) \cdot B$$
$$= -2\gamma B^*(\text{Re } S)B \tag{4.5}$$

It can easily be shown that the real part of S_{ij} is a positive definite matrix, so the total excitation probability of the system always decreases monotonically.

Since the matrix S is symmetric it can be diagonalized. If we write its eigenvalues as s_ℓ for $\ell = 1 \cdots n$, and the corresponding eigenvectors as $B^{(\ell)}$ we have

$$SB^{(\ell)} = s_\ell B^{(\ell)} \tag{4.6}$$

and the vectors $B^{(\ell)}(t)$ then denote excitation modes that decay exponentially

$$B^{(\ell)}(t) = B^{(\ell)}(0)e^{-\gamma s_\ell t}. \tag{4.7}$$

These are, in a sense, the normal modes of the collective radiative damping process.

The eigenvalues s_ℓ are in general complex. The real parts $\mathrm{Re}\, s_\ell$ can never be negative, and $\gamma\, \mathrm{Re}\, s_\ell$ will govern the rate of exponential decay of each mode. The imaginary part of the root s_ℓ will lead to a frequency shift characteristic of the mode. It will be a shift due explicitly to the presence of other atoms nearby. In the ℓ-th mode then the excitation amplitudes will oscillate with the frequency

$$\omega_0' + \gamma\, \mathrm{Im}\, s_\ell = \omega_0 + \delta\omega_0 - \delta\omega_g + \gamma\, \mathrm{Im}\, s_\ell. \tag{4.8}$$

The single atomic spectrum line is split in general into an n-fold multiplet, (many components of which may be very closely spaced).

The orthogonal transformation that diagonalizes the matrix S preserves its trace. Since all the diagonal elements of S_{ij} are equal to one, the trace is n, and the roots must obey the identity

$$\sum_\ell s_\ell = n. \tag{4.9}$$

The real and imaginary parts of this relation constitute two interesting sum rules. The first is

$$\sum_\ell \mathrm{Re}\, s_\ell = n, \tag{4.10}$$

which severely constrains the individual decay rates. If any mode, for example, has the maximal decay rate $n\gamma$, all the remaining decay rates must vanish. They characterize "dark" modes from which no radiation can escape. The second sum rule

$$\sum_\ell \mathrm{Im}\, s_\ell = 0, \tag{4.11}$$

implies that the sum of the frequency displacements in the n-fold multiplet is zero. The multiplet remains centered on $\omega_0 + \delta\omega_0 - \delta\omega_g$.

The exponential time dependence of the mode amplitudes $B^{(\ell)}(t)$ makes it easy to find the radiation amplitudes $\alpha_k(t)$. The result may be written compactly by recalling the position dependence of the coupling constants λ_{jk} given by Eq. (2.14) and gathering the phase factors $e^{i\mathbf{k}\cdot\mathbf{r}_j}$ into an n-component vector

$$V = (e^{i\mathbf{k}\cdot\mathbf{r}_1}, \ldots, e^{i\mathbf{k}\mathbf{r}_n}), \tag{4.12}$$

so that we can write

$$\sum_j e^{-i\mathbf{k}\cdot\mathbf{r}_j} \beta'_j(t) = V^* \cdot B(t). \tag{4.13}$$

If we then evaluate the integrand in Eq. (3.2) for the ℓ-th mode we find the radiation amplitude

$$\alpha_k(t) = \lambda_k^* e^{-i\omega_k t} \frac{\left(1 - e^{-[\gamma s_\ell - i(\omega_k - \omega'_0)]t}\right)}{\omega_k - \omega'_0 + i\gamma s_\ell} (V^* \cdot B^{(\ell)}) \tag{4.14}$$

for times $t > (\gamma \operatorname{Re} s_\ell)^{-1}$ then, the ℓ-th mode radiates a Lorentzian spectrum line centered at $\omega_k = \omega'_0 + \gamma \operatorname{Im} s_\ell$ with half-width $\gamma \operatorname{Re} s_\ell$. The mode vectors $B^{(\ell)}$ can be assumed to form a complete orthonormal set so an arbitrary initial excitation $B(0)$ may be expanded in terms of them as

$$B(0) = \sum_\ell B^{(\ell)}(B^{(\ell)} \cdot B(0)). \tag{4.15}$$

The ℓ-th line of the radiated multiplet then, if it is well enough separated from the others, will have an intensity proportional to

$$|V^* \cdot B^{(\ell)}|^2 |B^{(\ell)} \cdot B(0)|^2. \tag{4.16}$$

5. The Radiative Decay Matrix

To find the time-dependent atomic excitation modes we solve the equation (4.2). We shall have first to evaluate the elements S_{ij} of the matrix that describes the mutual induction of radiative decay processes by the individual atoms. These matrix elements are expressed by Eq. (2.21) as summations over all the excitation modes k of the field. For radiation in free space then, they are integrations over the space of plane wave propagation vectors $\mathbf{k}$ together with sums over the pair of transverse polarizations associated with each $\mathbf{k}$. To evaluate the S_{ij} then we shall need more explicit expressions for the coupling coefficients λ_{jk}, which include their polarization dependence and their dependence on the atomic positions $\mathbf{r_j}$.

The interaction of the atomic electric dipole moments with the electric field (in rationalized units) is characterized by the coupling constants

$$\lambda_{jk} = -i\sqrt{\frac{\omega_k}{2\hbar V}}\,\boldsymbol{\mu}_j \cdot \hat{\mathbf{e}}^{(p)}(\mathbf{k})e^{i\mathbf{k}\cdot\mathbf{r}_j}. \tag{5.1}$$

In this expression $\boldsymbol{\mu}_j$ is the transition matrix element of the dipole moment vector for the j-th atom, $\hat{\mathbf{e}}^{(p)}$ is one of the two polarization basis vectors (assumed real-valued) associated with the propagation vector, $\mathbf{k}$, and V is the quantization volume.

The expression for S_{ij} given by Eq. (2.21) then, as an integral over $\mathbf{k}$-space and a polarization sum is

$$S_{ij} = -\frac{i}{2\hbar\gamma}\int\frac{k\,d\mathbf{k}}{(2\pi)^3}\sum_p \boldsymbol{\mu}_i\cdot\hat{\mathbf{e}}^{(p)}\hat{\mathbf{e}}^{(p)}\cdot\boldsymbol{\mu}_j\left\{\frac{e^{i\mathbf{k}\cdot\mathbf{r}_{ij}}}{k-k_0-i\epsilon} + \frac{e^{-i\mathbf{k}\cdot\mathbf{r}_{ij}}}{k+k_0}\right\}, \tag{5.2}$$

where we have introduced the abbreviation $\mathbf{r}_{ij} = \mathbf{r}_i - \mathbf{r}_j$. We can carry out the polarization sum by noting the dyadic relation

$$\sum_{p=1,2}\hat{\mathbf{e}}^{(p)}(\mathbf{k})\hat{\mathbf{e}}^{(p)}(\mathbf{k}) = \mathbf{1} - \hat{\mathbf{k}}\hat{\mathbf{k}}, \tag{5.3}$$

in which $\mathbf{1}$ is the unit dyadic and $\hat{\mathbf{k}} = \mathbf{k}/k$ is a unit vector in the direction of $\mathbf{k}$. Then we have

$$\sum_p \boldsymbol{\mu}_i\cdot\hat{\mathbf{e}}^{(p)}\hat{\mathbf{e}}^{(p)}\cdot\boldsymbol{\mu}_j = \boldsymbol{\mu}_i\cdot\boldsymbol{\mu}_j - (\boldsymbol{\mu}_i\cdot\hat{\mathbf{k}})(\boldsymbol{\mu}_j\cdot\hat{\mathbf{k}}) \tag{5.4}$$

as an expression to include in the integrand of Eq. (5.2).

If, as in the preceding sections of this paper, we assume our two-level atoms all have the same dipole moment orientation, i.e., $\boldsymbol{\mu}_1 = \boldsymbol{\mu}_2 = \cdots = \boldsymbol{\mu}_n$, then the polarization sum reduces to

$$\mu^2 - (\boldsymbol{\mu}\cdot\hat{\mathbf{k}})^2, \tag{5.5}$$

where μ is the common value of the $\boldsymbol{\mu}_j$. We shall extend our model to include arbitrary and indepedent orientations of the $\boldsymbol{\mu}_j$ in the next section, but find it expedient meanwhile to separate that problem from the integration to be carried out in Eq. (5.2) by replacing the sum (5.4) by its value averaged over all directions of the dipole vector $\boldsymbol{\mu}$.

$$\langle\mu^2 - (\boldsymbol{\mu}\cdot\hat{\mathbf{k}})^2\rangle_{\mathrm{av}} = \frac{2}{3}\mu^2. \tag{5.6}$$

The polarization sum will be treated more fully in due course.

With the expression (5.6) substituted for the polarization sum in Eq. (5.2) we can write the averaged matrix element $\bar{S}_{ij}$ as

$$\bar{S}_{ij} = -\frac{i}{3\hbar\gamma}\frac{\mu^2}{(2\pi)^3}\int k^3 dk\, d\Omega\left\{\frac{1}{k-k_0-i\epsilon'}+\frac{1}{k+k_0}\right\}e^{i\mathbf{k}\cdot\mathbf{r}_{ij}} \quad (5.7)$$

in which $d\Omega$ is an element of solid angle in $\mathbf{k}$-space. We can write this integral as

$$\bar{S}_{ij} = -\frac{i}{3\hbar\gamma}\frac{\mu^2}{(2\pi)^3}\nabla^2\int k\, dk\, d\Omega\left\{\frac{1}{k-k_0-i\epsilon'}+\frac{1}{k+k_0}\right\}e^{i\mathbf{k}\cdot\mathbf{r}_{ij}} \quad (5.8)$$

where ∇^2 is the Laplacian differential operator in the space of $\mathbf{r}_{ij}$.

Let us now recall that in the first term in brackets in Eq. (5.8), the infinitesimal term $i\epsilon'$ has been added to k_0 to define the behavior of a fraction that is otherwise singular for $k = k_0$. The second fraction in the brackets, on the other hand, is not singular in the range of integration, so adding $i\epsilon'$ to k_0 in it makes no change at all in the limit $\epsilon' \to 0$. If indeed we add this term and also carry out the unrelated angular integration we find

$$\bar{S}_{ij} = \frac{i\mu^2}{6\pi^2\hbar\gamma}\nabla^2\frac{1}{r_{ij}}\int_0^\infty dk\left\{\frac{1}{k-k_0-i\epsilon'}+\frac{1}{k+k_0+i\epsilon'}\right\}\sin kr_{ij}. \quad (5.9)$$

The two integrals from 0 to ∞ can equally well be combined as a single integral from $-\infty$ to ∞, and evaluated as

$$\int_{-\infty}^{\infty}\frac{\sin kr}{k-k_0-i\epsilon'}dk = \frac{1}{2i}\int_{-\infty}^{\infty}\frac{e^{ikr}-e^{-ikr}}{k-k_0-i\epsilon'}dk$$
$$= \pi e^{ik_0 r}, \quad (5.10)$$

so that we have

$$\bar{S}_{ij} = \frac{i\mu^2}{6\pi\hbar\gamma}\nabla^2\frac{e^{ik_0 r_{ij}}}{r_{ij}}$$
$$= -\frac{i\mu^2 k_0^2}{6\pi\hbar\gamma}\frac{e^{ik_0 r_{ij}}}{r_{ij}}. \quad (5.11)$$

The coefficient simplifies as well. We can write the photon emission rate γ defined by Eq. (2.11) with λ_{jk} expressed by Eq. (5.1) as

$$\gamma = \frac{\mu^2\eta_0^3}{6\pi\hbar}, \quad (5.12)$$

so we are led to the simple result

$$\bar{S}_{ij} = -i\frac{e^{ik_0 r_{ij}}}{k_0 r_{ij}}. \quad (5.13)$$

It will be useful for the construction of recursion relations to note that this is just the spherical Hankel function[5]

$$\bar{S}_{ij} = h_0^{(1)}(k_0 r_{ij}).$$ (5.14)

6. A More General Atomic Model

Our two-state model for the atoms, as we have noted earlier, makes their behavior quite anisotropic. It provides for each atom a spatial direction along which its electric dipole moment oscillates. We have thus had to specify a spatial direction for each of the dipole matrix elements $\boldsymbol{\mu}_j$. Taking those directions all to be the same and averaging over them was, as noted, an oversimplification. Real atoms can have dipole moments that oscillate in any direction. They achieve that isotropy by having an abundance of rotational quantum states available. For atoms with the appropriate rotational symmetry we can no longer specify directions for the dipole moments as if they were classical variables. The dipole moments $\boldsymbol{\mu}_j$ then become quantum variables; they can fluctuate in direction. We can describe this fluctuation while still retaining the assumption that the atoms have just two energy levels by introducing rotational states that are degenerate in energy.

Let us assume, for example, that the atomic ground state has zero angular momentum and that there are three degenerate excited states with unit angular momentum. The S-state and three P-states then form a four-state model of the atom with the same two energy levels we have considered earlier. It will be convenient as a matter of notation to choose the three P-states to be the ones with zero component of angular momentum along the three coordinate axes. This orthogonal set has real-valued wave functions and transforms under rotations like the components of a 3-vector. An excited state of the j-th atom can then be any linear combination of these three states. We must regard the operators $\sigma_j^{\pm}$ defined by Eq. (1.1), which excite or de-excite these states, as forming vectors $\boldsymbol{\sigma}_j^{\pm}$. The coupling constants λ_{jk} must likewise be regarded as vectors $\boldsymbol{\lambda}_{jk}$ referring to the three-component excited states. The interaction Hamiltonian of Eq. (1.4) then becomes a sum of scalar products,

$$H_1 = \hbar \sum_{jk} (\boldsymbol{\sigma}_j^+ + \boldsymbol{\sigma}_j^-) \cdot (\boldsymbol{\lambda}_{jk} a_k + \boldsymbol{\lambda}_{jk}^* a_k^\dagger).$$ (6.1)

The coupling coefficient vectors analogous to the constants of Eq. (5.1) are then

$$\boldsymbol{\lambda}_{jk} = -i\sqrt{\frac{\omega_k}{2\hbar V}} \mu \hat{\mathbf{e}}^{(p)}(\mathbf{k}) \, e^{i\mathbf{k}\cdot\mathbf{r}_i},$$ (6.2)

in which μ is simply the magnitude of the matrix element for the P to S transition.

The excitation amplitude for the j-th atom, which we have written earlier as β_j will now be replaced by a set of three amplitudes which we can represent as the components of a 3-vector amplitude $\boldsymbol{\beta}_j$. There will be corresponding changes in the amplitudes ζ_{ijk} and $\eta_{jkk'}$ defined earlier in Eq. (1.5), but these are sufficiently straightforward that there is no need to detail them explicitly. Suffice it to say that repeating the steps we have gone through in Sections 2 and 3 leads to a set of coupled equations relating the vector excitation amplitudes $\boldsymbol{\beta}'_j = \boldsymbol{\beta}_j(t) \exp[i(\omega_0 + \delta\omega_0 - \delta\omega_g)t]$. These are

$$\dot{\boldsymbol{\beta}}'_i = -\gamma\boldsymbol{\beta}'_i - \gamma \sum_{j \pm i} \mathbf{S}_{ij} \cdot \boldsymbol{\beta}'_j, \qquad (6.3)$$

in which each of the matrix elements $\mathbf{S}_{ij}$ must be regarded as a dyadic

$$\mathbf{S}_{ij} = -\frac{i}{\gamma} \sum_k \left\{ \frac{\boldsymbol{\lambda}_{ik}\boldsymbol{\lambda}^*_{jk}}{k - k_0 - i\epsilon'} + \frac{\boldsymbol{\lambda}^*_{ik}\boldsymbol{\lambda}_{jk}}{k + k_0} \right\}. \qquad (6.4)$$

When the vector coupling coefficients of Eq. (6.2) are inserted in this expression it preserves much of the same form as the integral in Eq. (5.7), except that the factor $2/3$ that resulted from the directional average in Eq. (5.6) is replaced by the dyadic polarization sum

$$\sum_p \hat{\mathbf{e}}^{(p)}(\mathbf{k})\hat{\mathbf{e}}^{(p)}(\mathbf{k}). \qquad (6.5)$$

If we let $\mathbf{1}$ be the unit dyadic and introduce $\hat{\mathbf{k}} = \mathbf{k}/k$ as a unit vector in the direction of $\mathbf{k}$ we need only insert

$$\sum_p \hat{\mathbf{e}}^{(p)}(\mathbf{k})\hat{\mathbf{e}}^{(p)}(\mathbf{k}) = \mathbf{1} - \hat{\mathbf{k}}\hat{\mathbf{k}} \qquad (6.6)$$

into the integrand of Eq. (5.7) and remove the factor $2/3$ in order to have the value of $\mathbf{S}_{ij}$.

Then by following essentially the same steps as took us from Eq. (5.7) to Eq. (5.13) we are led to the result

$$\mathbf{S}_{ij} = \frac{i\mu^2}{4\pi\hbar\gamma}(\mathbf{1}\nabla^2 - \nabla\nabla)\frac{e^{ik_0 r_{ij}}}{r_{ij}}. \qquad (6.7)$$

By again using Eqs. (5.11) and (5.12) we find

$$\mathbf{S}_{ij} = \frac{3}{2}\left(1 + \frac{1}{k_0^2}\nabla\nabla\right)\left(-i\frac{e^{ik_0 r_{ij}}}{k_0 r_{ij}}\right)$$

$$= \frac{3}{2}\left(1 + \frac{1}{k_0^2}\nabla\nabla\right)h_0^{(1)}(k_0 r_{ij}). \tag{6.8}$$

There is now a certain convenience in identifying the spherical Hankel function $h_0^{(1)}$, since we can use familiar recursion relations[6] for the Hankel functions and their derivatives to evaluate the double gradient term in Eq. (6.8). If we let $\hat{\mathbf{r}}$ be the unit vector $\mathbf{r}/r$ in the direction of $\mathbf{r}_{ij}$ we find

$$\frac{1}{k_0^2}\nabla\nabla h_0^{(1)}(k r_{ij}) = -\frac{1}{3}(h_0^{(1)} + h_2^{(1)})\mathbf{1} + \hat{\mathbf{r}}\hat{\mathbf{r}}h_2^{(1)} \tag{6.9}$$

so that the dyadic matrix elements of Eq. (6.7) are finally

$$\mathbf{S}_{ij} = \left(h_0^{(1)} - \frac{1}{2}h_2^{(1)}\right)\mathbf{1} + \hat{\mathbf{r}}\hat{\mathbf{r}} \cdot \frac{3}{2}h_2^{(1)} \quad \text{for} \quad i \neq j. \tag{6.10}$$

If we take the diagonal element $\mathbf{S}_{ii}$ to be the unit dyadic, $\mathbf{1}$, we can write the equations that govern the coupled excitation amplitudes as

$$\dot{\boldsymbol{\beta}}_i' = -\gamma\sum \mathbf{S}_{ij} \cdot \boldsymbol{\beta}_j'. \tag{6.11}$$

The set of n vectors $\boldsymbol{\beta}_i$ now comprise $3n$ excitation amplitudes and the symmetric matrix of dyadics $\mathbf{S}$ will have $3n$ (generally complex) eigenvalues s_ℓ, and $3n$ eigenvectors associated with them. Together they define what we have called the eigenmodes of radiative decay. The real and imaginary parts of the s_ℓ will now obey the sum rules

$$\sum_{\ell=1}^{3n} \text{Re } s_\ell = 3n, \qquad \sum_{\ell=1}^{3n} \text{Im } s_\ell = 0. \tag{6.12}$$

We can easily illustrate the calculation of the eigenvalues s_ℓ and their eigenmodes for the interaction of two atoms since in that case the equations for the excitation amplitudes separate into uncoupled blocks corresponding to longitudinal and transverse polarizations. For this purpose we can express the dyadics $\mathbf{S}_{ij}$ given by Eq. (6.10) in terms of the longitudinal projection dyadic $\mathcal{P}_L = \hat{\mathbf{r}}\hat{\mathbf{r}}$ and the transverse projection dyadic $\mathcal{P}_T = \mathbf{1} - \hat{\mathbf{r}}\hat{\mathbf{r}}$ as

$$\mathbf{S}_{ij} = \left(h_0^{(1)} - \frac{1}{2}h_0^{(1)}\right)\mathcal{P}_T + \left(h_0^{(1)} + h_2^{(1)}\right)\mathcal{P}_L \quad (i \neq j). \tag{6.13}$$

The longitudinal projections $\mathcal{P}_L\beta'_j$ of the excitation amplitudes will then obey the matrix equation

$$\begin{pmatrix} \mathcal{P}_L\dot{\beta}'_1 \\ \mathcal{P}_L\dot{\beta}'_2 \end{pmatrix} = -\gamma \begin{pmatrix} 1 & h_0^{(1)} + h_2^{(1)} \\ h_0^{(1)} + h_2^{(1)} & 1 \end{pmatrix} \begin{pmatrix} \mathcal{P}_L\beta'_1 \\ \mathcal{P}_L\beta'_2 \end{pmatrix}. \tag{6.14}$$

The two eigenvalues are obviously $s_L^\pm = 1 \pm (h_0^{(1)} + h_2^{(1)})$, with s_L^+ corresponding to the symmetric longitudinal mode,

$$\mathcal{P}_L\beta'_1 = \mathcal{P}_L\beta'_2 \sim e^{-\gamma s_L^+ t} \tag{6.15}$$

and s_L^- corresponding to the antisymmetric mode

$$\mathcal{P}_L\beta'_1 = -\mathcal{P}_L\beta'_2 \sim e^{-\gamma s_L^- t}. \tag{6.16}$$

The transverse projections $\mathcal{P}_T\beta'_j$ can be further subdivided into two orthogonal polarizations which will provide pairs of degenerate modes, one pair with the eigenvalue $s_T^+ = 1 + h_0^{(1)} - \frac{1}{2}h_2^{(1)}$ corresponding to symmetric excitations and another with the eigenvalue $s_T^- = 1 - h_0^{(1)} + \frac{1}{2}h_2^{(1)}$ and antisymmetric excitations.

These six decay modes form a complete orthogonal set. Their decay constants are given by the real parts of the complex eigenvalues

$$\gamma \, \mathrm{Re}\, s_L^\pm = \gamma[1 \pm (j_0 + j_2)] \tag{6.17}$$

$$\gamma \, \mathrm{Re}\, s_T^\pm = \gamma \left[1 \pm \left(j_0 - \frac{1}{2}j_2\right)\right], \tag{6.18}$$

in which the functions $j_n(k_0 r_{12})$ are spherical Bessel functions.[5]

The excitation amplitudes α_k for the field radiated by the ℓ-th decay mode $\beta_j^{(\ell)}$ will still be given by an expression similar to that of Eq. (4.14). It is reached by replacing $\lambda_k^* V^* \cdot B^{(\ell)}$ in that expression by the factor $\sum_j \lambda_{kj}^* \cdot \beta_j^{(\ell)}$.

7. Three Colinear Atoms

Another example which can be analyzed in elementary terms is that of three atoms equally spaced along a line. If the neighboring atoms are a distance r apart we shall find it convenient to define the quantities

$$C = h_0^{(1)}(k_0 r) + h_2^{(1)}(k_0 r) \tag{7.1}$$

$$D = h_0^{(1)}(2k_0 r) + h_2^{(1)}(2k_0 r). \tag{7.2}$$

Then the equation that defines the longitudinal eigenmodes $\mathcal{P}_L\beta_j$ will be

$$\begin{pmatrix} 1 & C & D \\ C & 1 & C \\ D & C & 1 \end{pmatrix} \begin{pmatrix} \mathcal{P}_L\beta_1 \\ \mathcal{P}_L\beta_2 \\ \mathcal{P}_L\beta_3 \end{pmatrix} = s \begin{pmatrix} \mathcal{P}_L\beta_1 \\ \mathcal{P}_L\beta_2 \\ \mathcal{P}_L\beta_3 \end{pmatrix}, \tag{7.3}$$

where s is the corresponding eigenvalue.

An obvious choice for an eigen-mode is the antisymmetric one $\mathcal{P}_L\beta_2 = 0$, $\mathcal{P}_L\beta_3 = -\mathcal{P}_L\beta_1$ for which the eigenvalue is

$$s = 1 - D. \tag{7.4}$$

We can find the two remaining eigenvalues by factorizing the secular equation

$$\begin{vmatrix} 1-s & C & D \\ C & 1-s & C \\ D & C & 1-s \end{vmatrix} = (1 - D - s)\left[\left(1 + \frac{1}{2}D - s\right)^2 - \frac{1}{4}(D^2 + 8C^2)\right] = 0 \tag{7.5}$$

to reveal the two remaining roots

$$s^\pm = 1 + \frac{1}{2}D \pm \frac{1}{2}\sqrt{D^2 + 8C^2}. \tag{7.6}$$

These correspond to symmetrical patterns of longitudinal excitation.

To find the degenerate pairs of transverse excitation modes we need only redefine the quantities C and D of Eqs. (7.1) and (7.2), letting them be

$$C' = h_0^{(1)}(k_0 r) - \frac{1}{2}h_2^{(1)}(k_0 r) \tag{7.7}$$

$$D' = h_0^{(1)}(2k_0 r) - \frac{1}{2}h_2^{(1)}(2k_0 r). \tag{7.8}$$

The doubly degenerate roots for the transverse modes are then given by Eqs. (7.4) and (7.6) with C and D, replaced by C' and D', respectively.

Acknowledgement

The author is indebted to Sudhakar Prasad for discussion of this radiation problem.

References

1. M. J. Stephen, *J. Chem. Phys.* **40**, 669 (1964).
2. D. A. Hutchison and H. F. Hameka, *J. Chem. Phys.* **41**, 2006 (1964); R. H. Lehmberg, *Phys. Rev.* **A2**, 883, 889 (1970); H. Morawitz, *Phys. Rev.* **A7**, 1148 (1973); P. W. Milonni and P. L. Knight, *Phys. Rev.* **A10**, 1096 (1974).

3. R. Friedberg and J. T. Manassah, *Phys. Lett.* **A372**, 2514 (2008). These authors have also taken the counterrotating terms into account to treat radiation in a scalar analog version of electrodynamics.

4. V. Weisskopf and E. Wigner, *Z. Physik* **63**, 54 (1930); P. Meystre and M. Sargent III. *Elements of Quantum Optics* (Springer-Verlag, Berlin, Heidelberg, 1990), p. 351.

5. M. Abramowitz and I. Stegun, *Handbook of Mathematical Functions* (Dover Publications, New York, 1965), p. 437.

6. Ref. 5, p. 439.

COHERENT CONTROL OF ULTRACOLD MATTER: FRACTIONAL QUANTUM HALL PHYSICS AND LARGE-AREA ATOM INTERFEROMETRY

EDINA SARAJLIC,[1] NATHAN GEMELKE,[2] SHENG-WEY CHIOW,[1]
SVEN HERRMAN,[1] HOLGER MÜLLER,[1,3] and STEVEN CHU[1,3,*]

[1] *Physics Department, Stanford University, Stanford, CA 94305, USA,*
[2] *James Franck Institue, University of Chicago, Chicago, IL 60637*
[3] *Department of Physics, University of California, Berkeley, CA 94720, USA and
Lawrence Berkeley National Laboratory, One Cyclotron Road, Berekely, CA 94720*
* E-mail: schu@lbnl.gov

We describe our efforts to study the physics of the fractional quantum Hall effect using ultracold quantum gases in an optical lattice and to perform precision measurements using large-area atom interferometry.

Keywords: Fractional quantum Hall effect; coherent control; ultracold quantum gas; atom interferometry.

1. Highly Correlated States in the Fractional Quantum Hall Regime of a Rotating Bose Gas

Early after achievement of Bose-Einstein condensation of neutral atoms it was recognized that a key aspect of superfluidity in these gases was the presence of irrotational flow and quantized vorticity. It was quickly demonstrated in a series of experiments that long-lived vortices could be excited and observed in a variety of ways.[1,2] Large numbers (on order of 100) vortices were soon produced in individual trapped condensates, forming the expected Abrikosov lattice of vortex cores.[3] Much theoretical effort was eventually devoted to the case of extremely high vorticity, where the number of vortices is comparable to the total number of particles N in the gas, or equivalently the total angular momentum is of order $N(N-1)$. In this fractional quantum Hall (FQH) regime, the rotation rate of the trapped gas Ω approaches the centrifugal limit $\Omega \to \omega$, where ω is the harmonic trapping frequency, and the system becomes closely analogous to a two-dimensional electron gas in a magnetic field.[4] In this limit, the single-particle energy levels are organized into nearly degenerate Landau levels separated by twice the harmonic trap frequency, and the energy for the system confined to the lowest Landau level may be written as $\mathcal{H} = (\Omega - \omega)L + \mathcal{V}$. Here, L is the total angular momentum of the system, and $\mathcal{V}$ represents repulsive contact interactions bewteen the atoms. It was shown that many of the ground state wavefunctions discussed in the context of the FQH effect for electrons appear as ground states near the centrifugal limit of the rotating Bose gas, where the contribution to the total energy from interactions is sufficient to mix single particle states in the lowest Landau level. Similar to the situation for electrons, strong correlations are expected to occur as a result of this lifting of the large-scale single particle degeneracy. Additionally, excited states of the system have been shown to possess fractional statistical character,[5] owing to the presence of reduced dimensionality and the presence of a gauge potential. Unfortunately, reaching the fractional quantum Hall regime with a single gas consisting of an easily probed number of atoms of order $N = 10^4$ particles requires both temperatures and precision in trap manipulation beyond the reach of previous experiments, due primarily to the inverse dependence of excitation energy on particle number. To circumvent these limitations, we describe work performed with an ensemble of rotating gases confined near the potential minima of an optical lattice potential, each of which contains a small number of particles.

To produce an optical lattice of locally rotating potentials, three laser beams of equal intensity detuned far from atomic resonance are combined with their propagation directions evenly distributed on the surface of a cone with a small apex angle of $\theta = 8°$ (see Fig. 1(a)). The optical interference pattern created by these beams consists of a triangular lattice of intensity maxima, whose light shifts form a conservative trapping potential for atomic motion. Near to the minima, the potential is locally harmonic, approximately cylindrically symmetric, and may be described by $V(x, y) = -V_0 \sum_{j=1}^{3} \cos(\sqrt{(3)}k_e r_j)$, where $r_j = x \cos(2\pi j/3) + y \sin(2\pi j/3)$, with x, y coordinates in the plane of the lattice, and $k_e = k \sin \theta$ is the reduced wavevector caused by shallow intersection.

By choosing a small intersection angle, the spacing between lattice sites may be made large, in this case $3.5 \mu m$, which has the effect of reducing the tunneling rate of atoms between lattice sites in the 2D potential to a value negligible for the experiment timescale, and simultaneously making the trapping potential effectively more harmonic by reducing its vibration frequency at a fixed total depth. The three lattice beams are combined and focused to $150 \mu m$ by a single lens; each is derived from a common

1.5W, fiber-coupled beam, intensity stablized by an acousto-optic modulator and sourced from a 10W single-mode 1064nm Nd:YAG ring laser injection-locked to a stable non-planar ring oscillator (Lightwave NPRO). In order to produce a nominally cylindrically-symmetric potential near the bottom of each two-dimensional lattice site, center-of-mass vibration frequencies for loaded atoms are measured, and beam intensities adjusted to equalize these frequencies to a typical precision of 0.3%. To rotate the local potential, two electrooptic phase modulators are inserted into two of the beams forming the 2D lattice potential. By adiabatically manipulating the relative optical phase of the beams (see Fig. 1), the time-averaged potential near the lattice minima may be made to appoximate a rotating anisotropic harmonic oscillator.

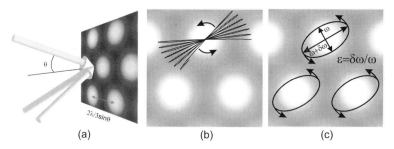

(a)   (b)   (c)

Fig. 1. A two-dimensional triangular optical lattice is formed by the intersection of three far-detuned laser beams (a). By manipulating the relative phases of the beams, an arbitrary translation of the lattice potential in two dimensions is possible; scanning the potential rapidly (500kHz) along a given direction effectively time-averages the potential, weakening the trap curvature along the axis of translation. By slowly pivoting this axis in time (b), the time-averaged local potential approximates an anisotropic harmonic oscillator whose principal axes rotate at the rate of pivot Ω (c).

In order to enhance the effect of interactions between atoms loaded into this potential, an additional optical potential is applied along the axis of rotation. For this purpose two additional beams, frequency-offset from those forming the 2D potential are added, counterpropagating along the normal to the plane of the 2D lattice potential. The final potential is then a three-dimensional array of highly oblate and strongly confining 'dot-like' traps, approximately harmonic with radial trapping frequencies of up to 3kHz and axial frequencies up to 30kHz. Interaction strength is characterized by the dimensionless ratio of scattering length a_s to oscillator length ℓ from confinement along the rotation direction, $\eta = a_s/\sqrt{2\pi}\ell$. For this experiment, $\eta \sim 0.07$.

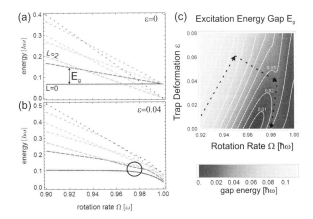

Fig. 2.   Energy levels for interacting few-atom system in a rotating trap. (a) Shows the levels for a cylindrically symmetric trap as viewed in a rotating coordinate frame for two particles interacting via a repulsive contact potential. Plot (b) shows the same spectrum after a small corotating asymmetry is added to the trap via a quadrupolar deformation ($\epsilon = 0.04$). In this case, the ground state level crossing (circled) between zero angular momentum (L=0) and the L=2 ($\frac{1}{2}$-Laughlin) state is made avoided by the perturbation to the trap, which couples levels whose angular momenta differ by $2\hbar$. In plot c, the energy gap between the lowest energy state and first excited is plotted as a function of trap deformation strength and rotation rate.

Atoms are loaded from a ^{87}Rb Bose-Einstein condensate of 10^5 particles in the $|F = 2, m_F = 2\rangle$ state at a temperature $T = 30$nK formed by evaporative cooling in a time-orbiting-potential magnetic trap with final average trap frequency $\bar{\omega} = 2\pi \times 46$Hz. After evaporation, the two-dimensional lattice potential is adiabatically increased from zero intensity to its full value of 0.5W per beam, loading the atoms at a peak linear density of approximately 300 atoms/μm per tube. In order to reduce the density, the TOP trap is deformed into a quadrupole trap whose center is pulled below the position of the atoms loaded into the tubelike 2D lattice potential, and the axial confinement of atoms trapped in the two-dimensional lattice potential is adiabatically released from 42Hz to 3Hz over a time of 1s by reducing the magnetic quadrupole field, during which time the size of the cloud increases from a Thomas-Fermi diameter of 20μm to a full-width half-max of 220μm. Following this, the axial standing wave intensity is increased to inhibit axial motion. In order to produce a well-defined mean occupancy in the full three-dimensional lattice potential, a tomographic technique is used to remove atoms far from the center of the lattice volume along the axial direction, creating a top-hat density profile. A weak magnetic field gradient is applied, and a microwave field is applied to transfer atoms from the $|2, 2\rangle$ state into $|1, 1\rangle$. By slowly sweeping the microwave frequency, atoms are adiabatically transferred between internal states at the edges of the cloud. Following this, a strong field gradient is applied to completely remove atoms in the state $|1, 1\rangle$. The two-dimensional lattice intensity is then slowly reduced to evaporate atoms from the center of the trap, until the desired mean occupancy is reached as inferred from absorption imaging performed transverse to the rotation axis. The lattice intensity is then returned to its full value in order to begin interrogating atoms in rotation.

In order to drive atoms from the non-rotating ground state into correlated states at nonzero angular momentum, an adiabatic pathway is followed in trap rotation rate and deformation strength. The deformation is characterized by $\epsilon = \delta\omega$, where ω and $\omega + \delta\omega$ are the minor and major axis vibration frequencies, respectively. A promising pathway can be inferred from a plot of the excitation energy from the lowest to first excited state as a function of these sweep parameters. This is shown in Figure 2, the result of a full numerical calculation for the interacting few-body system (in this case for four particles), taking into account single-particle basis functions up to a cutoff total angular momentum, here $L < 12\hbar$. It is important to note that as the particle number and angular momentum is increased, the energy of the first excited state in the centrifugal limit decreases roughly as $1/N$, suggesting the necessary ramp rate and trap precision to reach correlated ground states scale favorably only for small particle numbers. For the case of four particles, assuming an experimentally feasible interaction size of $\eta \sim 0.07$, the gap at the first ground state crossing (the four-particle Pfaffian state) is expected to be approximately $0.028\hbar\omega = h\,84\text{Hz}$ in an $\omega = 2\pi \times 3\text{kHz}$ trap. It is also important to note that the adiabatic transfer need not be sensitively tuned for a particular occupancy, provided in all cases one follows the ground state contour adiabatically. A representative pathway chosen for this experiment is illustrated in Figure 2c ; this pathway is translated horizontally (in rotation rate) by a variable amount to provide a control parameter to probe the onset of interparticle correlations shown in Figure 3.

Once the adiabatic transfer has been completed, short range correlation in the gas is probed by applying a brief pulse of light tuned to a photoassociative transition to an electronically excited molecular state. This transfers pairs of particles found at short range (determined by the extent of the excited molecule) into short-lived molecules, whose decay is accompanied by sufficient energy to remove the constituents from the lattice trap.

The rate of observed photoassociation loss is shown in Figure 3 as a function of the final frequency of the adiabatic rotation sequence, showing a strong depression near the centrifugal limit.

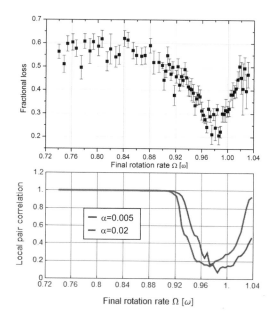

Fig. 3. Loss of atoms following a short photoassociation pulse probes local pair correlation as a function of final rotation rate (plotted here in units of the harmonic trap frequency ω) in the adiabatic pathway (a), showing strong suppression near the centrifugal limit. Average occupancy for this data set is $< N >= 5$ atoms per lattice site, as inferred from absorption imaging. Qualitative agreement can be found with full numeric evolution (b) of the four-body system, including effects of interaction, anharmonicity, and nonadiabaticity. In (b), two predicted responses are shown for differing degrees of anharmonicity, illustrating the lower-frequency downturn expected for this case. Here, α refers to the fractional deviation of the second single-particle vibrational splitting from the first; the parameters in part (a) correspond to $\alpha = 0.02$ at the center of the lattice volume.

A qualitative agreement can be found by comparing the measured loss to a zero-free-parameter numeric evolution (similar to that described previously by Popp, *et al*[6]) of the dynamic few-body system, accounting for nonadiabaticity and anharmonicity , as shown in figure 3. Proper inclusion of the effects of anharmonicity in the local lattice potential is necessary in order to account for the detailed downturn of the photoassociation rate, as well as observed time-of-flight momentum distributions; this strongly reduces the inferred fidelity of coupling to the strongly correlated states known from the fractional quantum Hall effect. However it is likely, given the strong modification of photoassociation rates, that these states exhibit reasonably strong correlations, as is to be expected from the strength of interactions and the relative proximity to the centrifugal limit.

2. Large-Area Atom Interferometry

Light-pulse atom interferometers[7] have been used for experiments of outstanding precision, like gravimeters,[8] gravity gradiometers,[9] gyroscopes,[10] measurements of Newton's gravitational constant G,[11,12] the fine-structure constant α,[13,14] or tests of gravitational theories.[15,16] They apply the momentum $\hbar k$ of photons to direct an atom on two (or more) paths which interfere when recombined. The sensitivity of atom interferometers increases with the phase shift between the arms. This depends linearly on the momentum splitting between the interferometer arms in gravimeters or gyroscopes — or even quadratically, like in measurements of α or certain gradiometers. However, many interferometers to date are limited to a splitting of $2\hbar k$ by the use of two-photon Raman transitions. Larger splittings of up to $6\hbar k$ have been achieved with multiple two-photon pulses or Bragg diffraction in atomic beam setups[17–19] and up to $12\hbar k$ using Bloch oscillations.[20]

We have recently made progress towards increased sensitivity in atom interferometry in several ways, that we will briefly discuss below.

2.1. Atom Interferometry with 24-Photon-Momentum-Transfer Bragg Beam Splitters

We have demonstrated the use of up to 24-photon Bragg diffraction[21] as a beam splitter in light-pulse atom interferometers, the largest splitting in momentum space so far. Relative to the 2-photon processes used in the most sensitive present interferometers, these large momentum transfer (LMT) beam splitters increase the phase shift 12-fold for Mach-Zehnder (MZ-) and 144-fold for Ramsey-Bordé (RB-) geometries. We achieve a high visibility of the interference fringes (up to 52% for MZ or 36% for RB) and long pulse separation times and superior control of systematic effects that are typical of atomic fountain setups. As the atom's internal state is not changed, important systematic effects can cancel. Figure 4 shows a gallery of interference fringes obtained in MZ and RB geometry at momentum transfers between 12-24$\hbar k$. More details will be found in Ref. 22.

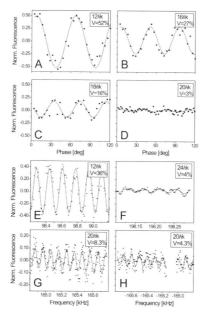

Fig. 4. A-D show MZ fringes with between 12 and 20$\hbar k$ momentum transfer; E and F are RB fringes with 12 and 24$\hbar k$. G and H show a conjugate 20$\hbar k$ RB-pair. Throughout, $T = 1\,\mathrm{ms}$, $T' = 2\,\mathrm{ms}$. Each data point is from a single launch (that takes 2 s), except for F, where 5-point adjacent averaging was used. The lines represent a sinewave fit.

2.2. *Noise-Immune, Recoil-Sensitive, Large-Area Atom Interferometers*

We have created a pair of simultaneous conjugate RB atom interferometers, see Fig. 5, left. Their sensitivity towards the photon recoil is similar, but the one towards inertial forces is reversed. That allows us to cancel the influence of gravity and, with simultaneous operation, noise.

Cancellation of vibrations between similar interferometers at separate locations has been demonstrated before.[9] In some important applications, however, the interferometers must be dissimilar so that a large differential signal can be picked up. Here, we present a method to cancel vibrational noise between dissimilar interferometers, with LMT beam splitters, see Fig. 5.

The cancellation of vibrations is based on the simultaneous application of the beam splitters for the conjugate interferometers. Our experimental setup is optimized to provide the laser radiation needed with an extremely tight phase relationship; any vibrationally-induced phase shifts are thus

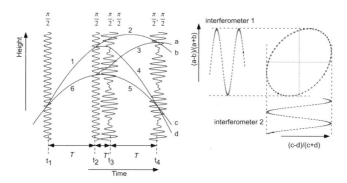

Fig. 5.   Correlating the fringes of two interferometers creates an ellipse whose shape (eccentricity and major axis) allows to determine the relative phase.

common mode and can be taken out in an ellipse-fitting analysis of the correlation. At short pulse separation times of 1 ms, a contrast of around 25-31% is achieved at momentum transfers between $(8-20)\hbar k$, see Fig. 6 for examples. This should be compared to the theoretical contrast of 50%. Also, it is evident that the strong dependence of the contrast upon the momentum transfer, observed in previous LMT interferometers,[22] is absent.

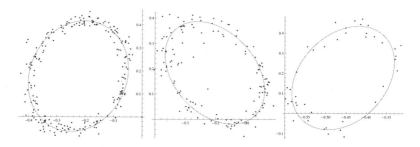

Fig. 6.   A: $12\hbar k$, 1ms, 25% contrast. B: $14\hbar k$, 1ms, 25% contrast; C: $20\hbar k$, 1ms, 27%

For $20\hbar k$ interferometers, about 10% contrast can be obtained at $T = 50\,\text{ms}$. Without simultaneous conjugate interferometers (SCIs), this is only possible at $T = 1\,\text{ms}$,[22] so the use of SCIs allows us to improve the pulse separation time T to 50 ms from 1 ms, without loss of contrast. This corresponds to a 2,500-fold increase in the enclosed area. At 70 ms, a contrast of 4.1% is still observable. and paves the path towards enhanced sensitivity in many cutting-edge applications. Examples include improved measurements of the photon recoil and the fine structure constant[13,14,23] and tests of the equivalence principle.[16]

To further confirm the applicability of our method, we have taken 15,000 pairs of data for a $10\hbar k$ interferometer with a pulse separation time of 50 ms over a 12-h period, see Fig. 7. By ellipse-specific fitting, we extract the differential phase to a resolution of 6.6 ppb. This is also the resolution to which the interferometers can determine $\hbar/M$; correspondingly, they are sensitive to the fine structure constant α via $\alpha^2 = (2R_\infty/c)(M/m_e)(h/M)$ to a resolution of 3.3 ppb.

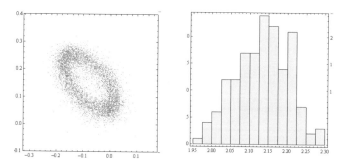

Fig. 7.   Left: 9958 Data pairs out of 15,000 that were taken during a 12-h session. Right: Histogram showing the distribution of ellipse fitting results.

2.3. *Very large area atom interferometers by differential optical acceleration*

The Bragg diffraction beam splitters used for LMT so far require extremely large laser power for increased momentum transfer. Even using our injection-locked 6.2 W Ti:sapph laser, which, we believe, is the strongest laser at a wavelength of 852 nm, we are limited to $20\hbar k$ for a reasonable contrast of the interference fringes. To increase the diffraction order, a further increase of the laser power would be required, which seems hard to achieve.

Adiabatic transfer[13] or Bloch oscillations of matter waves in an accelerated optical lattice[14] can been used to transfer a thousand $\hbar k$, but this affects the common momentum of the arms, not the splitting. Here, we have developed a method that can increase the momentum transfer without being limited by the laser power.

To do so, we have first demonstrated the differential acceleration of atomic samples by Bloch oscillations in two superimposed optical lattices.

Decoupling of the samples is due to their initial momentum separation, provided by a 4th order Bragg diffraction. A Bloch oscillation — Bragg diffraction — Bloch oscillation sequence forms a "BBB" beam splitter. Four BBB splitters make a RB atom interferometer, see Fig. 8. Two of them, running simultaneously to reject noise and systematic effects, show 15% contrast at 24-photon-momentum splitting each, see Fig. 9.

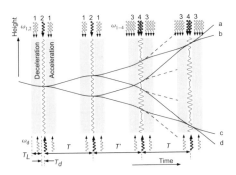

Fig. 8. Space-Time diagram of simultaneous conjugate Ramsey-Bordé BBB-Interferometers. 1: Dual optical lattice; 2: Single Bragg beam splitter; 3: Quadruple optical lattice; 4: Dual Bragg beam splitter; a-d: outputs. The dashed lines indicate trajectories that do not interfere.

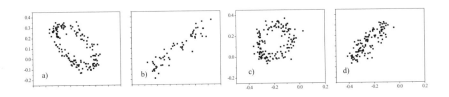

Fig. 9. Ellipses from simultaneous conjugate interferometers with Bragg-Bloch beam splitters. The x and y axes show the normalized flourescence of the upper and lower interferometer. (a) $\Delta p = 12\hbar k, C = 16.5\%$. (b) $\Delta p = 18\hbar k, C = 20.3\%$; (c) $\Delta p = 20\hbar k, C = 16.9\%$, (d) $\Delta p = 24\hbar k, C = 15.1\%$.

2.4. Towards fundamental physics measurements by atom interferometry

Taken together, the advances we just reported allow for a tremendous increase in the sensitivity of atom interferometers. We will soon apply them for a measurement of the fine structure constant α at the part per billion level of accuracy. By comparison to the value derived from the electron's anomalous gyromagnetic moment $g - 2$,[24] this will correspond to testing the theory of quantum electrodynamics at the highest precision ever. The influence of muons and hadrons on $g - 2$ will be revealed for the first time. Moreover, this measurement would provide a limit on low-energy dark matter candidates or supersymmetric particles, and serve as a probe for the internal structure of the electron. Indeed, a measurement to 0.1 ppb would correspond to a search for physics beyond the standard model on the TeV energy scale.

References

1. M. R. Matthews, B. P. Anderson, P. C. Haljan, D. S. Hall, C. E. Wieman and E. A. Cornell, *Phys. Rev. Lett.* **83**, 2498 (1999).
2. K. W. Madison, F. Chevy, W. Wohlleben and J. Dalibard, *Phys. Rev. Lett.* **84**, 806 (2000).
3. V. Schweikhard, I. Coddington, *et al.*, *Phys. Rev. Lett.* **92**, 040404-4 (2004).
4. N. K. Wilkin and J. M. F. Gunn, *Phys. Rev. Lett.* **84**, 6–9 (2000).
5. B. Paredes and P. Fedichev, *et al.*, *Phys. Rev. Lett.* **87**, 010402 (2001).
6. M. Popp and B. Paredes, *et al.*, *Phys. Rev. A* **70**, 053612 (2004).
7. A. D. Cronin, J. Schmiedmayer and D. E. Pritchard, arXiv:0712.3703v1.
8. A. Peters, K. Y. Chung and S. Chu, *Nature (London)* **400**, 849 (1999).
9. M. J. Snadden *et al.*, *Phys. Rev. Lett.* **81**, 971 (1998).
10. T. L. Gustavson, A. Landragin and M. A. Kasevich, *Class. Quantum Gravity* **17**, 2385 (2000).
11. J. B. Fixler *et al.*, *Science* **315**, 74 (2007).
12. G. Lamporesi *et al.*, *Phys. Rev. Lett.* **100**, 050801 (2008).
13. A. Wicht *et al.*, *Physica Scripta* **T102**, 82 (2002).
14. P. Cladé *et al.*, *Phys. Rev. Lett.* **96**, 033001 (2006); *Phys. Rev. A* **74**, 052109 (2006).
15. H. Müller *et al.*, *Phys. Rev. Lett.* **100**, 031101 (2008).
16. S. Dimopoulos *et al.*, *Phys. Rev. Lett.* **98**, 111102 (2007).
17. J. M. McGuirk, M. J. Snadden and M. A. Kasevich, *Phys. Rev. Lett.* **85**, 4498 (2000).
18. S. Gupta *et al.*, *Phys. Rev. Lett.* **89**, 140401 (2002).
19. D. M. Giltner, R. W. McGowan and S. A. Lee, *Phys. Rev. Lett.* **75**, 2638 (1995); *Phys. Rev. A* **52**, 3966 (1995).
20. J. Hecker Denschlag, *et al.*, *J. Phys. B* **35**, 3095 (2002).
21. H. Müller, S.-W. Chiow and S. Chu, *Phys. Rev. A* **77**, 023609 (2008).
22. H. Müller *et al.*, *Phys. Rev. Lett.* **100**, 180405 (2008).
23. H. Müller *et al.*, *Appl. Phys. B* **84**, 633 (2006).
24. G. Gabrielse *et al.*, *Phys. Rev. Lett.* **97**, 030802 (2006); *ibid.* **99**, 039902 (2007).

MORE ACCURATE MEASUREMENT OF THE ELECTRON MAGNETIC MOMENT AND THE FINE STRUCTURE CONSTANT

D. HANNEKE, S. FOGWELL, N. GUISE, J. DORR and G. GABRIELSE

Department of Physics, Harvard University,
Cambridge, MA 02138, USA

A measurement reported in 2008 uses a one-electron quantum cyclotron to determine the electron magnetic moment in Bohr magnetons, $g/2 = 1.001\,159\,652\,180\,73\,(28)\,[0.28\,\text{ppt}]$, with an uncertainty 2.7 and 15 times smaller than for previous measurements in 2006 and 1987. The electron is used as a magnetometer to allow lineshape statistics to accumulate, and its spontaneous emission rate determines the correction for its interaction with a cylindrical trap cavity. The new measurement and QED theory determine the fine structure constant, with $\alpha^{-1} = 137.035\,999\,084\,(51)\,[0.37\,\text{ppb}]$, and an uncertainty 20 times smaller than for any independent determination of α.

Keywords: Electron magnetic moment, electron g value, fine structure constant, quantum cyclotron.

1. New Measurement of the Electron $g/2$

A 2008 measurement[1] of the electron magnetic moment $\boldsymbol{\mu}$ determines $g/2$, which is the magnitude of $\boldsymbol{\mu}$ scaled by the Bohr magneton, $\mu_B = e\hbar/(2m)$. For an eigenstate of spin $\mathbf{S}$,

$$\boldsymbol{\mu} = -\frac{g}{2}\,\mu_B\,\frac{\mathbf{S}}{\hbar/2}. \tag{1}$$

This is one of the few measurable properties of one of the simplest of elementary particles – quantifying its interaction with the fluctuating QED vacuum, and probing for electron size or composite structure that has not yet been detected. For a point electron in the simplest renormalizable Dirac description, $g/2 = 1$. QED predicts that vacuum fluctuations and polarization slightly increase this value. Physics beyond the standard model of particle physics could make $g/2$ deviate from the Dirac/QED prediction, as internal quark-gluon substructure does for a proton.

The 1987 measurement that provided the accepted $g/2$ for nearly 20 years[2] was superceded in 2006 by a measurement that used a one-electron quantum cyclotron.[3] Key elements that made the measurement possible included quantum jump spectroscopy and quantum non-demolition (QND) measurements of the lowest cyclotron and spin levels,[4] a cylindrical Penning trap cavity[5] (Fig. 2(a)), inhibited spontaneous emission,[6] and a one-particle self-excited oscillator (SEO).[7] The 2008 measurement[1] has an uncertainty that is 2.7 and 15 times lower than the 2006 and 1987 measurements, respectively, and confirms a 1.8 standard deviation shift from the 1987 value (Fig. 1(a)).

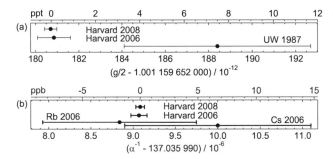

Fig. 1. (a) Most accurate measurements of the electron $g/2$, and (b) most accurate determinations of α.

2. Most Accurate Determination of α

The fine structure constant, $\alpha = e^2/(4\pi\epsilon_0\hbar c)$, is the fundamental measure of the strength of the electromagnetic interaction in the low energy limit. The fine structure constant is also a crucial ingredient of our system of fundamental constants.[8]

The new measurement of the electron $g/2$, with the help of recently updated QED theory,[9] determines α with an uncertainty nearly 20 times smaller than does any independent method (Fig. 1(b)). The uncertainty in α is now limited a bit more by the need for a higher-order QED calculation (underway[9]) than by the measurement uncertainty in $g/2$.

The standard model relates g and α by

$$\frac{g}{2} = 1 + C_2\left(\frac{\alpha}{\pi}\right) + C_4\left(\frac{\alpha}{\pi}\right)^2 + C_6\left(\frac{\alpha}{\pi}\right)^3 + C_8\left(\frac{\alpha}{\pi}\right)^4$$
$$+ C_{10}\left(\frac{\alpha}{\pi}\right)^5 + \ldots + a_{\mu\tau} + a_{\text{hadronic}} + a_{\text{weak}}, \tag{1}$$

with the asymptotic series and $a_{\mu\tau}$ coming from QED. Unambiguously prescribed QED calculations (recently summarized[10]) give exact C_2, C_4 and C_6 (all checked numerically), along with a numerical value and uncertainty for C_8, and a small $a_{\mu\tau}$. Very small hadronic and weak contributions are included, along with the assumption that there is no significant modification from electron substructure or other physics beyond the standard model.

The fine structure constant is determined by solving Eq. 1 for α in terms of the measured electron $g/2$:

$$\alpha^{-1} = 137.035\,999\,084\,(33)\,(39) \quad [0.24\text{ ppb}]\,[0.28\text{ ppb}],$$
$$= 137.035\,999\,084\,(51) \qquad [0.37\text{ ppb}]. \qquad (2)$$

The first line shows experimental (first) and theoretical (second) uncertainties that are nearly the same. The theory uncertainty contribution to α is divided as (12) and (37) for C_8 and C_{10}. It should decrease when a calculation underway[9] replaces the crude estimate $C_{10} = 0.0\,(4.6)$.[8,10] The α^{-1} of Eq. 2 will then shift by $2\alpha^3\pi^{-4}C_{10}$, which is $8.0\,C_{10} \times 10^{-9}$. A small change Δ_8 in the calculated $C_8 = -1.9144\,(35)$ would add $2\alpha^2\pi^{-3}\Delta_8$.

The total 0.37 ppb uncertainty in α is nearly 20 times smaller than for the next most precise independent methods (Fig. 1(b)). These so-called atom recoil methods[11,12] utilize measurements of transition frequencies and mass ratios, as well as either a Rb recoil velocity (in an optical lattice) or a Cs recoil velocity (in an atom interferometer). (A report in these proceedings may slightly decrease the reported uncertainty in the Rb measurement.)

3. Other Applications of the New Measurement

The accuracy of the new $g/2$ sets the stage for an improved CPT test with leptons. With a one-positron quantum cyclotron we hope to measure the positron $g/2$ at the same level of accuracy as we did for the electron. The goal is a CPT test with leptons that is much more than an order of magnitude more precise than any other.

Already the most precise test of QED comes from comparing our measured $g/2$ to what can be calculated using Eq. 1 using α from the atom recoil measurements.[10] The accuracy of the QED test is limited almost entirely by the uncertainties in the atom recoils measurements, and not by the much smaller uncertainties in the measured $g/2$ and the QED theory calculation.

Finally, a report[13] suggests that the the accurately measured electron $g/2$ will make possible the discovery of low-mass dark-matter particles, or will exclude of this possibility. An improved sensitivity requires the new $g/2$ along with a better independent measurement of α.

4. One Electron Quantum Cyclotron

Fig. 2(b) represents the lowest cyclotron and spin energy levels for an electron weakly confined in a vertical magnetic field $B\hat{z}$ and an electrostatic quadrupole potential. The latter is produced by biasing the trap electrodes of Fig. 2(a). The measured $g/2$ value is primarily determined by the cyclotron frequency $\bar{f}_c \approx 149$ GHz (blue in Fig. 2(b)) and the measured anomaly frequency $\bar{\nu}_a \approx 173$ MHz (red in Fig. 2(b)),[3]

$$\frac{g}{2} \simeq 1 + \frac{\bar{\nu}_a - \bar{\nu}_z^2/(2\bar{f}_c)}{\bar{f}_c + 3\delta/2 + \bar{\nu}_z^2/(2\bar{f}_c)} + \frac{\Delta g_{cav}}{2}. \tag{1}$$

Small adjustments are needed for the measured axial frequency, $\bar{\nu}_z \approx 200$ MHz, and for the relativistic shift, $\delta/\nu_c \equiv h\nu_c/(mc^2) \approx 10^{-9}$. A cavity shift $\Delta g_{cav}/2$ is the fractional shift of the cyclotron frequency caused by the interaction with radiation modes of the trap cavity. Small terms of higher order in $\bar{\nu}_z/\bar{f}_c$ are neglected. The Brown-Gabrielse invariance theorem[14] has been used to eliminate from Eq. 1 the effect of the lowest order imperfections of a real trap – quadratic distortions of the electrostatic potential and misalignments of the trap electrode axis with **B**.

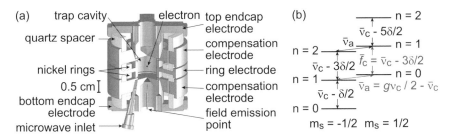

Fig. 2.   (a) Cylindrical Penning trap cavity used to confine a single electron and inhibit spontaneous emission, and (b) the cyclotron and spin levels of an electron confined within it.

Quantum jump spectroscopy determines $\bar{f}_c$ and $\bar{\nu}_a$. For each of many trials the system is prepared in the spin-up ground state, $|n = 0, m_s = 1/2\rangle$, after which the preparation drives and detection amplifier are turned off for 1 s. Either a cyclotron drive at a frequency near to $\bar{f}_c$, or an anomaly drive at frequency near $\bar{\nu}_a$, is then applied for 2 s. The amplifier and a feedback system are turned on to provide QND detection of either a one-quantum cyclotron excitation or a spin flip. Cavity-inhibited spontaneous emission makes the cyclotron excitation persist long enough to allow such detection. The fraction of the excitation trials resulting in excitations is measured.

The cyclotron drive is microwave field injected into the trap cavity through a cold attenuator that keeps black body photons from entering the trap. The anomaly drive is an oscillatory potential applied to electrodes at frequencies near $\bar{\nu}_a$ to drive off-resonant axial motion through the magnetic bottle gradient from two nickel rings (Fig. 2(a)). The electron, radially distributed as a cyclotron energy eigenstate, sees an oscillating magnetic field perpendicular to **B** as needed to flip its spin, with a gradient that allows a simultaneous cyclotron transition.[15] To ensure that the electron samples the same magnetic variations while $\bar{\nu}_a$ and $\bar{f}_c$ transitions are driven, both drives are kept on with one detuned slightly so that only the other causes transitions. Low drive strengths keep transition probabilities below 20% to avoid saturation effects.

QND detection of one-quantum changes in the cyclotron and spin energies takes place because the magnetic bottle shifts the oscillation frequency of the self-excited axial oscillation as $\Delta\bar{\nu}_z \approx 4\,(n + m_s)$ Hz. After a cyclotron excitation, cavity-inhibited spontaneous emission provides the time needed to turn on the electronic amplification and feedback, so the SEO can reach an oscillation amplitude at which the shift can be detected.[7] An anomaly transition is followed by a spontaneous decay to the spin-down ground state, $|n = 0, m_s = -1/2\rangle$, and the QND detection reveals the lowered spin energy.

5. Uncertainties and Corrections

Expected asymmetric lineshapes arise from the thermal axial motion of the electron through the magnetic bottle gradient. The axial motion is cooled by a resonant circuit in about 0.2 s to as low as $T_z = 230$ mK (from 5 K) when the detection amplifier is off. For the cyclotron motion these fluctuations are slow enough that the lineshape is essentially a Boltzmann distribution with a width proportional to T_z.[16] For the anomaly resonance, the fluctuations are effectively more rapid, leading to a resonance shifted in proportion to T_z.

The weighted averages of $\bar{\nu}_a$ and $\bar{f}_c$ from the lineshapes determine $g/2$ via Eq. 1. With saturation effects avoided, these pertain to the magnetic field averaged over the thermal motion. It is crucial that any additional fluctuations in B that are symmetric about a central value will broaden such lineshapes without changing the mean frequency.

To test this weighted mean method we compare maximum likelihood fits to lineshape models. The data fit well to a convolution of a Gaussian resolution function and a thermal-axial-motion lineshape.[16] The broadening may arise from vibrations of the trap and electron through the slightly inhomogeneous field of the external solenoid, or from fluctuations of the solenoid field itself. Because we have not yet identified its source we add a "lineshape" uncertainty based upon the discrepancy (beyond statistical uncertainty) between the $g/2$ values from the mean and fit for the four measurements. To be cautious we take the minimum discrepancy as a correlated uncertainty, and then add the rest as an uncorrelated uncertainty. An additional probe of the broadening comes from slowly increasing the microwave frequency until a one-quantum cyclotron excitation is seen. The distribution of excitations is consistent with the Gaussian resolution functions determined from the fits.

Drifts of B are reduced below 10^{-9}/hr by regulating five He and N_2 pressures in the solenoid and experiment cryostats, and the surrounding air temperature.[3] Remaining slow B drift is corrected based upon lineshapes measured once every three hours. Unlike the one-night-at-a-time analysis used in 2006, all data taken in four narrow ranges of B values (Table 1) are combined, giving a lineshape signal-to-noise that allows the systematic investigation of lineshape uncertainty.

Better measurement and understanding of the electron-cavity interaction removes cavity shifts as a major uncertainty. Cavity shifts are the downside of the cavity-inhibited spontaneous emission which usefully narrows resonance lines and gives the averaging time we need to turn on the SEO and determine the cyclotron state. The shifts arise when the cyclotron oscillator has its frequency pulled by the coupling to nearby radiation modes of the cavity. The cylindrical trap cavity was invented[5] and developed[17] to deliberately modify the density of states of the free space radiation modes in a controllable and understandable way (though not enough to require modified QED calculations[18]). Radiation mode frequencies must still be measured to determine the effective dimensions of a right-circular cylindrical cavity which has been imperfectly machined, which has been slit (so sections of the cavity can be separately biased trap electrodes), and whose dimensions change as the electrodes cool from 300 to 0.1 K.

To the synchronized-electrons method used in 2006, the 2008 measurement also adds a new method – using the electron itself to determine the

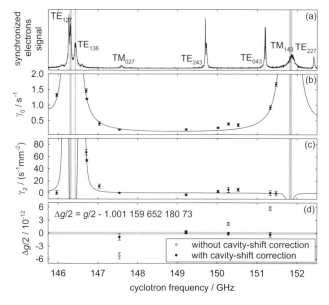

Fig. 3.   (a) Modes of the trap cavity are observed with synchronized electrons,[3] as well as (b) with a single electron damping rate γ_0 and (c) its amplitude dependence γ_2. (d) Offset of $g/2$ from our result in Eq. 1 without (open circle) and with (points) cavity-shift corrections, with an uncertainty band for the average.

cavity-electron interaction. The measured spontaneous emission rate for its cyclotron motion, $\gamma = \gamma_0 + \gamma_2 A^2$, depends upon the amplitude A of the axial oscillation through the standing waves of cavity radiation modes. The amplitude is varied by adjusting the SEO,[7] and it is measured by fitting to a cyclotron quantum-jump lineshape.[7,16] Fits of γ_0 and γ_2 (Fig. 3(b)–(c)) to a renormalized calculation of the coupling of the electron and cavity[19] determine the frequencies (with uncertainties represented by the vertical gray bands in Fig. 3(a)–(c)) and Q values of the nearest cavity modes, and the cavity-shift corrections for $g/2$ (Table 1). (Subtleties in applying this calculation to measurements will be reported.) Substantially different cavity-shift corrections bring the four $g/2$ measurements into good agreement (Fig. 3(d)).

6. Results

The measured values, shifts, and uncertainties for the four separate measurements of $g/2$ are in Table 1. The uncertainties are lower for measurements with smaller cavity shifts and smaller linewidths, as might be

Table 1. Measurements and shifts with uncertainties multiplied by 10^{12}. The cavity-shifted "$g/2$ raw" and corrected "$g/2$" are offset from our result in Eq. 1.

f_c	147.5 GHz	149.2 GHz	150.3 GHz	151.3 GHz
$g/2$ raw	-5.24 (0.39)	0.31 (0.17)	2.17 (0.17)	5.70 (0.24)
Cav. shift	4.36 (0.13)	-0.16 (0.06)	-2.25 (0.07)	-6.02 (0.28)
Lineshape				
correlated	(0.24)	(0.24)	(0.24)	(0.24)
uncorrelated	(0.56)	(0.00)	(0.15)	(0.30)
$g/2$	-0.88 (0.73)	0.15 (0.30)	-0.08 (0.34)	-0.32 (0.53)

expected. Uncertainties for variations of the power of the $\bar{\nu}_a$ and $\bar{f}_c$ drives are estimated to be too small to show up in the table. A weighted average of the four measurements, with uncorrelated and correlated errors combined appropriately, gives the electron magnetic moment in Bohr magnetons,

$$g/2 = 1.001\,159\,652\,180\,73\,(28) \quad [0.28 \text{ ppt}]. \quad (1)$$

The uncertainty is 2.7 and 15 times smaller than the 2006 and 1987 measurements, and 2300 times smaller than has been achieved for the heavier muon lepton.[20]

Items that warrant further study could lead to a future measurement of $g/2$ to higher precision. First is the broadening of the expected lineshapes which limits the splitting of the resonance lines. Second, a variation in measured axial temperatures, not understood, increases the uncertainty contributed by the wider lineshapes. Third, cavity sideband cooling could cool the axial motion to near its quantum ground state for a more controlled measurement. Fourth, a new apparatus should be much less sensitive to vibration and other variations in the laboratory environment.

7. Self-Excited Proton

The self-excited one-electron oscillator was a crucial ingredient of accurate measurements of the electron $g/2$. Fig. 4 shows one of the first electrical signals detected from a self-excited single proton. Our hope is to improve the sensitivity of this oscillator until non-destructive spin flips of a single trapped proton can be observed as a way to measure g for a proton, and then for an antiproton. If this approach is successful it may be possible to improve the accuracy with which the magnetic moment of the antiproton is measured by a factor of a million or more. An proton/antiproton spin flip is much harder to observe than that of an electron/positron because a nuclear magneton is 2000 times smaller than a Bohr magneton.

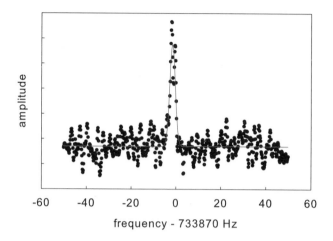

Fig. 4.   First example of the signal from a self-excited proton oscillator.

8. Directly Driven Electron Spin Flip

The two electron spin states could potentially be a very high fidelity q-bit. As one small step, Fig. 5 shows the first line shape for a electron driven directly near its spin frequency, rather than at the difference between the spin and cyclotron frequencies.

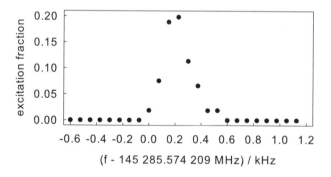

Fig. 5.   First example of the lineshape for electron spin flips driven directly at the spin frequency.

9. Conclusion and Acknowledgments

In conclusion, a new measurement of the electron $g/2$ is 15 times more accurate than the 1987 measurement that provided $g/2$ and α for nearly 20 years, and 2.7 times more accurate than the 2006 measurement that superseded it. Achieving the reported electron $g/2$ uncertainty with a positron seems feasible, and would make the most stringent lepton CPT test. With QED and the assumption of no new physics beyond the standard model of particle physics, the new measurement determines α almost 20 times more accurately than any independent method. The measured $g/2$ is accurate enough to allow testing QED, probing for electron size, and searching for a low mass dark matter particle if a more accurate independent measurement of α is realized.

More details are in the thesis of D. Hanneke, and being readied for publication. This work was supported by the NSF AMO program.

References

1. D. Hanneke, S. Fogwell and G. Gabrielse, *Phys. Rev. Lett.* **100**, 120801 (2008).
2. R. S. Van Dyck, Jr., P. B. Schwinberg and H. G. Dehmelt, *Phys. Rev. Lett.* **59**, 26 (1987).
3. B. Odom, D. Hanneke, B. D'Urso and G. Gabrielse, *Phys. Rev. Lett.* **97**, 030801 (2006).
4. S. Peil and G. Gabrielse, *Phys. Rev. Lett.* **83**, 1287 (1999).
5. G. Gabrielse and F. C. MacKintosh, *Intl. J. Mass Spec. Ion Proc.* **57**, 1 (1984).
6. G. Gabrielse and H. Dehmelt, *Phys. Rev. Lett.* **55**, 67 (1985).
7. B. D'Urso, R. Van Handel, B. Odom, D. Hanneke and G. Gabrielse, *Phys. Rev. Lett.* **94**, 113002 (2005).
8. P. J. Mohr and B. N. Taylor, *Rev. Mod. Phys.* **77**, 1 (2005).
9. T. Aoyama, M. Hayakawa, T. Kinoshita and M. Nio, *Phys. Rev. Lett.* **99**, 110406 (2007).
10. G. Gabrielse, D. Hanneke, T. Kinoshita, M. Nio and B. Odom, *Phys. Rev. Lett.* **97**, 030802 (2006), *ibid.* **99**, 039902 (2007).
11. P. Cladé, E. de Mirandes, M. Cadoret, S. Guellati-Khélifa, C. Schwob, F. Nez, L. Julien and F. Biraben, *Phys. Rev. A* **74**, 052109 (2006).
12. V. Gerginov, K. Calkins, C. E. Tanner, J. J. McFerran, S. Diddams, A. Bartels and L. Hollberg, *Phys. Rev. A* **73**, 032504 (2006).
13. C. Boehm and J. Silk, *Phys. Lett. B* **661**, 287 (2008).
14. L. S. Brown and G. Gabrielse, *Phys. Rev. A* **25**, 2423 (1982).
15. F. L. Palmer, *Phys. Rev. A* **47**, 2610 (1993).
16. L. S. Brown, *Ann. Phys. (NY)* **159**, 62 (1985).
17. J. N. Tan and G. Gabrielse, *Appl. Phys. Lett.* **55**, 2144 (1989).
18. D. G. Boulware, L. S. Brown and T. Lee, *Phys. Rev. D* **32**, 729 (1985).
19. L. S. Brown, G. Gabrielse, K. Helmerson and J. Tan, *Phys. Rev. Lett.* **55**, 44 (1985).
20. G. W. Bennett *et al.*, *Phys. Rev. D* **73**, 072003 (2006).

DETERMINATION OF THE FINE STRUCTURE CONSTANT WITH ATOM INTERFEROMETRY AND BLOCH OSCILLATIONS

M. CADORET, E. de MIRANDES, P. CLADÉ, C. SCHWOB, F. NEZ, L. JULIEN*
and F. BIRABEN

*Laboratoire Kastler Brossel, ENS, CNRS, UPMC, 4 place Jussieu,
75252 Paris CEDEX 05, France
*E-mail: biraben@spectro.jussieu.fr
www.spectro.jussieu.fr*

S. GUELLATI-KHÉLIFA

*Institut National de Métrologie, Conservatoire National des Arts et Métiers,
61 rue Landy, 93210 La Plaine Saint Denis, France*

We use Bloch oscillations to coherently transfer many photon momenta to atoms. Then we can measure accurately the recoil velocity $\hbar k/m$ and deduce the fine structure constant α. The velocity variation due to Bloch oscillations is measured using atom interferometry. This method yields a value of the fine structure constant $\alpha^{-1} = 137.035\,999\,45\,(62)$ with a relative uncertainty of about 4.5×10^{-9}.

Keywords: Fundamental constants, fine structure constant, Bloch oscillations, atom interferometry.

1. Introduction

The fine structure constant α is the fundamental physical constant characterizing the strength of the electromagnetic interaction. It is a dimensionless quantity, i.e. independent of the system of units used. It is defined as:

$$\alpha = \frac{e^2}{4\pi\epsilon_0\hbar c} \tag{1}$$

where ϵ_0 is the permittivity of vacuum, c is the speed of light, e is the electron charge and $\hbar = h/2\pi$ is the reduced Planck constant. The fine structure constant is a key part of the adjustment of the fundamental physical constants.[1,2] The different measurements of α are shown on Fig. 1. These values are obtained from experiments in different domains of physics, such as the quantum Hall effect and Josephson effect in solid state physics, or the measurement of the muonium hyperfine structure in atomic physics. The most precise determinations of α are deduced from the measurements of the electron anomaly a_e made in the eighties at the University of Washington[3] and recently at Harvard.[4-6] This last experiment and an impressive improvement of QED calculations[7,8] have lead to a new determination of α with a relative uncertainty of 3.7×10^{-10}. Nevertheless this last determination of α relies on very difficult QED calculations. To test it, other determinations of α are required, such as the values deduced from the measurements of h/m_{Cs}[9] and h/m_{Rb} (m_{Cs} and m_{Rb} are the masses of Cesium and Rubidium atoms) which are also indicated in Figure 1. In this paper, we present the measurements of h/m_{Rb} made in Paris in 2005 and 2008.

The principle of our experiment is the measurement of the recoil velocity v_r of a Rubidium atom absorbing or emitting a photon ($v_r = \hbar k/m$, where k is the wave vector of the photon absorbed by the atom of mass m). As the relative atomic masses A_r are measured very precisely, the measurement of h/m_{Rb} is a way to accurately determine α via the Rydberg constant R_∞:

$$\alpha^2 = \frac{2R_\infty}{c} \frac{A_r(\text{Rb})}{A_r(e)} \frac{h}{m_{Rb}} \tag{2}$$

In this equation, the relative atomic mass of the electron $A_r(e)$ and the Rubidium $A_r(\text{Rb})$ are known with the relative uncertainties of 4.4×10^{-10} and 2.0×10^{-10}, respectively.[10,11] As the fractional uncertainty of R_∞ is 7×10^{-12},[12,13] the factor limiting the accuracy of α is the ratio h/m_{Rb}.

2. Principle of the experiment

The principle of the experiment is to coherently transfer as many recoils as possible to the atoms (i.e. to accelerate them) and to measure the final velocity distribution. In our experiment, the atoms are efficiently accelerated by means of N Bloch oscillations (BO). The velocity selection and velocity measurement are done with Raman transitions. The experiment develops in three steps. i) Firstly, we select from a cold atomic cloud of ^{87}Rb a bunch of atoms with a very narrow velocity distribution. This selection is performed by a Doppler velocity sensitive counter-propagating Raman transition. In 2005, we have used a π-pulse to transfer the atoms from the $F = 2$ to the $F = 1$ hyperfine level of ^{87}Rb. In 2008, we have modified the experimental scheme to take advantage of Ramsey spectroscopy: we use a pair of $\pi/2$ pulses which produces a fringe pattern in velocity space. ii) Secondly, we

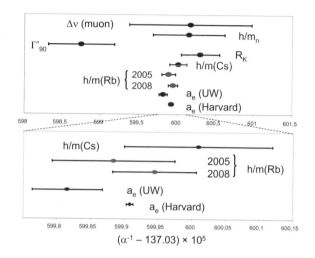

Fig. 1.   Determinations of the fine structure constant in different domains of physics. The most precise measurements are shown in the lower part of the figure. They are deduced from the anomaly of electron and from the ratios h/m_{Cs} and h/m_{Rb}. We have taken into account the most recent result of the group of Gabrielse.[6] The two values deduced for h/m_{Rb} are presented in this paper.

transfer to these selected atoms as many recoils as possible by means of Bloch oscillations as explained later. iii) Finally, we measure the final velocity of the atoms by a second Raman transition which transfers the atoms from the $F = 1$ to the $F = 2$ hyperfine level. In short, we have used two different pulse sequences, the $\pi - \mathrm{BO} - \pi$ and $\pi/2 - \pi/2 - \mathrm{BO} - \pi/2 - \pi/2$ configurations.

The accuracy of our measurement of the recoil velocity relies in the number of recoils $(2N)$ that we are able to transfer to the atoms. Indeed, if we measure the final velocity with an accuracy of σ_v, the accuracy on the recoil velocity measurement σ_{v_r} is:

$$\sigma_{v_r} = \frac{\sigma_v}{2N} \tag{3}$$

Bloch oscillations have been first observed in atomic physics by the groups of Salomon in Paris and Raizen in Austin.[14–16] In a simple way, Bloch oscillations can be seen as Raman transitions where the atom begins and ends in the same energy level, so that its internal state ($F = 1$ for $^{87}\mathrm{Rb}$) is unchanged while its velocity has increased by $2v_r$ per Bloch oscillation. This is illustrated on Fig. 2 which shows the atomic kinetic energy versus the atomic momentum. The velocity distribution obtained after the $\pi/2 - $

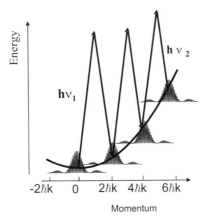

Fig. 2. Acceleration of cold atoms with a frequency chirped standing wave. The variation of energy versus momentum in the laboratory frame is given by a parabola. The energy of the atoms increases by the quantity $4(2j + 1)E_r$ in each cycle. The Ramsey fringe patterns represents the momenta distribution of the atoms in the $F = 1$ hyperfine level.

$\pi/2$ selection is also represented. Bloch oscillations are produced in a one dimensional optical lattice which is accelerated by linearly sweeping the relative frequency of two counter propagating laser beams (frequencies ν_1 and ν_2). The frequency difference $\Delta\nu$ is increased so that, because of the Doppler effect, the beams are periodically resonant with the atoms ($\Delta\nu = 4(2j + 1)E_r/h$, $j = 0, 1, 2, 3..$ where E_r/h is the recoil energy in frequency units and j the number of transitions). This leads to a succession of rapid adiabatic passages between momentum states differing by $2\hbar k$. In the solid-state physics approach, this phenomenon is known as Bloch oscillations in the fundamental energy band of a periodic optical potential. The atoms are subject to a constant inertial force obtained by the introduction of the tunable frequency difference $\Delta\nu$ between the two waves that create the optical potential.[14]

We now describe the acceleration process following the Bloch formalism. If, after the selection, the atom has a well defined momentum $\hbar q_0$ with $|q_0| < k$, the atomic wave function is modified when the optical potential is increased adiabatically (without acceleration) and becomes in the first energy band:

$$|\Psi_{0,q_0}\rangle = \sum_l \phi_0(q_0 + 2lk)|q_0 + 2lk\rangle \qquad (4)$$

with $l \in \mathbb{Z}$. Here $|q_0\rangle$ denotes the ket associated with a plane wave of momentum q_0 and the amplitudes ϕ_0 correspond to the Wannier function[17] in momentum space of the first band. When the potential depth is close to zero, the limit of the Wannier function ϕ_0 is 1 over the first Brillouin zone and zero outside. On the contrary, if the potential depth is large, the Wannier function selects several components in velocity space. When the optical lattice is accelerated adiabatically, the Wannier function is continuously shifted in momentum space following the relation:

$$|\Psi(t)\rangle = \sum_l \phi_0(q_0 + 2lk - mv(t)/\hbar)|q_0 + 2lk\rangle \qquad (5)$$

where $v(t)$ is the velocity of the optical lattice. The enveloping Wannier function ϕ_0 is shifted by $mv(t)$ in momentum space. After the acceleration, the potential depth is decreased adiabatically and, in equation 5, the Wannier function selects only one component of the velocity distribution. At the end, the wave function is $|\Psi\rangle = |q_0 + 2Nk\rangle$. If Δv is the velocity variation due to the acceleration, the number of Bloch oscillations N is such that $|\hbar q_0 + m\Delta v - 2N\hbar k| < \hbar k$. Consequently, if the initial atomic velocity distribution fits within the first Brillouin zone, it is exactly shifted by $2Nv_r$ without deformation, as shown in Fig. 2 for the velocity distribution produced by a pair of $\pi/2$ pulses.

3. Results in the $\pi - \mathrm{BO} - \pi$ configuration

Our experimental setup has been previously described in detail.[18,19] Briefly, we use a magneto-optical trap (MOT) and an optical molasses to cool the atoms to about 3 μK. The determination of the velocity distribution is performed using a $\pi - \pi$ pulse sequence of two vertical counter-propagating laser beams (Raman beams):[20] the first pulse with a fixed frequency δ_{sel}, transfers atoms from the $5S_{1/2}$, $|F = 2, m_F = 0\rangle$ state to the $5S_{1/2}$, $|F = 1, m_F = 0\rangle$ state, into a narrow velocity class (width of about $v_r/15$). Then a laser beam resonant with the $5S_{1/2}$ $(F = 2)$ to $5P_{3/2}$ $(F = 3)$ cycling transition pushes away the atoms remaining in the ground state $F = 2$. Atoms in the state $F = 1$ make N Bloch oscillations in a vertical accelerated optical lattice. We then perform the final velocity measurement using the second Raman π-pulse, whose frequency is δ_{meas}. The populations of the $F = 1$ and $F = 2$ levels are measured separately by using a time of flight technique. To plot the final velocity distribution we repeat this procedure by scanning the Raman beam frequency δ_{meas} of the second pulse.

To avoid spontaneous emission and to reduce other stray effects (light

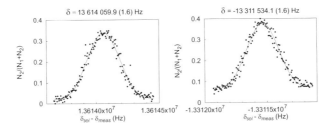

Fig. 3. Velocity spectra obtained in the $\pi - \mathrm{BO} - \pi$ configuration. Here N_1 and N_2 are respectively the number of atoms in $F = 1$ and $F = 2$ after the acceleration process. These two spectra are obtained by performing the Bloch acceleration upwards or downwards. The frequency difference between these spectra corresponds to 1780 recoil velocities.

shifts and refraction index), the Raman lasers and the optical lattice are blue detuned by ~ 1 THz and ~ 40 GHz respectively from the one photon transition. The delay between the two π-pulses is 12 ms and their duration 3.4 ms. The optical potential depth is 70 E_r. For an acceleration of about 2000 ms^{-2} we transfer about 900 recoil momenta in 3 ms with an efficiency greater than 99.97% per recoil. To avoid atoms from reaching the upper windows of the vacuum chamber, we use a double acceleration scheme: instead of selecting atoms at rest, we first accelerate them using Bloch oscillations and then we make the three step sequence: selection-acceleration-measurement. This way, the atomic velocity at the measurement step is close to zero. In order to eliminate the effect of gravity, we make a differential measurement by accelerating the atoms in opposite directions (up and down trajectories) keeping the same delay between the selection and measurement π-pulses. The ratio $\hbar/m$ can then be deduced from the formula:

$$\frac{\hbar}{m} = \frac{(\delta_{sel} - \delta_{meas})^{up} - (\delta_{sel} - \delta_{meas})^{down}}{2(N^{up} + N^{down})k_B(k_1 + k_2)} \tag{6}$$

where $(\delta_{meas} - \delta_{sel})^{up/down}$ corresponds respectively to the center of the final velocity distribution for the up and the down trajectories, $N^{up/down}$ are the number of Bloch oscillations in both opposite directions, k_B is the Bloch wave vector, and k_1 and k_2 are the wave vectors of the Raman beams. Moreover, the contribution of some systematic effects (energy level shifts) is inverted when the direction of the Raman beams are exchanged: for each up or down trajectory, the Raman beams directions are reversed and we record two velocity spectra. Finally, each determination of h/m_{Rb} and α is obtained from 4 velocity spectra. Fig. 3 shows two velocity spectra for the up and down trajectories.

The determinations of h/m_{Rb} and α have been derived from 72 experimental data point taken during four days. In these measurements, the number of Bloch oscillations were $N^{up} = 430$ and $N^{down} = 460$. Then, the effective recoil number is $2(N^{up} + N^{down})=1780$. The dispersion of these $n = 72$ measurements is $\chi^2/(n-1) = 1.3$ and the resulting statistical relative uncertainty in h/m_{Rb} is 8.8×10^{-9}. This corresponds to a relative statistical uncertainty in α of 4.4×10^{-9}. All systematic effects affecting the experimental measurement have been analyzed in detail in reference.[19] The total correction due to the systematic effects is $(10.98 \pm 10.0) \times 10^{-9}$ on the determination of h/m_{Rb}. With this correction, we obtain for α:

$$\alpha^{-1} = 137.035\,998\,84\,(91) \quad [6.7 \times 10^{-9}] \tag{7}$$

This value of the fine structure constant is labeled $h/m(\mathrm{Rb})2005$ in Fig. 1.

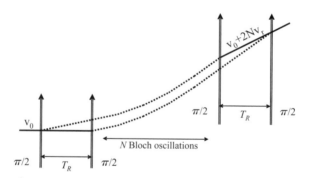

Fig. 4.   Scheme of the interferometer used for the measurement of h/m_{Rb}. The first pair of $\pi/2$ pulses produces a fringe pattern in the velocity distribution which is measured by the second pair of $\pi/2$ pulses. Between these two pairs of pulses, the atoms are accelerated upwards or downwards. The solid line corresponds to the atom in the $F = 2$ state, and the dashed line to the $F = 1$ state.

4. Measurement of the fine structure constant by atom interferometry

We describe in this section the results obtained in the $\pi/2-\pi/2-\mathrm{BO}-\pi/2-\pi/2$ configuration. The scheme of this interferometric method is shown in Fig. 4. The frequency resolution is now determined by the time T_R within each pair of pulses while the duration of each $\pi/2$ pulse determines the spectral width of the pulses and the number of atoms which contribute to the signal. This interferometer is similar to the one of reference,[9] except that effective Ramsey k-wavevectors point in the same direction. Consequently, this interferometer is not sensitive to the recoil energy, but only to the velocity variation due to Bloch oscillations which take place between the two sets of $\pi/2$ pulses.

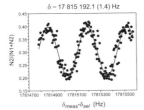 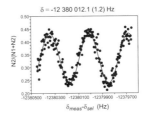

Fig. 5. Velocity spectra obtained in the $\pi/2 - \pi/2 - BO - \pi/2 - \pi/2$ configuration. Here N_1 and N_2 are respectively the number of atoms in $F = 1$ and $F = 2$ after the acceleration process. The spectrum on the left corresponds to the downwards acceleration (600 Bloch oscillations) and on the right to the up acceleration (400 Bloch oscillations). The frequency difference between these spectra corresponds 2000 recoil velocities.

As in the $\pi - BO - \pi$ configuration, a value of h/m_{Rb} is deduced from four spectra obtained with the upwards or downwards acceleration and by exchanging the directions of the Raman beams. An example of two spectra is shown in Fig. 5. In this case, the total number of Bloch oscillations is $N^{up} + N^{down} = 1000$, corresponding to 2000 recoil velocities between the up and down trajectories. The duration of each $\pi/2$ pulse is 400 μs and the time T_R is 2.6 ms (the total time of a pair of $\pi/2$ pulses is 3.4 ms). For these experiments, the blue detuning of the Raman lasers is 310 GHz. By comparison with the $\pi - BO - \pi$ configuration (see Fig. 3) the resolution is better: the half period of the fringes is about 160 Hz while the line width of the spectra of Fig. 3 was about 500 Hz. Nevertheless, the reduction of the uncertainties is not in the same ratio. This is due to the phase noise of the Raman laser which becomes more important. To lower this effect, we have set up an active anti vibration system: then the precision of each frequency measurement increases by about a factor of two. After this amelioration, to improve the resolution, it is tempting to increase the time T_R between the $\pi/2$ pulses. Nevertheless, another limitation appears, which is the size of the vacuum cell. Indeed, when we make the selection, the velocity of the atoms is close to $2Nv_r$ and, during the selection, the atom travels a distance of $2Nv_rT_R$. For example, in the case of the spectra of Fig. 5, the total displacement of the atoms is about 88 mm and 53 mm for the down and up trajectories (the upper window of the vacuum cell is 70 mm above the center of the cell): practically, with the parameters corresponding to the spectra of Fig. 5, we use all the size of the cell. Consequently, if we want to increase the delay T_R, we have to reduce the number N of Bloch oscillations and there is no benefit.

To surpass this limit, we have developed a method, called the *atomic elevator*, to better use the volume of the vacuum cell. The idea is to move the atoms to the top or the bottom of the cell before making the sequence described above. Then, we use the total size of the cell to accelerate and decelerate the atoms. To displace the atoms, we accelerate the atoms with 300 Bloch oscillations during 4 ms and, after a dead time of 13 ms, we decelerate the atoms with 300 Bloch oscillations. This sequence displaces the atoms by about 50 mm. With this technique, we have increased at the same time N to 800 and T_R to 5.7 ms. Fig. 6 shows two records obtained with these parameters. The visibility of the fringes is similar to the one of Fig. 5 and the half period of the fringe is about 90 Hz. Now the frequency difference between the two spectra corresponds to $3200v_r$. During the selection, the atomic velocity is about 10 m/s and the atom travels a distance close to 6 cm. This shows the limitation due to the size of the vacuum cell.

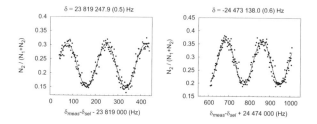

Fig. 6.   Velocity spectra obtained in the $\pi/2-\pi/2-\mathrm{BO}-\pi/2-\pi/2$ configuration with the atomic elevator. The spectrum on the left corresponds to the downwards acceleration (800 Bloch oscillations) and on the right to the upwards acceleration (800 Bloch oscillations). The frequency difference between these spectra corresponds 3200 recoil velocities.

We present now the result deduced from 221 measurements of h/m_{Rb}. For these measurements, we have used the two methods described previously, with and without the atomic elevator. The total number of Bloch oscillations $N^{up} + N^{down}$ varies from 200 to 1600. The dispersion of these $n = 221$ measurements is $\chi^2/(n-1) = 1.85$ and the resulting relative statistical uncertainty in α is 3×10^{-9}. The systematic effects are similar to

the ones described in Ref. 19. The two main effects are due to the geometry of the laser beams and to the second order Zeeman effect. To evaluate
the first effect, we have measured the wave front curvatures with a Shack-
Hartmann wave front analyzer (HASO 128 from Imagine Optics). From
these measurements, we have obtained a correction of $(-11.9 \pm 2.5) \times 10^{-9}$
in the determination of α. As explained above, the effect of parasitic level
shifts is eliminated by inverting the direction of the Raman beams. Nevertheless, this assumes that the measurements are made exactly at the same
position when the direction of the Raman beam is inverted. In fact, these
positions are not exactly the same because the directions of the recoils given
in the first Raman transition are opposite. For the timing used in our experiment, they differ by about $\delta_z = 300$ μm. We have precisely measured
the spatial magnetic field variations to control this effect. This correction
depends of the number of Bloch oscillations. For example, in the case of the
records of Fig. 6, its value is $(7 \pm 1) \times 10^{-9}$. Finally the relative uncertainty
due to the systematic effects is 3.4×10^{-9} and we obtain for α:

$$\alpha^{-1} = 137.035\,999\,45\,(62) \quad [4.5 \times 10^{-9}] \tag{8}$$

This value corresponds to the point labeled h/m(Rb)2008 in Fig. 1 and
is in agreement with our 2005 measurement. Our two results are also in
agreement with the most precise value deduced from the electron anomaly
(labeled a_e(Harvard) in Fig. 1).

5. Conclusion

We have presented two determinations of the fine structure constant α.
Depending on the Raman pulse arrangement ($\pi-$BO$-\pi$ or $\pi/2-\pi/2-$BO$-$
$\pi/2-\pi/2$ configurations), our experiment can run as an atom interferometer
or not. The comparison of the two resulting values, which are in good
agreement, provides an accurate test of these methods. The comparison
with the value extracted from the electron anomaly experiment[6] is either
a strong test of QED calculations or, assuming these calculations exact,
it gives a limit to test a possible internal structure of the electron. Our
goal is now to reduce the relative uncertainty of α. We are building a new
experimental setup with a larger vacuum chamber. With the new cell, we
plan to multiply the number of Bloch oscillations by a factor of three. Then,
it will be possible to reduce the uncertainty to the 10^{-9} level to obtain an
unprecedented test of the QED calculations.

Acknowledgments

This experiment is supported in part by the Laboratoire National de Métrologie et d'Essais (Ex. Bureau National de Métrologie) (Contract 033006), by IFRAF (Institut Francilien de Recherches sur les Atomes Froids) and by the Agence Nationale pour la Recherche, FISCOM Project-(ANR-06-BLAN-0192).

References

1. P. Mohr and B. N. Taylor, *Rev. Mod. Phys.* **77**, 1–107 (2005).
2. P. Mohr, B. N. Taylor and D. B. Newell, *Rev. Mod. Phys.* **80**, 633–730 (2008).
3. R.-S. Van Dick, P.-B. Schwinberg and H.-G. Dehmelt, *Phys. Rev. Lett.* **59**, 26–29 (1987).
4. B. Odom, D. Hanneke, B. D'Urso and G. Gabrielse, *Phys. Rev. Lett.* **97**, 030801 (2006).
5. G. Gabrielse, D. Hanneke, T. Kinoshita, M. Nio and B. Odom, *Phys. Rev. Lett.* **99**, 039902 (2007).
6. D. Hanneke, S. Fogwell and G. Gabrielse, *Phys. Rev. Lett.* **100**, 120801 (2008).
7. G. Gabrielse, D. Hanneke, T. Kinoshita, M. Nio and B. Odom, *Phys. Rev. Lett.* **97**, 030802 (2006).
8. T. Kinoshita and M. Nio, *Phys. Rev. D* **7**, 013003 (2006).
9. A. Wicht, J.-M. Hensley, E. Sarajilic and S. Chu, *Phys. Scr.* **T102**, 82–88 (2002).
10. T. Beir, H. Häffner, N. Hermansphan, S.-G. Karshenboim, H.-J. Kluge, W. Quint, S. Sthal, J. Verdú and G. Werth, *Phys. Rev. Lett.* **88**, 011603 (2002).
11. M.-P. Bradley, J.-V. Porto, S. Rainville, J.-K. Thompson and D.-E. Pritchard, *Phys. Rev. Lett.* **83**, 4510 (1999).
12. Th. Udem, A. Huber, B. Gross, J. Reichert, M. Prevedelli, M. Weitz and T.-W. Hänsch, *Phys. Rev. Lett.* **79**, 2646 (1997).
13. C. Schwob, L. Jozefowski, B. de Beauvoir, L. Hilico, F. Nez, L. Julien, F. Biraben, O. Acef and A. Clairon, *Phys. Rev. Lett.* **82**, 4960 (1999).
14. M. Ben Dahan, E. Peik, J. Reichel, Y. Castin and C. Salomon, *Phys. Rev. Lett.* **76**, 4508–4511 (1996) 4508.
15. E. Peik, M. B. Dahan, I. Bouchoule, Y. Castin and C. Salomon, *Phys. Rev. A* **55**, 2989–3001 (1997).
16. S. R. Wilkinson, C. F. Bharucha, K. W. Madison, Q. Niu and M. G. Raizen, *Phys. Rev. Lett.* **76**, 4512–4515 (1996).
17. G. H. Wannier, *Phys. Rev.* **52**, 191–197 (1937).
18. P. Cladé, E. de Mirandes, M. Cadoret, S. Guellati-Khélifa, C. Schwob, F. Nez, L. Julien and F. Biraben, *Phys. Rev. Lett.* **96**, 033001 (2006).
19. P. Cladé, E. de Mirandes, M. Cadoret, S. Guellati-Khélifa, C. Schwob, F. Nez, L. Julien and F. Biraben, *Phys. Rev. A* **102**, 052109 (2006).
20. P. Cladé, S. Guellati-Khélifa, C. Schwob, F. Nez, L. Julien and F. Biraben, *Eur. Phys. J. D* **33**, 173–179 (2005).

PRECISE MEASUREMENTS OF S-WAVE SCATTERING PHASE SHIFTS WITH A JUGGLING ATOMIC CLOCK

STEVEN GENSEMER, RUSSELL HART, ROSS MARTIN, XINYE XU,

RONALD LEGERE, and KURT GIBBLE

Department of Physics, The Pennsylvania State University,
104 Davey Laboratory, University Park, PA 16802, USA

Servaas Kokkelmans

Eindhoven University of Technology, P. O. Box 513,
5600MB Eindhoven, The Netherlands

We demonstrate an interferometric scattering technique that allows highly precise measurements of s-wave scattering phase shifts. We collide two clouds of cesium atoms in a juggling fountain clock. The atoms in one cloud are prepared in a coherent superposition of the two clock states and the atoms in the other cloud are prepared in one of the F,m_F ground states. When the two clouds collide, the clock states experience s-wave phase shifts as they scatter off of the atoms in the other cloud. We detect only the scattered part of the clock atom's wavefunction for which the relative phase of the clock coherence is shifted by the difference of the s-wave phase shifts. In this way, we unambiguously observe the differences of scattering phase shifts. These phase shifts are independent of the atomic density to lowest order, enabling measurements of scattering phase shifts with atomic clock accuracy. Recently, we have observed the changes in scattering phase shifts as a function of magnetic field over a range of values where Feshbach resonances may be expected and where inelastic scattering channels open and close. A number of these measurements will precisely test and tightly constrain our knowledge of cesium-cesium interactions. With such knowledge, future measurements may place stringent limits on the time variation of fundamental constants, such as the electron-proton mass ratio, by precisely probing phase shifts near a Feshbach resonance.

Keywords: Precise measurements of s-wave scattering phase shifts; juggling fountain clock.

1. Introduction

We have recently demonstrated a fundamentally new type of scattering experiment that allows us to directly observe scattering phase shifts with atomic-clock-like accuracy.[1] Our statistical uncertainty is currently as good as 6 mrad after 5 minutes of data collection. This precision will allows us to observe small changes in scattering phase shifts, for example as inelastic scattering channels open and close. We describe the experimental technique and demonstrate that we directly observe the scattering phase shift which behaves very differently than the cold collision frequency shift.[2] We also show some recent measurements of s-wave phase shifts as a function of magnetic field.

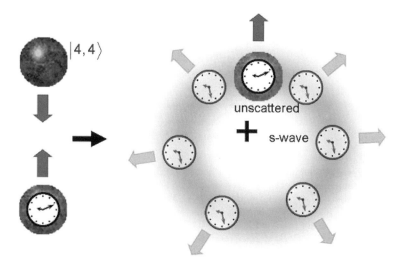

Fig. 1.   We collide an atom in a coherent superposition of the two cesium clock states with a cesium atom in a pure $|F, m\rangle$ state, such as $|4, 4\rangle$. When the clock states scatter, they experience different s-wave phase shifts, shifting the phase of the clock coherence by the difference of the s-wave phase shifts. By detecting only a scattered part of the clock atom's wavefunction, we directly observe the difference of the s-wave phase shifts as a large phase shift of the clock Ramsey fringes.

2. Description of the Experiment

The experiment is schematically illustrated in Fig. 1. We juggle atoms[3] in our cesium fountain clock by launching two clouds of atoms with a short time delay Δt so that the two clouds have a small relative velocity, of order $v_r = g\Delta t = 10$ cm/s. Both clouds of atoms pass through the microwave clock cavity which puts the atoms in the second cloud into a coherent superposition of the two Cs clock states $|Fm\rangle = |40\rangle$ and $|30\rangle$. These atoms collide with the atoms in the first cloud, which are prepared in a pure $|Fm\rangle$

state, such as $|44\rangle$. The spherically incoming s-wave component of the $|30\rangle$ state experiences an s-wave phase shift $\delta_{s,30,Fm}$, or simply δ_3. Similarly, the s-wave component of $|40\rangle$ acquires a phase shift of δ_4. The collision therefore shifts the relative phase of the s-wave scattered clock states by $\Phi = \delta_3 - \delta_4$. After colliding, the atoms return downwardly through the microwave cavity which drives a second $\pi/2$ pulse, converting the phase difference between the clock coherence and the microwave field into a population difference. This gives Ramsey fringes as in Fig. 2. By detecting only the scattered part of each clock atom's wavefunction, the Ramsey fringes we detect are shifted by $\delta_3 - \delta_4$, the difference of the s-wave phase shifts. Therefore we can directly observe the difference of scattering phase shifts with atomic-clock-like accuracy.

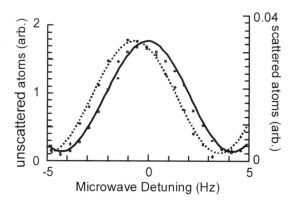

Fig. 2. Central Ramsey fringe for clock atoms that have s-wave scattered from $|33\rangle$ atoms at $90°$ (dashed) and unscattered clock atoms (solid). Here, the difference of the s-wave scattering phase shifts is $\Phi = \delta_3 - \delta_4 = 0.35$ for $v_r = 8.6$ cm/s. The fringes for the scattered atoms (dashed) have a much smaller amplitude than those for the unscattered atoms (solid).

A key feature of this technique is that the measurement is independent of the atomic density (Fig. 3(a)). It is independent of density to 0th order because each scattered clock atom that is detected experiences the two s-wave phase shifts. The density independence, combined with clock techniques, will allow highly precise measurements of the difference of scattering phase shifts which will allow unambiguous tests of cesium-cesium interactions and potentially precise tests of fundamental physics.

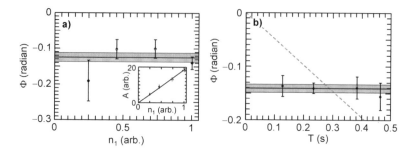

Fig. 3.   (a) The phase shift is independent of the density of cloud 1, n_1, in contrast to the usual cold-collision frequency shift which is proportional to density. The Ramsey fringe amplitude of scattered atoms A (inset) is proportional to the density as expected. (b) The phase shift Φ is independent of the free precession time T, as opposed to proportional to T for a frequency shift (dashed line). For the clock states scattering off of $|4, 4\rangle$, the best fit is $\Phi = 0.141(8)$.

Highly precise scattering measurements have not been possible because the atomic density n needs to be known to relate the measured number of scattered atoms N during some time interval Δt to the cross section, via $N = nv_r\sigma\Delta t$. To date, the best atomic density measurements do not approach even 1% accuracy.[4,5] Here, we can expect 100 microradian accuracy. Our first measurements have a statistical uncertainty of 7.6 mrad[1] and we expect improvements to better than 100 μrad accuracy. Our recent measurements have shown a statistical uncertainty of 6 mrad in 20 minutes of data collection. However, 75% of the data collection time was spent measuring three relatively small backgrounds. With optimal data collection, our current signal-to-noise reduces the averaging time to 5 minutes, or, for 100 μrad, about 2 weeks of data taking.

The phase shift of the scattered atoms' Ramsey fringes is also independent of the interrogation time T between the two Ramsey pulses (Fig. 3(b)). This is in contrast to the usual frequency shift in a clock[2] for which Φ increases linearly with the T (dashed line). As an example, the largest cold-collision frequency shift observed in a Cs clock was 5.5 mHz,[2] which is a very small phase shift of a 100 Hz Ramsey fringe (and just a fairly small phase shift of a 1 Hz fringe). These frequency shifts can be very large by comparison; for $T = 0.127$ s in Fig. 3(b), the frequency shift is 180 mHz for a relatively small phase shift of $\Phi = -0.14$. Near a Feshbach resonance,[6,7] the scattering phase shifts go through π and so the frequency shift of the Ramsey fringes will be several Hertz.

We now briefly describe the sequence to juggle and obtain Ramsey fringes. Our juggling fountain is based on a double magneto-optic trap (MOT)[8] in which the vapor cell MOT repeatedly loads the UHV MOT to collect the first cloud. We hold the first cloud while the second cloud is loaded in the vapor cell MOT below. The second cloud is then launched and, shortly before it reaches the UHV MOT, the first cloud is launched upward. Within 1 ms, a 9.2 GHz red-detuned moving-frame 3D optical lattice is switched on and, with degenerate Raman sideband cooling, the atoms are cooled to 300–500 nK[9] and optically pumped into $|33\rangle$. The second cloud is captured in the UHV MOT, trapped for as little as 1 ms, launched, and then cooled in the optical lattice. One of the experimental challenges is launching two clouds essentially on top of one another without heating the atoms by more than a photon recoil. For a 7 ms launch delay ($E = 20\ \mu$K), the two clouds are only separated by 1 cm, about their diameter, when launched. We "hide" the atoms in $|33\rangle$, carefully control the UHV and lattice beams,[3] and reduce the launch velocity of the second cloud[1] so that we can access collision energies as low as 16 μK ($v_r = 6.3$ cm/s). At the high end, our collision energies can exceed 200 μK ($v_r = 22.4$ cm/s).

The atoms are state prepared in four microwave cavities right above the UHV MOT & optical lattice.[1] When scattering off of $|44\rangle$, the first cloud is transferred from $|33\rangle$ to $|44\rangle$ and the second cloud goes from $|33\rangle$ to $|43\rangle$ to $|32\rangle$ to $|41\rangle$ with composite π pulses.[10] Finally, a velocity-selective two-photon Raman transition[11] transfers the atoms in cloud 2 from $|41\rangle$ to $|30\rangle$.

We isolate the scattered atoms by using the velocity-selection of another two-photon Raman transition. After the two clouds collide, we eliminate the $|44\rangle$ atoms in the first cloud and the $|40\rangle$ atoms in the second cloud with a pulse of a clearing beam. We detect the velocity distribution of the $|30\rangle$[3,8] atoms, with a second velocity-selective two-photon Raman transition to the $|40\rangle$ state, the solid "collisions" curves in Fig. 4(a) and (b). We also detect a "no-collisions" background (dashed curve) by clearing the first cloud right after the second cloud is transferred to the $|30\rangle$ state, well before the two clouds collide. The difference of the collision and no-collisions curves (solid – dashed) in Fig. 4(c) is the velocity distribution of the scattered atoms.

To observe the Ramsey fringes of the scattered atoms, we tune the two-photon Raman transition to select atoms at a particular scattering angle, such as 90° ($v_z = 0$ in Fig. 4(c)). We apply two $\pi/2$ pulses with the clock cavity and scan their frequency to obtain the Ramsey fringes in Fig. 2.

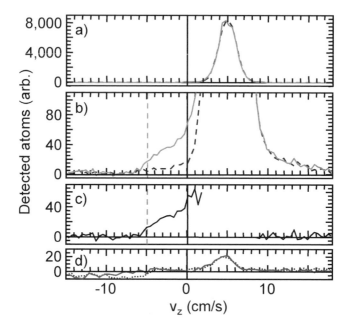

Fig. 4. (a) Velocity distribution of clock atoms in cloud 2, prepared in $|3,0\rangle$ and cloud 1 in $|4,4\rangle$, with $v_r = 9.92$ cm/s. The solid (dashed) curves show the velocity distribution of cloud 2 for 'collisions' ('no-collisions') when we clear cloud 1 from the fountain late (early). (b) Magnification of (a) by 100. In this center-of-mass frame, cloud 1 is centered at $v_{z1} = -4.96$ cm/s (dashed) and cloud 2 at $v_{z2} = 4.96$ cm/s. (c) The difference between the 'collisions' and 'no-collisions' curves represents scattered atoms, visible between $v_z = -5$ cm/s and 2 cm/s. (d) Background for 'collisions' (solid) and 'no-collisions' (dashed). The Ramsey fringes in Fig. 2 for scattered atoms (dashed) are taken at $v_z = 0$ and, for unscattered atoms (solid), at v_{z2}. About 0.1% of the atoms scatter into the 1.4 cm/s detected velocity width at 90°.

3. Brief Theory

Here we theoretically describe the scattering of an atom in a coherent super-position of two clock states off of an atom in a pure state $|Fm\rangle$. We relate the familiar cold collision clock shift and the interferometric scattering to the underlying s-wave phase shifts. In the center of mass frame, we take the wavefunction of the clock atom to be $\psi(\mathbf{r}) = (2v_r)^{-1/2}(|30\rangle + i|40\rangle)\exp(ikz)$ after its upward passage through the clock cavity where $v_r = \hbar k/\mu$ and μ is the reduced mass. The clock states acquire s-wave phase shifts as they scatter from atoms in the first cloud, giving:

$$\psi(\mathrm{r}) = \frac{1}{\sqrt{2v_r}} \left[e^{ikz} + e^{\delta_3} \sin(\delta_3) \frac{e^{ikr}}{kr} \right] |30\rangle$$

$$+ \frac{i}{\sqrt{2v_r}} \left[e^{ikz} + e^{\delta_4} \sin(\delta_4) \frac{e^{ikr}}{kr} \right] |40\rangle. \tag{1}$$

Here δ_3 and δ_4 are the s-wave phase shifts for states $|30\rangle$ and $|40\rangle$ scattering from $|Fm\rangle$.

After scattering, the atoms return back through the microwave cavity where the atoms experience a second $\pi/2$ pulse, but with a variable phase ϕ. Projecting the wavefunction onto the excited state $|40\rangle$, we get $\exp(i\phi/2)\langle 30|\psi(r)\rangle + exp(-i\phi/2)\langle 40|\psi(r)\rangle$.

The population of $|40\rangle$ has three contributions. One is from the unscattered part of each clock atom, one from the scattered part, and one from the interference between the scattered and unscattered waves. The unscattered probability current is

$$\vec{j}_{unsc} = \cos^2\left(\frac{\phi}{2}\right) \hat{z}, \tag{2}$$

which represents the usual Ramsey fringes. The interference term is

$$\vec{j}_{int} = \frac{\pi}{k^2} \left(2\cos^2\left(\frac{\phi}{2}\right) \left[\sin^2(\delta_3) + \sin^2(\delta_4)\right] \right.$$

$$\left. + \frac{1}{2}\sin(\phi)\left[\sin(2\delta_3) - \sin(2\delta_4)\right] \right) \hat{z}. \tag{3}$$

This term represents the cold collision frequency shift or mean field shift. In the limit of small phase shifts, it reduces to

$$\vec{j}_{int} = \frac{\pi}{k^2}(\delta_3 - \delta_4)\sin(\phi)\,\hat{z}, \tag{4}$$

which is an odd function of ϕ, producing a frequency shift of the Ramsey fringes. This acts as a phase shift of $\delta\Phi = nv_r\Delta t 2\pi(\delta_3 - \delta_4)/k^2$ of the Ramsey fringes for atoms detected in the forward direction.

The more interesting term that we detect here is the scattered probability current. By excluding the part of each atom's wavefunction in the forward direction, we detect

$$\vec{j}_{sc} = \frac{\pi}{k^2}\left[\sin^2(\delta_3) + \sin^2(\delta_4) + 2\sin(\delta_3)\sin(\delta_4)\cos(\phi + \delta_3 - \delta_4)\right]\hat{r}. \tag{5}$$

Here, the difference of the s-wave scattering phase shifts, $\delta_3 - \delta_4$ is directly observed as a phase shift of the Ramsey fringes. The probability flux in any direction is proportional to the density of scatterers in the first cloud and

$a_3^2 + a_4^2$ for small k, where a_i is the s-wave scattering length. If δ_3 and δ_4 are not equal, the fringe contrast is not 100%. But, because the Ramsey fringes represent an interference of the amplitudes of two states, fringes are visible for a wide range of δ_3/δ_4. Even for scattering cross-sections that differ by an order of magnitude, the contrast is more than 60%.[12] Similar expression can be written for higher partial waves, including the interference between different partial waves (e.g. the interference between s and p-waves).

4. Feshbach Resonances and Scattering Thresholds

The ability to precisely measure s-wave phase shifts will allow a number of scattering experiments. Several Feshbach resonances for the cesium clock states have been found at magnetic field values below 20 mG by measuring cold collision frequency shifts.[13] Using the cold collision shift is difficult because the phase shifts are so small, especially in comparison to the clock's quadratic Zeeman shift of 427 Hz/G^2. In comparison, for our interferometric technique, as the magnetic field is scanned through a Feshbach resonance, the s-wave phase shift will change by $\pm\pi/2$ so that the frequency shifts can be greater than 1 Hz.

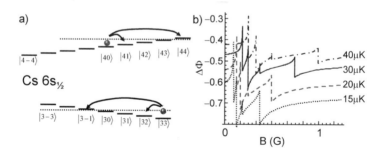

Fig. 5.   (a) Inelastic collisions between Cs hyperfine states $|40\rangle$ and $|33\rangle$ must conserve both z angular momentum (m_F) and energy. At $E_c = 30\ \mu K$ and $B = 0.25$ G, the dashed lines show the energy thresholds for $\Delta m = 4$ changing collisions which are energetically forbidden. (b) The predicted s-wave phase shifts Φ for the $|3,3\rangle$ state colliding with the clock states as a function of magnetic field for several collision energies. There are a series of thresholds between 0.1 and 0.4 G for collision energies $E_c = 15$ to 40 μK.

A second effect that we may observe are inelastic threshold effects as a function of magnetic field as depicted in Fig. 5(a). Considering the scattering of the $|40\rangle$ state off of $|33\rangle$, all inelastic scattering channels are energetically allowed for small magnetic fields. As the magnetic field increases above 0.23 G, the inelastic collision channel $|44\rangle|33\rangle \rightarrow |44\rangle|3-1\rangle$ closes and is no longer energetically allowed. As the field increase further, the $|44\rangle|33\rangle \rightarrow |43\rangle|30\rangle$ channel closes, then $|44\rangle|33\rangle \rightarrow |42\rangle|31\rangle$, and finally $|44\rangle|33\rangle \rightarrow |41\rangle|32\rangle$ closes at 0.89 G. In Fig. 5(b) we show a calculation of the s-wave phase shift as a function of magnetic field. The s-wave phase shift shows a threshold effect as each inelastic channel closes, often accompanied by a change in phase shift that can be as large as 100 mrad. The curves also show the expected threshold behavior that channels close at a higher magnetic field for higher collision energies.

Preliminary measurements of the phase shifts are shown in Fig. 6 for the clock states scattering off of the $|3m\rangle$ states. Over a 0.5 G range of magnetic fields, the s-wave phase shifts can change by more than 100 mrad. These variations could be caused by the opening and closing of inelastic scattering channels as in Fig. 5 or by Feshbach resonances.[14] Measurements are ongoing to extend the range of magnetic fields to greater than 1 G.

The quadratic Zeeman shift is an important systematic error. At 1 G, the sensitivity to changes and gradients is 0.854 Hz/mG, or 63 mrad/mG for our typical interrogation time of 0.23 s. So that we can easily increase the magnetic field, our fountain has an active control of the bias field and no passive shielding. We probe the magnitude of the field and its fluctuations over the atoms trajectories by detecting on each fountain launch both the scattered and unscattered atoms. The unscattered atoms serve as a probe of the magnetic field for that fountain cycle. In addition, on some fountain cycles we scatter the clock atoms off of atoms prepared in the $|3-3\rangle$ state to probe the gradients in the fountain region. The $|3-3\rangle$ collision channels should have no scattering thresholds and no Feshbach resonances are predicted. After scattering off of $|3-3\rangle$, the clock atoms follow the same trajectories and therefore sample the same gradients as when they scatter off of atoms in the other $|3m\rangle$ states. For the data in Fig. 6, we observe a linear variation of the phase shift for scattering off of $|3-3\rangle$ of -0.13 rad which is removed from the data in Fig. 6. With field shimming, we should be able to reduce the field gradients by a factor of 20.

5. Conclusions

We have demonstrated a new scattering technique in which we interferometrically observe the difference of s-wave phase shifts. With the high precision that atomic clock techniques enable, we can see small variations in the scattering phase shifts due to Feshbach resonances and the opening

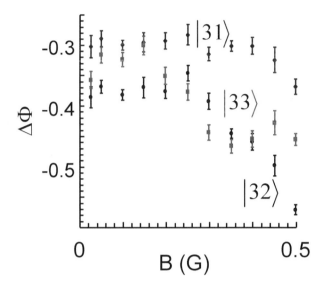

Fig. 6. Preliminary measurements of the difference of s-wave phase shifts $\Delta\Phi$ for the $|3m\rangle$ state colliding with the clock states as a function of magnetic field at 30 μK. The first and second cloud temperatures are 300 and 500 nK which produces a thermal broadening of ± 10 μK. There are a series of thresholds as shown in Fig. 5 in this region and there may also be Feshbach resonances in this range of magnetic fields. Future measurements will extend to higher fields.

and closing of inelastic scattering channels. Precise measurements near Feshbach resonances[15–17] will highly constrain the interatomic interactions[18] and can also place stringent limits on the time variation of fundamental constants, such as the electron to proton mass ratio.[19]

Acknowledgments

We acknowledge stimulating discussions with Boudewijn Verhaar and Eite Tiesinga and financial support from the NSF, NASA, and ONR.

References

1. R. A. Hart, X. Xu, R. Legere and K. Gibble, *Nature* **446**, 892 (2007).
2. K. Gibble and S. Chu, *Phys. Rev. Lett.* **70**, 1771 (1993).
3. R. Legere and K. Gibble, *Phys. Rev. Lett.* **81**, 5780 (1998).
4. J. R. Ensher, D. S. Jin, M. R. Matthews, C. E. Wieman and E. A. Cornell, *Phys. Rev. Lett.* **77**, 4984 (1996).

5. M. T. DePue, C. McCormick, S. L. Winoto, S. Oliver and D. S. Weiss, *Phys. Rev. Lett.* **82**, 2262 (1999).
6. H. Feshbach, *Ann. Phys.* **5**, 357 (1958).
7. E. Tiesinga, B. J. Verhaar and H. T. C. Stoof, *Phys. Rev. A* **47**, 4114 (1993).
8. K. Gibble, S. Chang and R. Legere, *Phys. Rev. Lett.* **75**, 2666 (1995).
9. P. Treutlein, K. Chung and S. Chu, *Phys. Rev. A* **63**, 051401 (2001).
10. M. H. Levitt, *Prog. Nucl. Mag. Res. Spectrosc.* **18**, 61 (1986).
11. M. Kasevich, D. S. Weiss, E. Riis, K. Moler, S. Kasapi and S. Chu, *Phys. Rev. Lett.* **66**, 2297 (1991).
12. K. Gibble and B. J. Verhaar, *Phys. Rev. A* **52**, 3370 (1995).
13. H. Marion, *et al.*, arXiv:physics/0407064 (2006).
14. Eite Tiesinga, private communication (2008).
15. C. Chin, V. Vuletic, A. J. Kerman and S. Chu, *Phys. Rev. Lett.* **85**, 2717 (2000).
16. P. J. Leo, C. J. Williams and P. S. Julienne, *Phys. Rev. Lett.* **85**, 2721 (2000).
17. C. Chin, V. Vuletic, A. J. Kerman, S. Chu, E. Tiesinga, P. J. Leo and C. J. Williams, *Phys. Rev. A* **70**, 032701 (2004).
18. A. Widera, F. Gerbier, S. Fölling, T. Gericke, O. Mandel and I. Bloch, *New J. Phys.* **8**, 152 (2006).
19. C. Chin and V. V. Flambaum, *Phys. Rev. Lett.* **96**, 230801 (2006).

QUANTUM CONTROL OF SPINS AND PHOTONS AT NANOSCALES

P. CAPPELLARO[1,2], J. M. MAZE[1], L. CHILDRESS[3], M. V. G. DUTT[4],

J. S. HODGES[1,5], S. HONG[1], L. JIANG[1], P. L. STANWIX[2], J. M. TAYLOR[6],

E. TOGAN[1], A. S. ZIBROV[1], P. HAMMER[7] A. YACOBY[1], R. L. WALSWORTH[1,2]

and M. D. LUKIN[1]

[1]*Department of Physics, Harvard University, Cambridge, MA 02138 USA.*
[2]*Harvard-Smithsonian Center for Astrophysics, Cambridge, MA 02138 USA.*
[3]*Department of Physics, Bates College, Lewiston, ME 04240 USA.*
[4]*Department of Physics and Astronomy, University of Pittsburgh, Pittsburgh, PA 15260, USA.*
[5]*Department of Nuclear Science and Engineering,* [6]*Department of Physics, Massachusetts Institute of Technology, Cambridge, MA 02139 USA.*
[7]*Department of Electrical and Computer Engineering, Texas A&M University, College Station, TX 77843 USA.*

The detection of weak magnetic fields with high spatial resolution is an outstanding problem in diverse areas ranging from fundamental physics and material science to data storage and bio-imaging. Here we describe a new approach to magnetometry that takes advantage of recently developed techniques for coherent control of solid-state spin qubits. We experimentally demonstrate this novel magnetometer employing an individual electronic spin associated with a Nitrogen-Vacancy (NV) center in diamond. Using an ultra-pure diamond sample, we achieve shot-noise-limited detection of nanotesla magnetic fields at kHz frequencies after 100 seconds of averaging. In addition, we demonstrate 0.5 microtesla/$\sqrt{Hz}$ sensitivity for a diamond nanocrystal with a volume of $(30 \text{ nm})^3$. This magnetic sensor provides an unprecedented combination of high sensitivity and spatial resolution – potentially allowing for the detection of a single nuclear spin's precession within one second.

Keywords: Quantum control, solid-state qubits, magnetometry.

1. Introduction

Magnetic field sensing has historically been achieved by using atomic/molecular systems or solid-state devices with distinctly different underlying physics. Precision measurement techniques in atomic and molecular systems,[1,2] which are also widely used to implement ultra-stable atomic clocks,[3–5] rely on monitoring the precession of angular momentum via the Zeeman effect for magnetic field sensing.[6,7] Sensitive solid-state magnetometers are generally based on many-body macroscopic phenomena such as the Josephson effect in SQUIDs[8,9] or the Hall effect in semiconductors.[10] However, even state-of-the-art techniques have difficulty detecting weak fields in small regions of space. Of particular interest would be the detection and localization of the magnetic field produced by a single electronic or nuclear spin. Some intriguing techniques — such as magnetic force microscopy[11,12] — could potentially yield better spatial resolution. Here we investigate a novel approach to the detection of weak magnetic fields using systems currently explored as qubits: isolated electronic spins in diamond.[13–15] Our approach to magnetic sensing[16–18] combines the coherent manipulation of individual electronic spin qubits embedded in a solid-state environment with optical read-out, to yield an unprecedented combination of high sensitivity and spatial resolution.

2. Concept of a magnetic sensor based on a single spin

As illustrated in Figure 1(a), the electronic spin of an individual NV impurity in diamond can be polarized via optical pumping and measured through spin state-selective fluorescence. Conventional ESR techniques are used to coherently manipulate the spin angular momentum via microwave fields. To achieve magnetic sensing, we monitor the electronic spin precession, which depends on external magnetic fields through the Zeeman effect. This method is directly analogous to precision magnetometry techniques in atomic and molecular systems.

Ultimately, sensitivity is determined by the spin coherence time and by spin projection-noise. Although solid-state electronic spins have shorter coherence times than gaseous atoms, quantum control techniques can decouple them from the local environment, leading to a substantial improvement in their sensitivity to external, time-varying magnetic fields. Even if this is less sensitive than for state-of-the-art macroscopic magnetometers, a key feature of our sensor is that it can be localized within a region of about 10 nm, either in direct proximity to a diamond surface or within a nano-sized diamond crystal, yielding high spatial resolution.

The canonical approach to detecting a Zeeman shift uses a Ramsey-type sequence. A $\pi/2$-pulse creates a superposition of two Zeeman levels, which acquire a relative phase $\phi = \delta\omega\,\tau \propto \frac{g\mu_B}{\hbar}B\tau$ from the external field B during the free evolution interval τ (here μ_B is the Bohr magneton and $g \approx 2$ for NV centers). Another $\pi/2$-pulse transforms the relative phase into a popula-

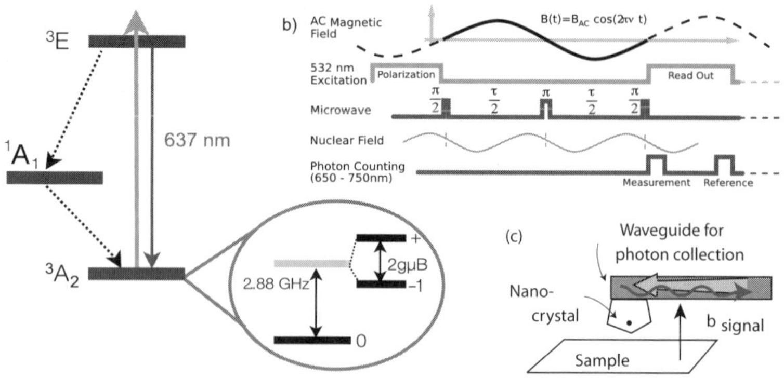

Fig. 1.   Principles of the individual NV electronic spin diamond magnetic sensor. (a) Energy levels for a single NV impurity. (b) Example of the pulse sequence structure of the experimental approach. (c) Scanning tip setup for high-spatial resolution magnetometry.

tion difference, which is measured optically. For small ϕ, the magnetometer signal $\mathcal{S}$ (proportional to the induced population difference) depends linearly on the magnetic field: $\mathcal{S} \approx \frac{g\mu_B}{\hbar} B\tau$. During the total averaging interval T, T/τ measurements can be made, yielding a shot-noise-limited sensitivity δB given by the minimum detectable field, $B_{min} \equiv \delta B/\sqrt{T} = \frac{\hbar}{g\mu_B} \frac{1}{\sqrt{\tau T}}$.

Increasing the interrogation time τ improves the sensitivity until interactions with the environment lead to signal decay. For solid-state spin systems, the coherence is limited by interactions with nearby lattice nuclei and paramagnetic impurities, resulting in an ensemble dephasing time $T_2^* \sim 1$ μs. Coherent control techniques can improve the sensitivity for AC fields. Due to the long correlation times characteristic of dipolar interactions between nuclear spins — the principal source of dephasing — spin echo techniques can dramatically extend the coherence time. Specifically, by adding an additional microwave π-pulse to the Ramsey sequence at time $\tau/2$, the Hahn echo sequence[19] removes the effect of environmental perturbations whose correlation time is long compared to τ. Thus a signal field $B(t)$ oscillating in-phase with the pulse sequence produces an overall additive phase shift, leading to a total phase accumulation, $\delta\phi = \frac{g\mu_B}{\hbar} [\int_0^{\tau/2} B(t)dt - \int_{\tau/2}^{\tau} B(t)dt]$. Correspondingly, the probability of the spin being in the $m_s = 0$ state at the end of the sequence is $P_0(\tau) = [1 + F(\tau)\cos(\delta\phi)]/2$, where $F(\tau)$ is the amplitude of the spin-echo signal envelope in the absence of a time varying field (Fig. 2(a)). For maximal response to CW signals $B_{AC}\sin(\nu t + \varphi_0)$ with known frequency and phase (assuming small B_{AC}), we find $\tau = 2\pi/\nu$

and $\varphi_0 = 0$ to be optimal. The resulting sensitivity per averaging time is

$$\delta B_{AC} \sim \frac{\pi\hbar}{2g\mu_B C\sqrt{T_2}}. \tag{1}$$

Here we introduced the parameter $C \leq 1$, which describes photon shot noise and a finite contrast to the Ramsey fringes.

The optimum sensitivity is achieved only for fields oscillating near $\nu \sim 1/T_2$. These results can be extended to higher frequency signals, by using composite pulse sequences such as CPMG[20] that may provide an even longer coherence time at the expense of a reduced bandwidth.

3. Implementation with Nitrogen-Vacancy centers

3.1. *Regimes of operation and achievable sensitivity*

In our work we use NV centers for magnetic sensing. The fine structure of the electronic ground state of a NV center, shown in Fig. 1(a), is a spin triplet. The crystal field splits the $m_s = \pm1$ Zeeman sublevels from the $m_s = 0$ sublevel by $\Delta = 2\pi \times 2.87$ GHz, setting the spin quantization axis parallel to the nitrogen-to-vacancy direction. There are two possible regimes of operations for the magnetometer, each providing a different compromise between the ease of control and achievable sensitivity.

In the presence of a low static magnetic field $B_{DC} \leq 10$ mT, it is preferable to use the $m_s = \pm1$ manifold which provides a better sensitivity. It has twice the energy splitting of the 0-1 manifold and is less affected by nuclear spin-induced decoherence at low fields, since inter-nuclear interactions are suppressed by the large hyperfine field.[21] For diamond where natural abundance (1.1%) Carbon-13 nuclei are the principal cause of decoherence, the signal decays as $F(\tau) = \exp[-(\tau/T_2)^3]$ with $T_2 \sim 300\mu$s.[22] We can optimize the sensitivity as a function of τ, $\delta B_{AC} = \frac{\pi\hbar}{2g\mu_B}e^{(\tau/T_2)^3}\sqrt{\tau + t_m}/C\tau$, obtaining a sensitivity of $\delta B_{AC} \approx 18$ nT Hz$^{-1/2}$ for a single NV center under current experimental conditions ($C \approx 0.05$ and measurement time $t_m \leq 2$ μs). Improved collection efficiencies ($C = 0.3$) would yield $\delta B_{AC} = 3$ nT/Hz$^{-1/2}$.

At high magnetic fields, it becomes more convenient to address the $m_s = \{0, 1\}$ manifold. The dynamics imposed on the electronic spin by the nuclear spin bath is however more complex. Figure 2(a) shows a typical spin-echo signal observed from an individual NV center. The periodic modulation of the echo is caused by a bath of ^{13}C nuclear spins, which create an effective precessing magnetic field of a few microtesla at the NV center. The precession of the nuclear spins around B_{DC} causes the NV spin-echo

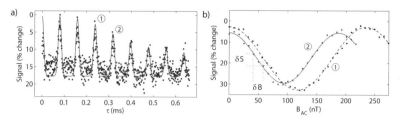

Fig. 2. Spin-echo-based magnetometry with an individual NV electronic spin in a bulk diamond sample. (a) Example electronic spin-echo measurement (dots) and fitting (solid line, see text). (b) Spin-echo signal as a function of applied AC magnetic field amplitude for two operating frequencies $\nu_1 = 3.15$ kHz (dashed line) and $\nu_2 = 4.21$ kHz, corresponding to revivals 1 and 2 in Fig. 2(a).

signal to collapse and revive[15] at half the rate of the ^{13}C Larmor frequency, $\omega_L = \gamma_{^{13}C} B_{DC}$. Note that substantial spin-echo revivals exist even after a free evolution of 0.6 ms. The decay of the echo signal envelope does not follow a simple exponential decay associated with typical ESR on bulk samples. This can be understood by noting that echo dynamics of a single NV center near the revivals is likely determined by a few ^{13}C that interact strongly with the electronic spin,[13–15,22,23] yielding multiple characteristic time scales for the echo decay. The envelope of the spin echo signal in Figure 2(a) has been modeled with an exponential decay $F(\tau) \propto \exp(-(\tau/T_2)^4)$ modulated by a pair of strongly interacting ^{13}C. The sensitivity is then

$$\delta B_{AC} = \frac{\pi\hbar}{g\mu_B} F(\tau)\sqrt{\tau + t_m}/C\tau. \tag{1}$$

With $T_2 \sim 600\ \mu s$, the predicted optimal sensitivity is $\delta B_{AC} \approx 4\ \text{nT}/\text{Hz}^{-1/2}$ for an ideal spin readout, while we expect a sensitivity $\gtrsim 25\ \text{nT}/\text{Hz}^{-1/2}$ with current collection efficiencies, corresponding to $C \sim 0.05$.

3.2. First experimental realization

To establish the sensitivity limits of our magnetometer, we performed a series of proof-of-principle experiments involving single NV centers in bulk ultra-pure single crystal diamond and in commercially available diamond nanocrystals. Our experimental methodology is depicted in Figure 1. Single NV centers are imaged and localized with ~ 170 nm resolution using confocal microscopy. The position of the focal point is moved near the sample surface using a galvanometer mounted mirror to change the beam path and a piezo-driven objective mount. A 20 micron diameter wire generates microwave pulses to manipulate the electronic spin states (see Figure 1(b)). A pair of Helmholtz coils are used to provide both AC and DC magnetic fields. An individual center is first polarized into the $m_S = 0$ sublevel. In Figure 1(b), a coherent superposition between the states $m_S = 0$ and $m_S = 1$ is created by applying a $\pi/2$ pulse tuned to this transition. The system freely evolves for a period of time $\tau/2$, followed by a π refocusing pulse. After a second $\tau/2$ evolution period, the electronic spin state is projected onto the $m_S = \{0, 1\}$ basis by a final $\pi/2$ pulse, at which point the ground state population is detected optically via spin-dependent fluorescence.

We consider first a single crystal diamond bulk sample, operating in the $m_s = \{0, 1\}$ manifold. To achieve the highest sensitivity, the revival rate of the spin-echo signal is adjusted by varying the strength of the applied DC magnetic field B_{DC}, such that the frequency of the echo revival peaks coincides with multiples of the AC field frequency ν to be detected. As shown in Figure 2(b), the observed peak of the spin-echo signal varies periodically as the amplitude of the external AC field (B_{AC}) is increased. The signal variation results from the accumulated phase due to the AC magnetic field, which is converted into a spin population difference, leading to variations in the detected fluorescence signal. Maximal signal in Figure 2 corresponds to an average number of photons $\bar{n} = 0.03$ detected during the 324 ns photon counting window of a single experimental run. In Figure 2(b), each displayed point is a result of $N = 7 \times 10^5$ averages of spin-echo sequences. The magnetometer is most sensitive to variations in the AC magnetic field amplitude (δB) at the point of maximum slope, with the sensitivity being limited by the uncertainty in the spin-echo signal measurement (δS).

Figure 3(a) shows the measured sensitivity δB after one second of averaging as a function of the AC magnetic field frequency $\nu = 1/\tau$. At high frequencies or short times, $F(1/\nu) \to 1$, and the sensitivity scales as $\sqrt{\nu}$, while at low frequencies decoherence degrades the sensitivity. For comparison we plot Eq. (1), with $F(\tau) = \exp(-(\tau/T_2)^4)(1 - \frac{(a^2 - b^2)}{a^2} \sin^2 a\tau \sin^2 b\tau)$,[23] with the parameters found from the fitting of the echo envelope ($T_2 = 676\mu s$, $b = 478$Hz and $a = 626$Hz). The absolute sensitivity depends on the signal to noise ratio in the readout of the NV electronic spin state. In our case, this is limited by photon collection efficiency $\approx 0.1\%$. The resulting photon shot noise[7,16] results in a degradation of the ideal magnetometer sensitivity given by Eq. (1). Our theoretical prediction of magnetometer sensitivity (solid curve in Figure 3(a)) combines the NV coherence properties shown in Figure 2(a) with the noise due to photon counting statistics and imperfect collection efficiency.[16,17] This prediction is in excellent agreement with the experimental results, indicating that our magnetometer is photon-shot-

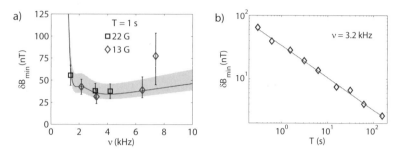

Fig. 3.   Magnetometer sensitivity characterization. (a) Measured sensitivity of a single NV spin magnetometer in bulk diamond after one second averaging. Error bars represent standard deviation for a sample size of 30. Also shown is the theoretically predicted sensitivity (solid line), with the shaded region representing uncertainty due to variations in photon collection efficiency. Measurements were carried out at two DC fields, $B_{DC} = 13$ (diamond) and 22 G (square). (b) Minimum measurable AC magnetic field as a function of averaging time, for AC field frequency $\nu = 3.2$ kHz and $B_{DC} = 13$ G (diamonds). Fit to the data shows that the sensitivity improves as the square root of averaging time, in agreement with theoretical estimates based on photon shot-noise limited detection.

noise limited. Figure 3(b) shows sensitivity for a fixed AC magnetic field frequency ν as a function of measurement time T. The solid line is a fit to $B_{min} \propto T^{-\alpha}$, where $\alpha = 0.5 \pm 0.01$. This indicates that magnetic fields as small as few nanotesla are resolvable after 100 seconds of averaging.

We also performed similar experiments with single NV centers in diamond nanocrystals (30 nm diameter) to demonstrate magnetic sensing within a nanoscale detection volume. The available nanocrystals contain a large number of impurities (probably paramagnetic substitutional nitrogen atoms containing unpaired electron spins) that shorten the electronic spin coherence time[29] to values ranging from 4 to 10 μs. Sensitive detection of AC magnetic fields is still possible as demonstrated experimentally in Figure 4. A magnetometer sensitivity of $\delta B \sim 0.5 \pm 0.1~\mu\mathrm{T}/\sqrt{Hz}$ is achieved for this nanocrystal at $\nu = 380$ kHz. In Figure 4(b), the maximum signal corresponds to an average number of photons $\bar{n} = 0.02$ counted during a 324 ns photon counting window; $N = 2 \times 10^6$ averages of spin-echo sequences were used.

4. Outlook

The high magnetic field sensitivity in a small volume offered by solid state spin-qubits such as NV centers in diamond can find a wide range of applications, from fundamental physics tests or quantum computing applications to detection of NMR signals, surface physics and material science, and med-

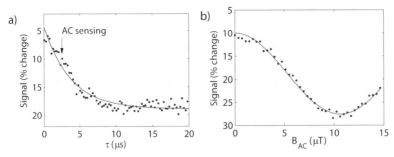

Fig. 4. Example of magnetic sensing with a single NV electronic spin in a diamond nanocrystal. (a) Spin-echo signal from a single NV center contained in a diamond nanocrystal with diameter of 34 ± 12 nm as determined by AFM. The arrow indicates the time at which magnetic sensing is performed in Fig. 4b. (b) Spin-echo signal as a function of the applied AC magnetic field amplitude at a frequency of $\nu = 380$ kHz. The resulting standard deviation yields a magnetometer sensitivity of 0.5 ± 0.1 μT/Hz$^{1/2}$.

ical imaging and biomagnetism. In particular, this robust technology could be invaluable both in nanoscale magnetic field imaging and in macroscopic field detection scenarios, such as low-field MRI.

For example, one of the outstanding challenges in magnetic sensing is the detection and real space imaging of single electronic and nuclear dipoles. Since the magnetic field from a single dipole decreases with distance as $\sim 1/r^3$, a magnetometer that can be brought into close proximity of the dipole offers a clear advantage. This can be achieved by using a diamond nanocrystal. For example, a ~ 25 nm diameter nanocrystal attached to the end of an optical fiber or plasmonic waveguide,[24] would provide a spatial resolution ~ 25 nm, while achieving orders of magnitude higher magnetic field sensitivity than magnetic force microscopy.[12] Provided the waveguide can yield high collection efficiency ($C \gtrsim 0.3$), a sensitivity better than 3 nT Hz$^{-1/2}$ could be achieved using echo-based techniques. This surpasses the sensitivity of Hall-bar[25] or SQUID[8] based microscopes by more than an order of magnitude, with 10 times better spatial resolution. For example, the magnetic field from a single proton is ~ 3 nT at 10-nm separation, which an NV nanocrystal magnetometer would be able to detect within one second. The ultimate limits to miniaturization of such nanocrystals, which are likely due to surface effects, are not yet well understood, but experiments have already demonstrated control of single NV centers in sub 50-nm nanocrystals,[27–29] as well as the use of such nanocrystals in scanning probe setups.[30]

Further improvements can be obtained by using multiple pulses. The narrow bandwidth associated with such an approach can also be exploited for a frequency selective measurement, ranging from tens of kHz up to MHz. This will enable distinguishing different isotopes, due to their unique gyro-magnetic ratios, and could improve the spatial resolution when used in combination with a strong magnetic field gradient. Much longer coherence and interrogation times should be possible by using isotopically pure diamond with low concentrations of both ^{13}C and nitrogen electron spin impurities. The signal-to-noise ratio may also be increased by improving the measurement readout efficiency. Near single-shot readout of an electronic spin in diamond has been achieved with cryogenic cooling using resonant excitation.[26] Photon collection efficiency at room temperature can also be substantially improved using either conventional far-field optics or evanescent, near-field coupling to optical waveguides.[24] Finally, another way to improve the magnetometer sensitivity is to use many sensing spins,[16] where we can take advantage of the relatively high achievable density of spins in the solid-state ($\sim 10^{17}$ cm^{-3}) compared to atomic magnetometers($\sim 10^{13}$ cm^{-3}).[7] Further extensions could include the use of non-classical spin states, such as squeezed states induced by the spin-spin coupling.

These considerations indicate that coherent control of electronic spins in diamond can be used to create a magnetic field sensor with an unprecedented combination of sensitivity and spatial resolution in a small, robust device. On a more general level, these ideas could apply to a variety of paramagnetic systems or even qubits sensitive to other perturbations of their environment. The vast range of potential applications for sensitive, spatially resolved measurements warrants a re-examination of solid-state quantum devices from the perspective of metrology.

Acknowledgments

We gratefully acknowledge conversations with D. Awschalom, A. Cohen, J. Doyle, D. Budker and M. P. Ledbetter. This work was supported by the NSF, DARPA, MURI and David and Lucile Packard Foundation.

References

1. D. Budker, D. F. Kimball and D. P. DeMille, *Atomic Physics: An Exploration Through Problems and Solutions* (Oxford University Press, 2004).
2. N. F. Ramsey, *Molecular Beams* (Oxford University Press, 1990).
3. A. D. Ludlow, *et al.*, *Science* **319**, 1805 (2008).
4. T. Rosenband, *et al.*, *Science* **319**, 1808 (2008).
5. D. J. Wineland, J. J. Bollinger, W. M. Itano, F. L. Moore and D. J. Heinzen, *Phys. Rev. A* **46**, R6797 (1992).

6. K. Kominis, T. W. Kornack, J. C. Allred and M. V. Romalis, *Nature* **422**, 596 (2003).

7. D. Budker and M. Romalis, *Nat. Phys.* **3**, 227 (April 2007).

8. S. J. Bending, *Adv. Phys.* **48**, 449 (1999).

9. R. Kleiner, D. Koelle, F. Ludwig and J. Clarke, *Proc. IEEE* **92**, 1534 (2004).

10. C. N. Owston, *J. Sci. Instrum.* **44**, 798 (1967).

11. D. Rugar, R. Budakian, H. J. Mamin and B. W. Chui, *Nature* **430**, 329 (2004).

12. H. J. Mamin, M. Poggio, C. L. Degen and D. Rugar, *Nat. Nano* **2**, 301 (2007).

13. F. Jelezko, T. Gaebel, I. Popa, A. Gruber and J. Wrachtrup, *Phys. Rev. Lett.* **92**, 076401 (2004).

14. F. Jelezko, *et al.*, *Phys. Rev. Lett.* **93**, 130501 (2004).

15. L. Childress, *et al.*, *Science* **314**, 281 (2006).

16. J. M. Taylor, *et al.*, *Nature Phys.* **4**, 810 (2008).

17. J. R. Maze, *et al.*, *Nature* **455**, 644 (2008).

18. G. Balasubramanian, *et al.*, *Nature* **455**, 648 (2008).

19. E. L. Hahn, *Phys. Rev.* **80**, 580 (1950).

20. S. Meiboom and D. Gill, *Rev. Sci. Instrum.* **29**, 688 (1958).

21. G. R. Khutsishvili, *Sov. Phys. Uspekhi* **8**, 743 (1966).

22. M. V. G. Dutt, *et al.*, *Science* **316**, 1312 (2007).

23. J. R. Maze, J. M. Taylor and M. D. Lukin, arXiv:0805.0327 (2008).

24. D. E. Chang, A. S. Sorensen, P. R. Hemmer and M. D. Lukin, *Phys. Rev. Lett.* **97**, 053002 (2006).

25. A. M. Chang, *et al.*, *Appl. Phys. Lett.* **61**, 1974 (1992).

26. J. Wrachtrup and F. Jelezko, *J. Phys. Cond. Matt.* **18**, S807 (2006).

27. T. Gaebel, *et al.*, *Nat. Phys.* **2**, 408 (2006).

28. J. R. Rabeau, *et al.*, *Appl. Phys. Lett.* **88**, p. 023113 (2006).

29. J. R. Rabeau, *et al.*, *Nano Lett.* **7**, 3433 (2007).

30. S. Kühn, C. Hettich, C. Schmitt, J.-Ph. Poizat and V. Sandoghdar, *J. Microsc.* **202**, 2 (2001).

ATOMIC ENSEMBLE QUANTUM MEMORIES

S.-Y. LAN*, R. ZHAO*, S. D. JENKINS[†], O. A. COLLINS*, Y. O. DUDIN*,

A. G. RADNAEV*, C. J. CAMPBELL*, D. N. MATSUKEVICH[‡],

T. A. B. KENNEDY* and A. KUZMICH*

*School of Physics, Georgia Institute of Technology, Atlanta, Georgia 30332-0430
[†]CNR-INFM, Dipartimento di Fisica e Matematica,
Università degli Studi dell'Insubria, Via Valleggio 11 22100 Como, Italy
[‡]Department of Physics, University of Maryland, College Park, Maryland 20742

A key ingredient for a practical quantum repeater is a long memory coherence time. We describe a quantum memory using the magnetically-insensitive clock transition in atomic rubidium confined in a 1D optical lattice. We observe quantum lifetimes exceeding 6 milliseconds. We also demonstrate a dozen independent quantum memory elements within a single cold sample, and describe matter-light entanglement generation involving arbitrary pairs of these elements.

Keywords: Quantum memory; quantum repeater; matter-light entanglement.

1. Introduction

Quantum mechanics provides a mechanism for absolutely secure communication between remote parties, see for example Ref. 1. For distances greater than about a 100 kilometers direct quantum communication via optical fiber is difficult, due to fiber losses. To overcome this difficulty, intermediate storage of the quantum information along the transmission channel using a quantum repeater protocol has been suggested.[2] The optically thick atomic ensemble has emerged as an attractive medium for quantum storage, matter-light qubit entanglement generation and distribution.[3] Efficient quantum state transfer between ensemble-based qubits and single photons can be achieved in free space by utilizing a very weak interaction at a single photon/single atom level. The realization of coherent quantum state transfer from a matter qubit to a photon qubit was achieved using cold rubidium at Georgia Tech in 2004,[4] followed by the first light-matter qubit conversion and entanglement of remote atomic qubits in 2005.[5] A scheme to achieve long-distance quantum communication at the absorption minimum of optical fibers, employing atomic cascade transitions, has been proposed and its critical elements experimentally verified.[6] In order to boost communication rates under conditions of limited quantum memory time, a modified quantum repeater based on dynamic allocation of quantum resources, multiplexed quantum repeater, has been proposed.[7]

Here we would like to report our recent progress on long-lived storage and retrieval of single quantum excitations, including a two order of magnitude increase in the quantum memory lifetime,[8] and the realization of multiple memory elements within a single cold atomic sample.

2. Long-lived quantum memory

Protocols for quantum communication are typically based on remote parties sharing and storing an entangled quantum state. The generation of such remote entanglement must necessarily be done locally and distributed by light transmission over optical fiber links or through free space.[16] For the distribution of entanglement over a length L the characteristic timescale for storage is the light travel time L/c, where c is the speed of light in the medium. For $L = 1000$ km, $L/c \approx 5$ ms for an optical fiber.

In recent advances involving atomic ensembles,[4–6,8–15] the quantum memory lifetime, was limited by residual magnetic fields, with the longest measured time of 32 μs.[8] To circumvent this limit one can use the ground-state hyperfine coherence of the $m = 0$ Zeeman levels as the basis of quantum storage. This so-called clock transition is only second-order sensitive to external magnetic fields, leading to a memory coherence time limit

$$\tau = [4\pi \cdot 575[\text{Hz/G}^2]B_0 B' l]^{-1}.$$

Under our experimental conditions, with bias magnetic field $B_0 \sim 0.5$ G, gradient $B' \ll 30$ mG/cm, and sample length $l \sim 1$ mm, we find $\tau \gg 100$ ms.

Ballistic expansion of the freely falling gas provides a memory time limitation which can be estimated from the time $\tau = \Lambda/(2\pi v)$ it takes an atomic spin grating to dephase by atomic motion. For representative MOT parameters, grating wavelength $\Lambda = 50$ μm, atomic velocity $v = \sqrt{k_B T/M} \simeq 8$ cm/s for $T = 70$ μK, we find $\tau \sim 100$ μs. These estimates indicate that in order to demonstrate quantum memory lifetimes of many milliseconds we must suppress atomic motion and use a magnetically-insensitive atomic coherence as the basis of the quantum memory.

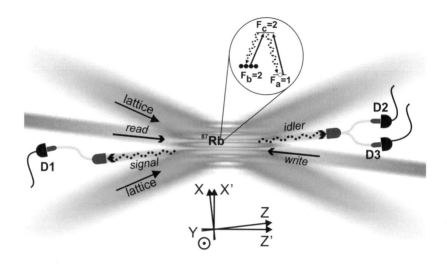

Fig. 1.   Essential elements of the experimental set-up. Between 10^5 and 10^6 sub-Doppler cooled ^{87}Rb atoms are loaded into an optical lattice, and detection of the signal field generated by Raman scattering of the *write* laser pulse heralds the presence of a *write* spin wave excitation. A *read*/control field converts the surviving atomic excitation into an idler field after a storage period T_s. The inset shows the atomic level scheme of ^{87}Rb with levels a and b the hyperfine components of the ground $5S_{1/2}$ level, and level c a hyperfine component of the excited $5P_{1/2}$ level. The *write* laser excites the $b \leftrightarrow c$ transition, with Raman emission of the signal field on $c \to a$. The *read* laser excites the $a \leftrightarrow c$ transition, with Raman emission of the idler field on $c \to b$.

2.1. Description of the system

In order to suppress atomic motion, we load an atomic cloud of ^{87}Rb into a one-dimensional optical lattice, as shown in Fig. 1. The ground hyperfine levels a and b of ^{87}Rb have angular momenta $F_a = 1$ and $F_b = 2$, and the upper and lower clock states are written as $|+\rangle \equiv |b, m = 0\rangle$ and $|-\rangle \equiv |a, m = 0\rangle$, respectively. If the atoms are prepared in the upper clock state by optical pumping, the $|+\rangle$ and $|-\rangle$ states can be coupled by Raman scattering of a weak linearly polarized *write* laser field into an orthogonally polarized signal field detected in the near-forward direction. Ideally, after a controllable storage period, the *read* pulse converts atomic spin excitations into an idler field propagating along the quantization axis z, and linearly polarized in the x-direction. Under these conditions the medium exhibits electromagnetically induced transparency with susceptibility for x-polarized idler field controlled by the read laser intensity

$$\chi_x(\Delta) = -\frac{1}{\omega_{cb}} \frac{n|\kappa|^2 \Delta}{\Delta(\Delta + i\Gamma_c/2) - \frac{1}{2}|\Omega_r|^2},$$

where ω_{cb} is the transition frequency between levels c and b, Δ and Γ_c are the read laser detuning and spontaneous decay rate, Ω_r is the read laser Rabi frequency and $\sqrt{n}|\kappa|$ is the collective Rabi frequency, where n is the atomic density. The corresponding group velocity of the idler field is

$$v_g = c \frac{|\Omega_c|^2}{|\Omega_c|^2 + n|\kappa|^2}.$$

In order to maximize the retrieval efficiency, the signal and idler spatial mode functions should be matched and the condition $\mathbf{k_i} = \mathbf{k_w} - \mathbf{k_s} + \mathbf{k_r}$ satisfied, where $\mathbf{k_w}$, $\mathbf{k_s}$, $\mathbf{k_i}$ and *read* fields, respectively.

The detection of the signal photon after a write pulse implies a momentum change $\hbar(\mathbf{k_w} - \mathbf{k_s})$ of the atoms (along the x'-axis). The excitation amplitude for an atom at position $\mathbf{r}$ is proportional to $e^{-i(\mathbf{k_w}-\mathbf{k_s})\cdot\mathbf{r}}$. Since the period of the lattice, 25 μm, is shorter than the spin grating wavelength $\Lambda \simeq \lambda/\theta \approx 50$ μm, determined by the angle $\theta \approx 0.9°$ between the write and signal fields of wavelength $\lambda = 795$ nm, optical confinement helps to preserve the spin wave coherence by suppressing atomic motion along the $\mathbf{k_w} - \mathbf{k_s}$ direction.

2.2. *Retrieval of single quantum excitations*

We characterize how well the retrieved idler field compares to a single photon state by measuring the α-parameter of Grangier et al.,[18] which is defined by

$$\alpha = \frac{p_1 p_{123}}{p_{12} p_{13}}.$$

Here p_1, p_2, p_3 are the photoelectric detection probabilities on the three detectors, D1-3, respectively, Fig. 1. A field in a single-photon state incident on a beamsplitter is either transmitted or reflected, and the joint photoelectric detection probability vanishes, Fig. 2. The measured idler field is gated by detection of the signal field by D1.

In Table 1 we give the measured values of α, demonstrating quantum memory for storage times up to 6 ms. The value $\alpha = 0$ corresponds to an ideal, heralded single-photon state, whereas for classical fields $\alpha \geq 1$.

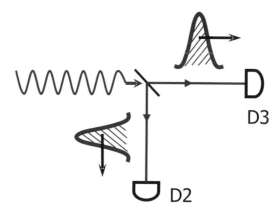

Fig. 2.   A single photon incident on a beamsplitter takes one of two paths leading to anti-correlated photoelectron counting events at D2 and D3. This forms the basis of the α-parameter measurement scheme.[18]

Table 1.   Measured values of α and intrinsic efficiency η_{int}.

storage time, ms	α	η_{int}
0.0012	0.02 ± 0.01	0.25
1	0.12 ± 0.04	0.11
4	0.17 ± 0.07	0.05
6	0.10 ± 0.10	0.045

Table 2.   Measured values of $g_D^{(2)}$ and intrinsic source efficiency ϵ_{int}.

protocol duration, ms	$g_D^{(2)}$	ϵ_{int}
4	0.06 ± 0.04	0.08
5	0 ± 0.06	0.06

An important, immediate application of this long quantum memory is the realization of a deterministic single photon source based on quantum measurement and feedback, as proposed in Ref. 8. As the protocol's success is based on long memory times, we are now able to significantly improve the quality of the single-photon source. It is demonstrated by measuring sub-Poissonian photoelectron statistics of the second-order coherence function $0 \le g_D^{(2)} < 1$, which is defined by

$$g_D^{(2)} = p_{23}/(p_2 p_3).$$

The source efficiency, defined as the probability ϵ to detect a photoelectric event per trial, is the second important figure of merit, ideally, $g_D^{(2)} = 0$

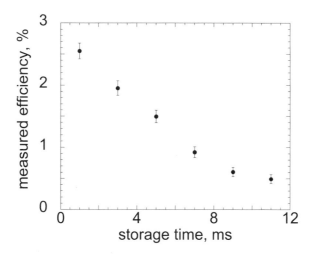

Fig. 3. Retrieval efficiency as a function of storage time for optically pumped atoms in an optical lattice with depth $U_0 = 40$ μK.

and $\epsilon = 1$. The measured passive losses from the atomic sample to the detector in the idler channel produce an efficiency factor of $0.25 \pm 10\%$. The measured values of $g_D^{(2)}$ and ϵ normalized by passive losses are given in Table 2.2.

In Fig. 3 we show the behavior of the measured retrieval efficiency — not normalized by passive losses — on the millisecond time-scale. The decay time is consistent with atomic motion in the lattice potential accompanied by differential light shifts of the clock states.[17]

3. Multiplexed quantum memory

The presence of multiple memory elements per node in a quantum repeater allows dynamic reallocation of resources improving the rate of quantum communication for short memory times.[7] Here we describe such a multiplexed quantum node. Individual addressing of memory elements within a single cold sample is achieved by means of 1D scanning with acoustic-optical deflectors (AODs). This allows us to demonstrate matter-light entanglement using an arbitrary pair of memory elements in the array.

3.1. *Quantum memory array*

Our experimental setup is illustrated in Fig. 4. Two AODs are used to scan *write* and *read* beams, respectively. Each mode from the *write* AOD

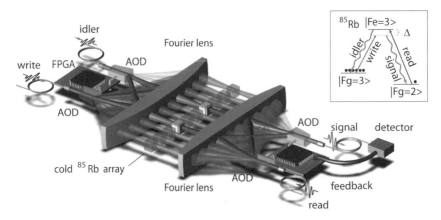

Fig. 4.   Schematic illustration of the experimental setup for the multiplexed memory array. The inset shows the relevant ^{85}Rb energy levels.

is matched to the *read* AOD. Another two AODs are employed for the collection of the signal and idler fields. The phase matching condition $\mathbf{k}_w^{(j)} + \mathbf{k}_r^{(j)} = \mathbf{k}_s^{(j)} + \mathbf{k}_i^{(j)}$ is satisfied at each memory element address $j = 1 - 12$.

A field-programmable gate array (FPGA) controls addressing of the memory elements via a digital-to-analog converter. Sequential pulses generated by the control logic with different voltage levels are fed into a voltage-controlled oscillator (VCO) which converts them into rf pulses with different frequencies. After amplification these are directed into the write AOD, which produces *write* pulses into a set of spatial modes. These pulses enable individual addressing of a localized sub-region of the atomic cloud that forms a memory element.

Using a MOT, ^{85}Rb atoms are prepared in the $|5S_{1/2}, F = 3\rangle$ ground level. The protocol begins when the atoms are released from the trap. By electronic control the driving frequency of the *write* AOD is changed and within 1 μs the deflector points to the desired memory element. A 300 ns optical pulse, red detuned from the $|5S_{1/2}, F = 3\rangle \rightarrow |5P_{1/2}, F = 3\rangle$ transition by 10 MHz (we use an additional acousto-optical modulator to compensate the frequency shift of the write AOD), is then sent to the memory element. Synchronously the signal AOD directs the scattered signal field from the memory element to the single photon detector. In this way, a 12 pulse train scans the atomic array in temporal order with a time interval of 1.3 μs. The detection of the signal field in a specific gate interval of 250 ns indicates the origin of the signal field.

3.2. *Matter-light entanglement with a quantum memory array*

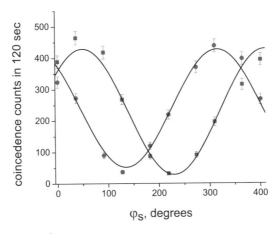

Fig. 5. Measured photoelectric coincidences for elements j=5 and 8 as a function of ϕ_s for $\phi_i = 0$, squares and $\phi_i = \pi/4$, circles. The solid curves are sinusoidal fits.

The AODs can be used as dynamic beam splitters, allowing to realize matter qubits based on pairs of elements of the memory array.

The write AOD aligned in +1 Bragg diffraction order is provided with two different rf frequencies, f_j and f_k ($f_j > f_k$), to generate two spatially distinct pulses. These two *write* pulses, red detuned from the $|5S_{1/2}, F = 3\rangle \rightarrow |5P_{1/2}, F = 3\rangle$ transition by -10 MHz and $-10+(f_j\text{-}f_k)$ MHz, respectively, illuminate two different elements simultaneously. The signal fields are collected by the signal AOD, which is aligned and modulated, at frequencies f_j and f_k, in order to combine the signal fields into a common spatial mode with a relative phase ϕ_s, coupled to the optical fiber for detection.

After a 150 ns delay, two 200 ns long *read* pulses generated by the *read* AOD aligned in +1 Bragg order with frequencies f_j and f_k are sent through the two elements to convert the atomic excitations to the idler fields. The idler AOD, driven at at frequencies f_j and f_k, combines the idler frequency components into a single spatial mode, with a relative phase ϕ_i.

The rf phase shifters on the signal and idler AODs allow to vary ϕ_s and ϕ_i. The photoelectric coincidence counts of signal and idler fields as a function of these phases is shown in Fig. 5. We have measured violation of

Bell's inequality ($|S| \leq 2$), with $S_{exp} = 2.38 \pm 0.03$. Table 3.2 shows the interference fringe visibility V for different combinations of elements.

Table 3. Measured interference visibility with different combinations of quantum memory elements.

j	k	fringe visibility
5	8	$88 \pm 1\%$
7	8	$86 \pm 2\%$
7	10	$79 \pm 1\%$
5	10	$81 \pm 2\%$
1	12	$73 \pm 3\%$

4. Conclusion

We have observed a quantum memory with a lifetime in excess of six milliseconds and used it to implement a high-quality deterministic single photon source. We have also demonstrated matter-light entanglement with a quantum memory array in a single cold atomic sample.

This work was supported by the National Science Foundation, A. P. Sloan Foundation, Office of Naval Research, and the Army Research Office through the Georgia Tech Quantum Institute.

References

1. M. A. Nielsen and I. L. Chuang, *Quantum Computation and Quantum Information* (Cambridge University Press, 2000).
2. H.-J. Briegel, W. Dür, J. I. Cirac and P. Zoller, *Phys. Rev. Lett.* **81**, 5932 (1998).
3. L.-M. Duan, M. D. Lukin, J. I. Cirac and P. Zoller, *Nature* **414**, 413 (2001).
4. D. N. Matsukevich and A. Kuzmich, *Science* **306**, 663-666 (2004).
5. D. N. Matsukevich *et al.*, *Phys. Rev. Lett.* **96**, 030405 (2006)
6. T. Chanelière *et al.*, *Phys. Rev. Lett.* **96**, 093604 (2006).
7. O. A. Collins, S. D. Jenkins, A. Kuzmich and T. A. B. Kennedy, *Phys. Rev. Lett.* **98**, 060502 (2007).
8. D. N. Matsukevich *et al.*, *Phys. Rev. Lett.* **97**, 013601 (2006).
9. T. Chanelière *et al.*, *Nature* **438**, 833 (2005).
10. D. N. Matsukevich *et al.*, *Phys. Rev. Lett.* **95**, 040405 (2005).
11. D. N. Matsukevich *et al.*, *Phys. Rev. Lett.* **96** 033601 (2006).
12. T. Chanelière, D. N. Matsukevich, S. D. Jenkins, S.-Y. Lan, R. Zhao, T. A. B. Kennedy and A. Kuzmich, *Phys. Rev. Lett.* **98**, 113602 (2007).
13. Y. A. Chen *et al.*, *Nat. Phys.* **4** 103-107 (2007).

14. J. Simon, H. Tanji, S. Ghosh and V. Vuletic, *Nat. Phys.* **3** 765 (2007).
15. J. Laurat *et al.*, *Phys. Rev. Lett.* **99**, 180504 (2007).
16. M. Aspelmeyer *et al.*, *Science* **301**, 621 (2003).
17. S. Kuhr *et al.*, *Phys. Rev. A* **72**, 023406 (2005).
18. P. Grangier, G. Roger and A. Aspect, *Europhys. Lett.* **1**, 173 (1986).

QUANTUM NON-DEMOLITION PHOTON COUNTING AND TIME-RESOLVED RECONSTRUCTION OF NON-CLASSICAL FIELD STATES IN A CAVITY

S. HAROCHE[1,2]*, S. DELEGLISE[1], C. SAYRIN[1], J. BERNU[1]
S. GLEYZES[1], C. GUERLIN[1], S. KUHR[1], I. DOTSENKO[1,2]
M. BRUNE[1] and J. M. RAIMOND[1]

[1] *Laboratoire Kastler Brossel, ENS, CNRS, UMPC*
24 rue Lhomond, F-75005 Paris
[2] *Collège de France, 11 pl. M. Berthelot, 75005 Paris*
* E-mail: haroche@lkb.ens.fr ; www.cqed.org

We describe Cavity QED experiments in which a beam of circular Rydberg atoms is used to manipulate and probe non-destructively microwave photons trapped in a very high-Q superconducting cavity. We realize an ideal quantum non-demolition (QND) measurement of light, observe the radiation quantum jumps due to cavity relaxation and prepare non-classical fields such as Fock and Schrödinger cat states. Combining QND photon counting with a homodyne mixing method, we reconstruct the Wigner functions of these non-classical states and, by taking snapshots of these functions at increasing times, obtain movies of the decoherence process in the cavity.

Keywords: Cavity QED, QND measurements, Zeno effect, quantum state reconstruction

1. A weak dissipation photon trap to look at light in a new way

In atom or ion trap experiments, small particle ensembles held in a localized region of space are studied and probed by light beams. A wide range of experiments has been realized on these systems, demonstrating various kinds of multi-atom state engineering and tomographic reconstruction processes.[1] We have realized the opposite situation in which microwave quanta confined in a cavity are manipulated and interrogated by an atomic beam. By extracting information from the trapped field, we can count photons without destroying them, follow single realizations of the field exhibiting photon number quantum jumps, engineer various non-classical states of light, fully reconstruct them and observe how these states evolve under the effect of decoherence. In these experiments, the radiation field becomes "an object of investigation", which can be tailored and repeatedly looked upon, as is done with material particles in ion trap studies.

To perform these experiments, we use circular Rydberg atoms, very sensitive to microwave radiation, which cross the cavity one at a time.[2] The main challenge has been to assemble a cavity storing the field without losing a single photon while hundreds of atoms interact with it. We have realized such a cavity,[3] made of two mirrors in a Fabry Perot configuration, between which light bounces more than a billion time before decaying (Figure 1). The field lifetime, $T_c = 0.13$ s, is three to four orders of magnitude larger than the time each atom spends in the cavity. This exceptional long damping time has been obtained with superconducting-Niobium-coated copper mirrors machined to a very high precision. The combination of extremely small surface roughness minimizing scattering losses with the near-zero resistance of the superconducting layer yields an unprecedented finesse of $4.2 \, 10^9$, about four orders of magnitude larger than that of the best optical Fabry Perot resonators.

Fig. 1. The Fabry Perot superconducting cavity trapping microwave photons (mirror diameter: 5 cm, the mirror distance – 2.7 cm in the real set-up – is here exaggerated for clarity).

2. Quantum non-demolition photon counting

Usual photon counting methods are generally destructive. Light quanta are absorbed by photosensitive materials and transformed into electrical signals, so that the photons disappear while being counted. The principle of non-destructive light intensity detection (of the quantum non-demolition or QND type) has been proposed in the 1970s and demonstrated in the 1990s with signal laser beams propagating in transparent dispersive media whose refractive index was changed by the light irradiation.[4] This change affects the phase of a probe beam co-propagating with the signal in the medium and an interferometric detection of this phase change is used to measure in a non-destructive way the intensity of the signal beam. These experiments, relying on non-linear optical processes, require relatively intense fields and do not have the sensitivity of single photon detection. In our Cavity QED experiment, we implement a different kind of QND method, proposed by our group in 1990,[5] which uses as probes circular Rydberg atoms exhibiting a transition between two nearby excited states e and g that is slightly off-resonant with the field to be measured. (e and g are the circular Rydberg states with principal quantum numbers 51 and 50 respectively.) We take advantage of the extremely large coupling of Rydberg atoms to microwave radiation in order to induce on this transition light shifts sensitive to single photons. Measuring the frequency of the atomic transition amounts to counting the photons without destroying them, since the non-resonant atoms cannot absorb radiation.

The experimental set-up is a Ramsey interferometer with two auxiliary cavities R_1 and R_2 between which the high-Q cavity C containing the field to be measured is inserted (Figure 2). In short, it is an atomic clock with trapped photons inside. The atoms, which cross the apparatus one by one, are prepared in a symmetric superposition of e and g by a microwave pulse in R_1 and the evolution of this superposition after it has crossed C is analysed with a second pulse applied in R_2. The final detection of the atoms in e or g by the field ionization detector D provides the information required to determine the atomic clock's phase. This phase is affected by the field in C and the resulting delay of the clock allows us to count non-destructively the light quanta.

We have started by studying the simple situation where the cavity stores a weak thermal field, fluctuating between 0 and 1 photon.[6] The light-shift is adjusted to produce, for a single photon, a π-phase shift of the atomic state superposition. The QND measurement then yields a telegraphic signal. A time interval during which hundreds of atoms are mostly detected in level

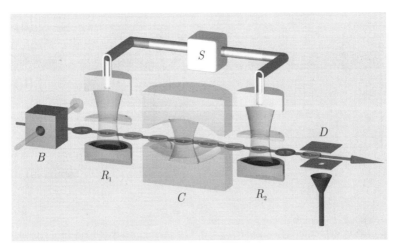

Fig. 2.   Scheme of experimental set-up. Atoms are prepared in the circular Rydberg state g in box B. They are brought to the symmetric superposition of e and g by a microwave pulse applied in the auxiliary microwave cavity R_1, then cross the superconducting cavity C which stores the photons to be counted. They are finally subjected to an analyzing microwave pulse in cavity R_2, before being counted by the detector D. The classical source S feeds the Ramsey cavities R_1 and R_2.

g, signalling that the cavity is empty, suddenly switches to a long sequence of atoms mostly detected in e, signalling that one photon has appeared between the mirrors. The sequences reverts back to level g when this photon subsequently disappears (Figure 3). These signals reveal for the first time the quantum jumps of light associated to the gain or the loss of a single light quantum exchanged with the cavity mirrors.

The method was then generalized to deal with stored fields containing on average several photons.[7] The field is produced by a coherent microwave source coupled to the cavity mode by diffraction on the mirror edges (source not shown in Figure 2). The field prepared in this way exhibits a poissonian statistics of its photon number. Each atom provides a partial information about this number. The phase shift per photon is adjusted to a fraction $2\pi/(n_m+1)$ where n_m is the maximum number of photons to be measured. We have considered fields with up to $n_m = 7$ photons. After about a hundred atoms has been detected, there is enough information to pin down precisely the atomic state superposition phase and hence the field intensity. The experiment realizes in this way a progressive projection of the initial coherent state onto a Fock state with a well-defined photon number. If the experiment is resumed under the same conditions after preparing anew the

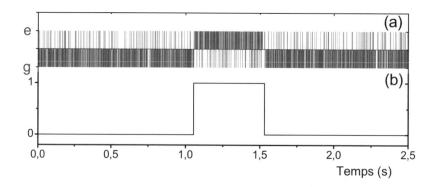

Fig. 3.   QND detection of one photon. (a) The atoms mostly found in g (lower bars) signal that the cavity is in vacuum (0 photon) while the atoms detected mostly in e (upper bars) signal that the cavity stores 1 photon. (b) Photon number determined by a majority vote involving 8 consecutive atomic detections. The quantum jumps from 0 to 1 and back are clearly visible. The photon detected on this trace survives about 0.5 s, i.e. four cavity damping times. Reproduced with permission of MacMillan publishers ltd: Nature **446**, 297 (2007).

same initial state, we again obtain a Fock state, generally different from the previous one, according to a random process. By accumulating enough statistics on a large number of realizations, we reconstruct the photon number distribution of the initially prepared state. This experiment verifies all the postulates of a quantum measurement of light. It also provides a practical way of generating Fock states with large photon numbers. Such states are very hard to generate by other means.

Once the field has been projected onto a Fock state, we can go on measuring its energy with a longer atomic sequence and observe in this way the relaxing field in the cavity. We then observe a staircase-like evolution of the field energy. It reveals the random times at which the field, evolving irreversibly towards vacuum, jumps from one photon number to the next (Figure 4). These signals provide a quantum picture of the field evolution quite different from that given by classical physics which predicts a smooth exponential decay. This continuous behaviour is recovered by averaging a large number of staircase like signals.

The QND recording of the photon number can have strange consequences if one tries to observe continuously the evolution of a field de-

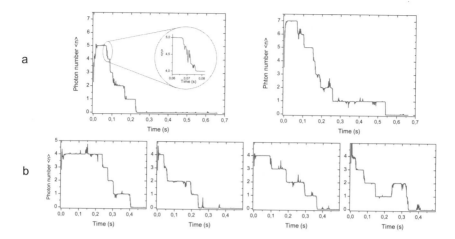

Fig. 4. Evolution of the mean photon number in the cavity along some realizations of the field. The traces (a) correspond to an initial projection of the field in the Fock state $n = 5$ (left) and $n = 7$ (right) respectively. The inset expands the time scale and shows that a quantum jump requires several milliseconds to be detected (time interval corresponding to the crossing of a few atoms). The traces (b) exhibit four examples of realizations corresponding to the initial projection of the field in the Fock state $n = 4$. Reproduced with permission of MacMillan publishers ltd: Nature **448**, 889 (2007).

scribed by a unitary coherent process. In this case, the mere observation of the photon number prevents the field from evolving. This is the quantum Zeno effect.[8] We have demonstrated it in an experiment in which we have tried to build a coherent field in the cavity by coupling it to a periodic sequence of radiation pulses produced by a classical source.[9] If the field is not measured between the pulses, its amplitude increases proportionally to the number of pulses and the average photon number starts to build up as the square of the elapsed time. If the field is instead measured in a QND way between the pulses, it is repeatedly projected back onto vacuum and the field growth is practically frozen. This experiment demonstrates for the first time the Zeno effect on an harmonic oscillator (here a field mode). It can also be interpreted as a manifestation of the backaction of the QND measurement of the field intensity on its phase. As the phase of the field injected by each pulse is periodically scrambled by the QND measurement, its complex amplitude undergoes a two dimensional Brownian motion which keeps the field near the phase space origin.

3. Reconstructing the quantum state of a non-classical field and observing its decoherence

The QND measurement of the field intensity allows us, by repeating the procedure on a large sample of realizations, to reconstruct the photon number distribution of the initial field. This quantity provides only a partial information about the field's quantum state. It does not tell us anything about the coherences between Fock states. This additional information can however be obtained by a homodyne mixing method. We translate the field in its phase space (by mixing it with a reference field of known amplitude and phase) and we subsequently measure in a QND way the photon number distribution of the translated field. By resuming the procedure on many realizations, with translations corresponding to various amplitudes and phases, we obtain enough information to reconstruct completely the state of the field. It is equivalently represented either by a density operator or by a Wigner function.[10] The latter is a real distribution in the phase plane which looks like a three dimension geographical map. Each of its points, defined by its distance to the plane origin and its direction, is associated to a value of the light field amplitude and phase. Usual radiation (emitted by heated bodies, by lasers or by a combination of such sources) is quite generally described by a landscape of positive peaks centred at the points corresponding to the most probable values of the field.

Non-classical fields have much less intuitive features. Their Wigner functions present oscillations exhibiting negative values in some areas of the phase plane. By measuring the field with QND Rydberg atom probes, we have reconstructed the maps of these strange states.[11] Fields with well-defined photon numbers have Wigner functions presenting concentric oscillations. So called Schrödinger cat states, which are quantum superpositions of classical states with different phases, are described by Wigner functions with two positive peaks corresponding to their classical components and, in between, a landscape of alternating positive ridges and negative valleys. We have initially prepared these states in the cavity (using a first atom dispersively interacting with a coherent field[12]) and then reconstructed these states at increasing times, performing field translations and using subsequent probe atoms serving as QND probes of the translated field intensity. We have obtained in this way snapshots of Schrödinger cat Wigner functions and observed the progressive vanishing of their negative features (Figure 5). This experiment reveals the fragility of the non-classical states which rapidly evolve under the effect of decoherence into classical states represented by strictly positive Wigner functions. Note that non-classical

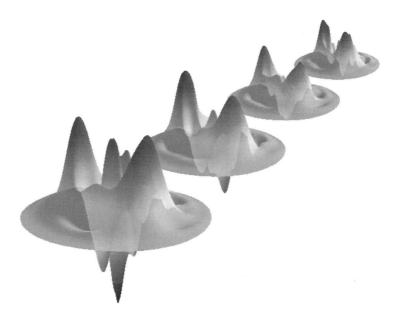

Fig. 5. Reconstructed Wigner function of a Schrödinger cat state at four successive times (3.3, 15.0, 32.2 and 46.3 ms after state preparation). Initially (lower left), the Wigner map exhibits two positive peaks corresponding to the cat's classical components. Between them, an oscillatory interference structure reveals the non-classical nature of the cat state. The interference pattern progressively vanishes as time goes. Finally (upper right) only the positive classical components remain.

field states with Wigner functions having features similar to those of the Schrödinger cat states described here have been recently reconstructed on propagating optical fields.[13] Decoherence was not observed in this experiment.

Being able to reconstruct in details the dynamics of the phenomenon that is at the heart of the quantum to classical boundary opens the way to the manipulation and control of decoherence. Procedures of quantum-feedback[14] become possible, in which atoms will be used to maintain in real time the non-classical features of a light field, and thus preserve the quantum properties which are essential for the realization of quantum information operations with light.

Acknowledgements

This work was supported by the Agence Nationale pour la Recherche (ANR), by the Japan Science and Technology Agency (JST), and by the

EU under the IP project SCALA. S.D. is funded by the Délégation Générale à l'Armement (DGA).

References

1. D. Leibfried, R. Blatt, C. Monroe and D. Wineland, *Rev. Mod. Phys.* **75**, 281 (2003).
2. J. M. Raimond, M. Brune and S. Haroche, *Rev. Mod. Phys* **73**, 565 (2001).
3. S. Kuhr, S. Gleyzes, C. Guerlin, J. Bernu, U. Busk Hoff, S. Deléglise, S. Osnaghi, M. Brune, J. M. Raimond, S. Haroche, E. Jacques, P. Bosland and B. Visentin, *Appl. Phys. Lett.* **90**, 164101 (2007).
4. P. Grangier, J. A. Levenson and J. P. Poizat, *Nature* **396**, 537 (1998).
5. M. Brune, S. Haroche, V. Lefèvre, J. M. Raimond and N. Zagury, *Phys. Rev. Lett.* **65**, 976 (1990).
6. S. Gleyzes, S. Kuhr, C. Guerlin, J. Bernu, S. Deléglise, U. Busk Hoff, M. Brune, J. M. Raimond and S. Haroche, *Nature* **446**, 297 (2007).
7. C. Guerlin, J. Bernu, S. Deléglise, C. Sayrin, S. Gleyzes, S. Kuhr, M. Brune, J. M. Raimond and S. Haroche, *Nature* **448**, 889 (2007).
8. B. Misra and E.C.G. Sudarshan, *J. Math. Phys. Sci.* **18**, 756 (1977).
9. J. Bernu, S. Deléglise, C. Sayrin, S. Kuhr, I. Dotsenko, M. Brune, J. M. Raimond and S. Haroche, *Phys. Rev. Lett.*, to be published (2008); arXiv 0809.4388 (2008).
10. S. Haroche and J.M.Raimond, *Exploring the quantum, atoms cavities and photons* (Oxford University Press, 2006).
11. S. Deléglise, I. Dotsenko, C. Sayrin, J. Bernu, M. Brune, J. M. Raimond and S. Haroche, *Nature* **455**, 510 (2008).
12. M. Brune, E. Hagley, J. Dreyer, X. Maître, A. Maali, C. Wunderlich, J. M. Raimond and S. Haroche, *Phys. Rev. Lett.* **77**, 4887 (1996).
13. A. Ourjoumtsev, H. Jeong, R. Tualle-Brouri and P. Grangier, *Nature* **448**, 784 (2007).
14. S. Zippilli, D. Vitali, P. Tombesi and J. M. Raimond, *Phys. Rev. A* **67**, 052101 (2003).

SPIN SQUEEZING ON AN ATOMIC-CLOCK TRANSITION

MONIKA H. SCHLEIER-SMITH, IAN D. LEROUX, and VLADAN VULETIĆ

Department of Physics, MIT-Harvard Center for Ultracold Atoms, and Research Laboratory of Electronics
Massachusetts Institute of Technology, Cambridge, Massachusetts 02139, USA
* E-mail: vuletic@mit.edu, http://www.rle.mit.edu/eap/

We generate input states with reduced quantum uncertainty (spin-squeezed states) for a hyperfine atomic clock by collectively coupling an ensemble of laser-cooled and trapped ^{87}Rb atoms to an optical resonator. A quantum non-demolition measurement of the population difference between the two clock states with far-detuned light produces an entangled state whose projection noise is reduced by as much as 9.4(8) dB below the standard quantum limit (SQL) for uncorrelated atoms. When the observed decoherence is taken into account, we attain 4.2(8) dB of spin squeezing, confirming entanglement, and 3.2(8) dB of improvement in clock precision over the SQL. The method holds promise for improving the performance of optical-frequency clocks.

Keywords: Spin squeezing; quantum noise; atomic clock.

1. Introduction: Projection Noise and the Standard Quantum Limit

In an atomic clock[1-3] or an atom interferometer,[4-6] the energy difference between two states is measured as a quantum mechanical phase accumulated in a given time, and the result read out as a population difference between the two states. An elegant and insightful description of the signal and noise[7,8] uses the angular-momentum formalism, where each individual atom i is formally associated with a spin $s_i = \frac{1}{2}$ system, while the ensemble is described by the total spin vector $\mathbf{S} = \sum_i \mathbf{s}_i$. Symmetric states of the ensemble of N_0 particles are then characterized by an ensemble spin quantum number S given by $S = \frac{1}{2}N_0$, while non-symmetric states correspond to a smaller quantum number, $S < \frac{1}{2}N_0$. An arbitrary symmetric state of N_0 uncorrelated particles (coherent spin state, or CSS) is described by an ensemble spin vector with maximal projection $S_1 = S$ along some direction $\mathbf{e}_1$ (see Fig. 1). Note that the length of the spin vector, $\sqrt{\langle S^2 \rangle} = \sqrt{S(S+1)}$

Fig. 1.   Illustration of a coherent spin state (CSS). For N_0 atoms, the state is represented by a circle of radius $\sqrt{S}$ on a Bloch sphere of radius $\sqrt{S(S+1)}$, where $S = N_0/2$.

(in units of $\hbar$) is larger than S, due to the fact that quantum mechanics imposes non-vanishing expectation values $\langle S_2^2 \rangle$, $\langle S_3^2 \rangle$ for the transverse spin components S_2, S_3. Graphically, the CSS thus corresponds to the circular intersection of a sphere of radius $\sqrt{S(S+1)}$ with the plane perpendicular to $\mathbf{e}_1$ at distance S from the origin. The finite radius $\sqrt{S}$ of the circle represents the angular momentum uncertainties $\Delta S_2 = \Delta S_3 = \sqrt{S/2}$. The possible measurement outcomes along any direction correspond to planes slicing the sphere at positions $M = -S, -S+1, \ldots S$ relative to the origin. For a CSS in the xy equatorial plane, which is the final state of a Ramsey clock sequence, the binomial distribution of possible $M = S_z$ values associated with the statistically independent measurement outcomes for the individual particles constitutes a fundamental source of noise that limits the precision of the measurement[7-9] at the standard quantum limit (SQL).

The SQL is the fundamental limit for measurements with ensembles of uncorrelated particles. However, quantum mechanics allows one to redistribute the quantum noise between different degrees of freedom by entangling the atoms in the ensemble. In Fig. 2(c) we represent the state of the system by a quasiprobability distribution of the noncommuting angular momentum components. The projection noise can be suppressed by reducing the quantum uncertainty in the variable of interest S_z at the expense of another variable, e.g. S_y, that is not directly affecting the experiment precision;[7,8] this corresponds to squeezing the circular uncertainty region of the CSS into an elliptical one. The redistribution of quantum noise for a system with a finite number of discrete states is referred to as "spin squeezing".[10] A state with reduced quantum uncertainty S_z is called "number squeezed". A state along x with reduced S_y is called "phase squeezed" (Fig. 2(c) iii, iv). The two states can be converted into each other by a common rotation of all individual spins.

Note that to demonstrate spin squeezing, it is necessary not only to measure the spin noise along some direction, but also to determine the length of the spin vector S, since processes that differently affect the individual spins $\mathbf{s}_i$ reduce the ensemble spin vector $\mathbf{S} = \sum \mathbf{s}_i$. The ensemble spin can be measured by determining the visibility of Rabi or Ramsey oscillations.[7,8] For an ensemble spin vector $\mathbf{S}$ oriented along the x axis, a state is number squeezed or phase squeezed[10–13] if $(\Delta S_z)^2 < |\langle S_x \rangle|/2$ or $(\Delta S_y)^2 < |\langle S_x \rangle|/2$, respectively.

Spin squeezing requires a Hamiltonian that is at least quadratic in the spin components, or equivalently, some form of interaction between the particles. While it is possible to use interatomic collisions in a Bose-Einstein condensate (BEC) for that purpose,[14,15] these density-dependent interactions are difficult to control in the setting of a precision measurement. An alternative proposal is to use the collective interaction of an atomic ensemble with a mode of an electromagnetic field.[16] In this approach, the ensemble interacts with a far-detuned light field, resulting in an entanglement between the ensemble spin S_z and the phase or amplitude of the light field. A subsequent near-quantum-limited measurement of the light results in a conditionally spin-squeezed state of the ensemble. The word "conditionally" signifies here that the particular spin-squeezed state that is created depends on the outcome of the measurement on the light field. If one were to ignore (trace over) the state of the light, no entanglement would be evident in the atomic state.

Nevertheless, even conditionally spin-squeezed input states can improve the sensitivity of an atomic clock,[17] since one can use the outcome of the measurement of the light field to determine the clock phase with improved precision compared to the SQL. A perhaps even more attractive possibility is to use the information gained *during* the measurement of the light field to steer the atomic quantum state to a desired location,[12,13,18] thus converting the conditional into unconditional spin squeezing.

In atomic Bose-Einstein condensates, interaction-induced spin-noise reduction below the projection noise limit has been inferred from an increased noise in another spin component,[19] and from a lengthening in coherence time in a system with atom-number-dependent mean-field energy.[15] In room-temperature vapor, spin squeezing[20] has been achieved by absorption of squeezed light,[21] and two-mode squeezing has been attained by a quantum non-demolition (QND) measurement on a light beam that has interacted with two ensembles.[22] A QND measurement[16] has been used to reduce the noise of a rotating spin in a room-temperature vapor below the projection noise limit, $(\Delta S_z)^2 < S_0/2$, but the length of the spin vector $|\langle S_x \rangle|$ was not measured.[23] The papers by Geremia *et al.* reporting spin squeezing for atoms with $s > \frac{1}{2}$ using a similar QND approach for cold atoms were recently retracted.[24] Light-induced squeezing within individual atoms of large spin $s = 3$, without squeezing the ensemble spin, has recently been demonstrated.[25]

2. Spin Squeezing by Optical Quantum Non-Demolition Measurement

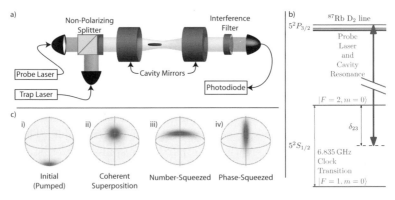

Fig. 2.   **Measurement-induced pseudo-spin squeezing on an atomic clock transition. (a) Setup.** A laser-cooled ensemble of ^{87}Rb atoms is loaded into a far-detuned optical dipole trap inside an optical resonator. A population difference N between hyperfine clock states $|1\rangle , |2\rangle$ produces a resonator frequency shift that is measured with a probe laser. **(b) Atomic level structure.** The resonator is tuned such that atoms in the two clock states produce equal and opposite resonator frequency shifts via the state-dependent atomic index of refraction. **(c) Preparing a squeezed input state for an atomic clock.** A number-squeezed state (iii) can be generated from a CSS along x (ii) by measurement of N. It can then be rotated by a microwave pulse into a phase-squeezed state (iv), allowing a more precise determination of the phase acquired in the free-evolution time of the atomic clock.

To prepare a spin-squeezed input state to an atomic clock, we adapt the proposal by Kuzmich, Bigelow, and Mandel[16] for a QND measurement of S_z with far off-resonant light.[22,23] By using the interaction of an optically thick ensemble with a single electromagnetic mode, the number of atoms in each of the clock states can be established beyond the projection noise limit without substantially reducing the system's coherence. For an optical depth exceeding unity, an accurate measurement of the atomic index of refraction, which can be viewed as a homodyne measurement of the forward-scattered field, with the directly transmitted field acting as the local oscillator, can be performed faster than the scattering of photons into free space reveals the states of the individual atoms and destroys the coherence. The attainable squeezing, in terms of variances, improves as the square root of the optical depth, which is why we use an optical resonator whose finesse $\mathcal{F} = 5600$ increases the optical depth by a factor of $\mathcal{F}/\pi \approx 1800$.

An ensemble of up to 5×10^4 laser-cooled ^{87}Rb atoms is trapped in a far-detuned optical dipole trap inside the optical resonator (Fig. 2). One resonator mode is tuned such that the state-dependent atomic index of refraction produces a mode frequency shift ω that is proportional to the population difference $N = N_2 - N_1 = 2S_z$ between the hyperfine clock states $|1\rangle = |5^2S_{1/2}, F = 1, m_F = 0\rangle$ and $|2\rangle = |5^2S_{1/2}, F = 2, m_F = 0\rangle$. The frequency shift is determined from our accurately measured resonator parameters as $d\omega/dN = 48(2)$ Hz/atom. This value is confirmed experimentally by measurement of the dual effect, namely the energy shift of the atomic levels by the intracavity light, that results in a phase shift between the clock levels of $\phi_{12} = 250(20)$ μrad per probe photon sent through the resonator. Given $d\omega/dN$, the average spin $\langle S_z \rangle$ and variance $(\delta S_z)^2$ are calculated from typically 50 repeated transmission measurements of a probe pulse tuned to the slope of the resonator mode. Light pulses of duration $T = 50$ μs, much longer than the resonator decay time of $\tau = \kappa^{-1} = 158$ ns, containing 10^5 to 10^6 photons traverse the atom-resonator system and are detected with an overall quantum efficiency of $Q_e = 0.43(4)$. A frequency stabilization system for probe laser and resonator ensures that the probe transmission noise is close to the photocurrent shot-noise limit. One of the experimental challenges is to stabilize the resonator length sufficiently well to resolve the mode shift due to atomic projection noise, typically a few kHz out of a 1 MHz resonator linewidth, while using light levels that lead only to a modest decoherence between the clock states.

We verify experimentally the projection noise level for the coherent spin state (CSS) of an uncorrelated ensemble[7,8,23] by measuring probe transmission for $p = 5 \times 10^5$ photons transmitted on average through the resonator. To reduce the effect of trap loading fluctuations, we perform a CSS preparation and measurement sequence (consisting of optical pumping into state $|1\rangle$, $\pi/2$ pulse, and measurement of S_z) twice with the same loaded atoms and determine the variance $(\delta S_z)^2$ between the two measurements. As a function of (effective) atom number N_0, projection noise is characterized by a variance $(\delta S_z)^2 \propto N_0$, while for technical noise $(\delta S_z)^2 \propto N_0^2$. (In

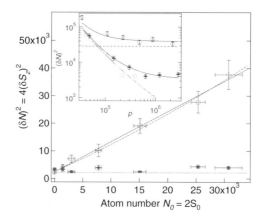

Fig. 3. Projection noise limit and spin noise reduction. The measured spin noise for an uncorrelated state (CSS, open circles) agrees with the theoretical prediction $(\Delta S_z)^2 = S_0/2$, with negligible technical noise (solid and dashed lines). Our measurement of S_z at photon number $p = 5 \times 10^5$ has an uncertainty $(\delta S_z)^2$ (solid diamonds) substantially below the SQL. **Inset:** Dependence of spin measurement variance $(\delta S_z)^2 = (\delta N)^2/4$ on probe photon number p for $N_0 = 3 \times 10^4$. With increasing photon number, the measurement uncertainty (solid diamonds) drops below the projection noise level $(\Delta S_z)^2_{\rm CSS} = aS_0/2$ (dashed line), while the variance measured for independently prepared CSSs (open circles) approaches $(\Delta S_z)^2_{\rm CSS}$. Also shown is the technical noise without atoms, expressed as an equivalent spin noise (open squares).

a standing-wave resonator with spatially-modulated atom-cavity coupling, we define the effective atom number $N_0 = \frac{4}{3}N_{tot}$ as the ideal projection noise variance for N_{tot} atoms evenly distributed along the cavity axis.) Unlike other experiments,[20,22,23] we have a reliable and accurate absolute calibration of the atom number via the resonator shift and can not only test the linear dependence $(\delta S_z)^2 = aN_0$ but also compare the slope a to a calculated value that takes into account the spatially inhomogeneous coupling between the trapped atoms and the probe light. Fig. 3 shows the dependence of variance $(\delta N)^2 = 4(\delta S_z)^2$ on atom number $N_0 = 2S_0$ (open circles). The fitted slope $a_f = 1.1(1)$ is slightly higher than the calculated value $a_c = 0.93(1)$ due to technical noise at large atom number. If we fix $a = a_c = 0.93$ and fit this quadratic technical noise, we find a small contribution $(\delta S_z)^2_{\rm tech} = 6(4) \times 10^{-6} N_0^2 \ll N_0$ (dashed curve in Fig. 3). This confirms that we have a system dominated by projection noise, and quantitatively establishes the SQL.

We prepare a state with conditionally reduced noise $(\Delta S_z)^2$ simply by measuring S_z for a CSS along x with a photon number sufficiently large to resolve S_z beyond the CSS variance $(\Delta S_z)^2_{\rm CSS} = S_0/2$. This measurement with variance $(\delta S_z)^2$ prepares a state with a random but known value of S_z whose quantum uncertainty is $(\Delta S_z)^2 = (\Delta S_z)^2_{CSS} (\delta S_z)^2 / ((\Delta S_z)^2_{CSS} + (\delta S_z)^2)$. (Throughout this report, δS_z refers to a measured standard deviation, while ΔS_z denotes a quantum uncertainty for the pure or mixed state that we are preparing. ΔS_z differs from δS_z because it includes the prior knowledge that the state is initially prepared as a coherent state along x. The distinction has little effect for strong squeezing, but for weak squeezing ensures that the initial quantum uncertainty is taken into account correctly.[26]) The faithfulness of the state preparation is verified with a second measurement, and we plot the variance of the two measurements $(\delta N)^2 = 4 (\delta S_z)^2$ vs. atom number N_0 in Fig. 3 (solid diamonds). While at low atom number the measurement noise exceeds the SQL due to photon shot noise and some technical noise (dash-dotted line in Fig. 3), at higher atom number $N_0 = 3 \times 10^4$ we achieve a 9.4(8) dB suppression of spin noise below the SQL.

The inset to Fig. 3 shows $(\delta N)^2$ vs. average transmitted photon number p at fixed $N_0 = 3 \times 10^4$ for the CSS as well as for the reduced-uncertainty state. At low p, photon shot noise prevents observation of the spin projection noise level (dashed line). For large p the observed noise for the CSS (open circles) reaches a plateau that corresponds to spin projection noise, while the squeezing measurement localizes the value of S_z to better than the projection noise (solid diamonds). For photon numbers $p \leq 5 \times 10^5$ the squeezing measurement is close to the technical noise without atoms (open squares).

Having established that we can prepare states with spin noise ΔS_z below the projection limit, we need to verify whether the system remains sufficiently coherent to guarantee entanglement. The prepared state is spin squeezed, and thereby entangled,[10] if $\zeta_{KU} = 2 (\Delta S_z)^2 / (a |\langle \widetilde{S} \rangle|) < 1$, where $\widetilde{S}$ is the ensemble spin in the xy-plane.[10] Fig. 4 shows, as a function of photon number in the preparation pulse, the normalized spin-noise $(\Delta S_z)^2 / (\Delta S_z)^2_{CSS}$ (open diamonds), and the measured clock contrast $C = |\langle \widetilde{S} \rangle| / S_0$ (open squares). Shown also is the squeezing parameter ζ_{KU} obtained by dividing the observed spin-noise reduction by C, demonstrating that we have achieved 4.2(8) dB of spin squeezing for $p = 3 \times 10^5$. We emphasize that in this analysis we use the full observed noise, including photon shot noise and all technical noise, and all contrast reduction, including contrast loss due to the resonator locking light (evident as finite contrast $C_{in} = 0.7$ for no probe pulse ($p = 0$) in Fig. 4). We find that C_{in} can be improved compared to Fig. 4 by choosing a larger detuning from atomic

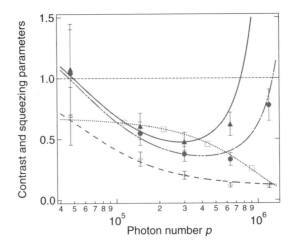

Fig. 4. Spin noise reduction, loss of contrast, and spin squeezing. The reduction of normalized spin noise $(\Delta S_z)^2 / (\Delta S_z)^2_{\text{CSS}}$ (open diamonds and dashed curve) below unity is accompanied by a loss of coherence observable as a reduced contrast C (open squares and dotted curve) in a Ramsey clock sequence. From these two measurements, we can deduce two squeezing parameters (see text), $\zeta_{\text{KU}} = 2\,(\Delta S_z)^2 /(a|\langle \tilde{S} \rangle|)$ (solid circles and dash-dotted curve), which characterizes the entanglement of the squeezed state, and $\zeta_{\text{W}} = 2\,(\Delta S_z)^2\, S_{\text{in}}/(a|\langle \tilde{S} \rangle|^2)$ (solid triangles and solid curve), which characterizes the squeezing-induced improvement in clock performance.

resonance for the lock light. (In Fig. 4 that detuning is $\sim 14\,\text{GHz}$.) The contrast reduction due to the probe light is probably due to a motion-induced fluctuation of the differential light shift between the clock states, and can be reduced by cooling the atoms further. The fundamental lower limit for contrast loss, set by the scattering of photons into free space, should allow the squeezing parameter ζ_{KU} to approach the 9 dB spin noise reduction observed at our highest probe photon numbers $p > 1 \times 10^6$. If technical noise can be reduced further, the fundamental limit associated with scattering is set by the optical depth OD of the sample[27] and for our present parameters ($OD = 5 \times 10^3$) amounts to ~ 18 dB of spin squeezing.

The usefulness of the state for precision measurements is quantified by the more stringent parameter[7,8] $\zeta_{\text{W}} = 2\,(\Delta S_z)^2\, S_{\text{in}}/(a|\langle \tilde{S} \rangle|^2) < 1$. This expression is easily understood as a reduction of the squared noise-to-signal ratio $(\Delta S_z)^2 /|\langle \tilde{S} \rangle|^2$ relative to its value in the unsqueezed coherent state $a/(2S_{\text{in}})$. For our system, the ensemble without squeezing has $S_{\text{in}} = S_0 C_{\text{in}}$, yielding $\zeta_{\text{W}} = \zeta_{\text{KU}} C_{\text{in}}/C$. This parameter, also plotted in Fig. 4, shows an improvement in clock precision of 3.2(8) dB.

3. Outlook

We have verified that the prepared number-squeezed state can be converted into a phase-squeezed state by a $\pi/2$ microwave pulse about $\langle S \rangle$, and used as an input state to a Ramsey type atomic clock. Note that the spin vector precesses through many revolutions in a typical atomic clock. Therefore in an optical-transition atomic-ensemble clock,[2,3] fractional frequency accuracies of 10^{-16} can be achieved with fairly modest absolute phase accuracies[2] of $\Delta\phi \sim 10^{-2}$, which can readily be improved by the squeezing technique investigated here. It should also be possible to apply this squeezing technique to atom interferometers[6] and other precision experiments with atomic ensembles. We believe that most of the technical limitations in the current experiment, such as remaining technical transmission noise due to imperfect laser-resonator frequency stabilization, and contrast loss due to spatially inhomogeneous light shifts, can be overcome in the near future, allowing for squeezing near the fundamental limit set by the sample's optical depth. Since even the latter can be improved by simply loading more atoms into the trap, we believe that 15 to 20 dB of spin squeezing should not represent an unrealistic goal for the near future.

Acknowledgments

We thank J. K. Thompson, M. D. Lukin, D. Stamper-Kurn, I. Teper, M. Kasevich, and E. Polzik for interesting discussions. This work was supported in part by the NSF, DARPA, and the NSF Center for Ultracold Atoms. M. Schleier-Smith acknowledges support from the Hertz Foundation Daniel Stroock Fellowship and NSF. I. D. Leroux acknowledges support from NSERC.

The group of E. Polzik has independently and simultaneously achieved results[28] similar to ours[29] in a Mach-Zehnder interferometer. In this meeting, M. Oberthaler and coworkers report spin squeezing in a Bose-Einstein condensate by atomic interactions in a multiple-well potential.[30]

References

1. G. Santarelli, P. Laurent, P. Lemonde, A. Clairon, A. G. Mann, S. Chang, A. N. Luiten and C. Salomon, *Phys. Rev. Lett.* **82**, 4619 (1999).
2. M. Takamoto, F.-L. Hong, R. Higashi and H. Katori, *Nature* **435**, 321 (2005).
3. R. Santra, E. Arimondo, T. Ido, C. H. Greene and J. Ye, *Phys. Rev. Lett.* **94**, 173002 (2005).
4. M. Kasevich and S. Chu, *Phys. Rev. Lett.* **67**, 181 (1991).
5. D. S. Weiss, B. C. Young and S. Chu, *Phys. Rev. Lett.* **70**, 2706 (1993).

6. J. B. Fixler, G. T. Foster, J. M. McGuirk and M. A. Kasevich, *Science* **315**, 74 (2007).

7. D. J. Wineland, J. J. Bollinger, W. M. Itano, F. L. Moore and D. J. Heinzen, *Phys. Rev. A* **46**, R6797 (1992).

8. D. J. Wineland, J. J. Bollinger, W. M. Itano and D. J. Heinzen, *Phys. Rev. A* **50**, R67 (1994).

9. J. L. Sørensen, J. Hald and E. S. Polzik, *Phys. Rev. Lett.* **80**, 3487 (1998).

10. M. Kitagawa and M. Ueda, *Phys. Rev. A* **47**, 5138 (1993).

11. A. S. Sørensen and K. Mølmer, *Phys. Rev. Lett.* **86**, 4431 (2001).

12. L. K. Thomsen, S. Mancini and H. M. Wiseman, *J. Phys. B: At., Mol. Opt. Phys.* **35**, 4937 (2002).

13. L. K. Thomsen, S. Mancini and H. M. Wiseman, *Phys. Rev. A* **65**, 061801 (2002).

14. A. Sørensen, L.-M. Duan, J. I. Cirac and P. Zoller, *Nature* **409**, 63 (2001).

15. G.-B. Jo, Y. Shin, S. Will, T. Pasquini, M. Saba, W. Ketterle and D. Pritchard, *Phys. Rev. Lett.* **98**, 030407 (2007).

16. A. Kuzmich, N. P. Bigelow and L. Mandel, *Europhys. Lett.* **42**, 481 (1998).

17. A. André, A. S. Sørensen and M. D. Lukin, *Phys. Rev. Lett.* **92**, 230801 (2004).

18. H. M. Wiseman, *Phys. Rev. A* **49**, 2133 (1994).

19. C. Orzel, A. K. Tuchman, M. L. Fenselau, M. Yasuda and M. A. Kasevich, *Science* **291**, 2386 (2001).

20. A. Kuzmich, K. Mølmer and E. S. Polzik, *Phys. Rev. Lett.* **79**, 4782 (1997).

21. L.-A. Wu, H. J. Kimble, J. L. Hall and H. Wu, *Phys. Rev. Lett.* **57**, 2520 (1986).

22. B. Julsgaard, A. Kozhekin and E. S. Polzik, *Nature* **413**, 400 (2001).

23. A. Kuzmich, L. Mandel and N. P. Bigelow, *Phys. Rev. Lett.* **85**, 1594 (2000).

24. J. M. Geremia, J. K. Stockton and H. Mabuchi, *Science* **304**, 270 (2004) and J. M. Geremia, J. K. Stockton and H. Mabuchi, *Phys. Rev. Lett.* **94**, 203002 (2005). Retractions available at J. M. Geremia, J. K. Stockton and H. Mabuchi, *Science* **321** 489 (2008) and J. M. Geremia, J. K. Stockton and H. Mabuchi, *Phys. Rev. Lett.* **101**, 039902(E) (2008).

25. S. Chaudhury, G. A. Smith, K. Schulz and P. S. Jessen, *Phys. Rev. Lett.* **96**, 043001 (2006).

26. K. Hammerer, K. Mølmer, E. S. Polzik and J. I. Cirac, *Phys. Rev. A* **70**, 044304 (2004).

27. I. Bouchoule and K. Mølmer, *Phys. Rev. A* **66**, 043811 (2002).

28. J. Appel, P. Windpassinger, D. Oblak, U. Hoff, N. Kjaergaard and E. S. Polzik, submitted to *Science* (2008).

29. M. H. Schleier-Smith, I. D. Leroux and V. Vuletić, submitted to *Science* (2008).

30. M. Oberthaler, *XXI International Conference on Atomic Physics*, Storrs (2008).

QUANTUM MICRO-MECHANICS WITH ULTRACOLD ATOMS

THIERRY BOTTER[1], DANIEL BROOKS[1], SUBHADEEP GUPTA[2],
ZHAO-YUAN MA[1], KEVIN L. MOORE[1], KATER W. MURCH[1],
TOM P. PURDY[1] and DAN M. STAMPER-KURN[1,3,*]

[1] *Department of Physics, University of California, Berkeley, CA 94720, USA*
[2] *Department of Physics, University of Washington, Seattle, WA 98195, USA*
[3] *Materials Sciences Division, Lawrence Berkeley National Laboratory, Berkeley, CA 94720, USA*
* E-mail: dmsk@berkeley.edu
physics.berkeley.edu/research/ultracold

In many experiments, isolated atoms and ions have been inserted into high-finesse optical resonators for fundamental studies of quantum optics and quantum information. Here, we introduce another application of such a system, as the realization of cavity optomechanics where the collective motion of an atomic ensemble serves the role of a moveable optical element in an optical resonator. Compared with other optomechanical systems, such as those incorporating nanofabricated cantilevers or the large cavity mirrors of gravitational observatories, our cold-atom realization offers direct access to the quantum regime. We describe experimental investigations of optomechanical effects, such as the bistability of collective atomic motion and the first quantification of measurement backaction for a macroscopic object, and discuss future directions for this nascent field.

Keywords: Quantum micro-mechanics; ultracold atoms; optomechanical systems

1. Introduction

Cavity opto-mechanics describes a paradigmatic system for quantum metrology: a massive object with mechanical degrees of freedom is coupled to and measured by a bosonic field. Interest in this generic system is motivated by several considerations. For one, the system allows one to explore and address basic questions about quantum limits to measurement. In this context, quantum limits to quadrature specific and non-specific measurements, both for those performed directly on the mechanical object and also those performed through the mediation of an amplifier have been derived.[1]

Second, as a detectors of weak forces, cavity opto-mechanical systems in the quantum regime may yield improvements in applications ranging from the nanoscale (e.g. for atomic or magnetic force microscopies) to the macroscale (e.g. in ground- or space-based gravity wave observatories). Finally, such systems, constructed with ever-larger mechanical objects, may allow one to test the validity of quantum mechanics for massive macroscopic objects. Striking developments in this field were presented at ICAP 2008 by Harris and Kippenberg.

Our contribution to this developing field is the realization that a cavity opto-mechanical system can be constructed using a large gas of ultracold atoms as the mechanical object. Having developed an apparatus that allows quantum gases to be trapped within the optical mode of a high-finesse Fabry-Perot optical resonator, we are now able to investigate basic properties of opto-mechanical systems. Several of these investigations are described below. The atoms-based mechanical oscillator may be considered small by some, with a mass ($\simeq 10^{-17}$ g) lying geometrically halfway between the single-atom limit explored at the quantum regime in ion and atom traps (10^{-22} g) [2,3, for example], and the small ($\simeq 10^{-12}$ g) nanofabricated systems now approaching quantum limits.[4,5] Nevertheless, our system offers the advantages of immediate access to the quantum mechanical regime, of the *ab initio* theoretical basis derived directly from quantum optics and atomic physics, and of the tunability and amenability to broad new probing methods that are standard in ultracold atomic physics. Our motivation for probing cavity opto-mechanics with our setup is not just to poach the outstanding milestones of this field (e.g. reaching the motional ground state or observing measurement backaction and quantum fluctuations of radiation pressure with a macroscopic object[6]). Rather, we hope to contribute to the development of macroscopic quantum devices by clarifying experimental requirements and the role of and limits to technical noise, developing optimal approaches to signal analysis and system control, exploring the operation and uses of multi-mode quantum devices, and defining different physical regimes for such systems. Also, our opto-mechanical system may have direct application as part of an atom-based precision (perhaps interferometric) sensor.

2. Collective modes of an intracavity atomic ensemble

The theoretical reasoning for considering a trapped atomic gas within a high-finesse optical resonator as a macroscopic cavity opto-mechanical system is laid out in recent work.[6] Recapping that discussion, we consider the dispersive coupling of an ensemble of N identical two-level atoms to a single standing-wave mode of a Fabry-Perot cavity, obtaining the spectrum of "bright" eigenstates of the atoms-cavity system according to the following Hamiltonian:

$$\mathcal{H} = \hbar\omega_c\hat{n} + \sum_i \frac{\hbar g^2(z_i)}{\Delta_{ca}} + \mathcal{H}_a + \mathcal{H}_{in/out}.$$ (1)

Here $\hat{n}$ is the cavity photon number operator, $\Delta_{ca} = \omega_c - \omega_a$ is the difference between the empty-cavity and atomic resonance frequencies, and $g(z_i) = g_0 \sin(k_p z)$ is the spatially dependent atom-cavity coupling frequency with z_i being the position of atom i and k_p being the wavevector at the cavity resonance. The term $\mathcal{H}_a$ describes the energetics of atomic motion while $\mathcal{H}_{in/out}$ describes the electromagnetic modes outside the cavity. Note that this expression already treats the atom-cavity coupling to second order in g. Repeating this analysis starting from the first-order term does not change our conclusions substantially.

Now, let us assume that all the atoms are trapped in harmonic potentials with "mechanical" trap frequency ω_z and neglect motion along directions other than the cavity axis. Further, we treat the atomic motion only to first order in atomic displacements, δz_i, from their equilibrium positions, $\bar{z}_i$; i.e. we assume atoms to be confined in the Lamb-Dicke regime with $k_p \delta z_i \ll 1$. We now obtain the canonical cavity opto-mechanical Hamitonian[7] as

$$\mathcal{H} = \hbar\omega_c'\hat{n} + \hbar\omega_z\hat{a}^\dagger\hat{a} - F\hat{Z}\hat{n} + \mathcal{H}_a' + \mathcal{H}_{in/out}.$$ (2)

We make several steps to arrive at this expression. First, we allow the cavity resonance frequency to be modified as $\omega_c' = \omega_c + \sum_i g^2(\bar{z}_i)/\Delta_{ca}$, accounting for the cavity resonance shift due to the atoms at their equilibrium positions. Second, we introduce the collective position variable $\hat{Z} = N_{\text{eff}}^{-1}\sum_i \sin(2k_p\bar{z}_i)\delta z_i$ that, along with a weighted sum $\hat{P} = \sum_i \sin(2k_p\bar{z}_i)p_i$ of the atomic momenta p_i, describes the one collective motion within the atomic ensemble that is coupled to the cavity-optical field. The operators $\hat{a}$ and $\hat{a}^\dagger$ are defined conventionally for this mode. In our treatment, without the presence of light within the optical cavity, this mode is harmonic, oscillating at the mechanical frequency ω_z, and endowed with a mass M equal to that of $N_{\text{eff}} = \sum_i \sin^2(2k_p\bar{z}_i)$ atoms. Third, we summarize the opto-mechanical coupling by the per-photon force $F = N_{\text{eff}}\hbar k g_0^2/\Delta_{ca}$ that acts on the collective mechanical mode. Finally, we lump all the remaining atomic degrees of freedom, and also the neglected higher order atom-cavity couplings, into the term $\mathcal{H}_a'$.

With this expression in hand, we may turn immediately to the literature on cavity opto-mechanical systems to identify the phenomenology expected for our atoms-cavity system. Several such phenomena are best described by referring to the opto-mechanical force on the collective atomic mode, given as

$$\hat{\mathcal{F}}_{\text{opto}} = -M\omega_z^2\hat{Z} + F\hat{n}. \tag{3}$$

We consider the following effects:

- If we allow the state of the cavity to follow the atomic motion adiabatically ($\omega_z \ll \kappa$), neglect quantum-optical fluctuations of the cavity field, and assume the collective atomic displacement remains small, the linear variation of $\langle\hat{n}\rangle$ with $\hat{Z}$ modifies the vibration frequency of the collective atomic motion. Here, κ is the cavity half-linewidth. This modification, known as the "optical spring," has been observed in various opto-mechanical systems and has been used to trap macroscopic objects optically.[8–10] We have made preliminary observations of the optical spring effect in our system as well.

- For larger atomic displacements, the opto-mechanical force may become notably anharmonic, and even, under suitable conditions, bistable.[11] Our observations of the resulting opto-mechanical bistability[12] are discussed in Sec. 4.

- When the cavity field no longer follows the atomic motion adiabatically, the opto-mechanical potential is no longer conservative. The dramatic effects of such non-adiabaticity are the cavity-induced damping or coherent amplification of the mechanical motion.[13] Such effects of dynamical backaction have been detected in several micro-mechanical systems[14–16] and also for single[17] or multiple atoms[18] trapped within a cavity.

- Finally, we consider also the effects of quantum-optical fluctuations of the intracavity photon number and, thereby, of the optical forces on the atomic ensemble. It can be shown that these force fluctuations represent the backaction of quantum measurements of the collective atomic position,[6] as described in Sec. 5.

3. Collective atomic modes in various regimes

The theoretical treatment described above is suitable in the Lamb-Dicke regime of atomic confinement and under the condition that the linear opto-mechanical coupling term ($F\hat{Z}\hat{n}$) is dominant (i.e. that the intracavity atomic gas is not tuned to positions of exclusively quadratic sensitivity). These conditions are met in our experiments at Berkeley, where an ultra-cold gas of about 10^5 atoms of ^{87}Rb is transported into the mode volume of a high-finesse Fabry-Perot optical resonator. The resonator length is tuned so that the resonator supports one TEM$_{00}$ mode with wavelength $\lambda_T = 850$ nm (trapping light) and another within a given detuning Δ_{ca} (in the range of 100's of GHz) of the D2 atomic resonance line (probe light). Laser light with wavelength λ_T is sent through the cavity to generate a 1D optical lattice potential in which the cold atomic gas is trapped (Fig. 1). The gas is strewn across over > 100 contiguous sites in this 1D optical lattice. Within each well, atoms are brought by evaporative cooling to a tempera-ture $T \sim 700$ nK. At this temperature, the atoms lie predominantly in the ground state of motion along the cavity axis, with $\hbar\omega_z/k_B \simeq 2\,\mu\mathrm{K} \gg T$, and the Lamb-Dicke condition is satisfied with respect to the wavevector of probe light ($k_p \simeq 2\pi/(780\,\mathrm{nm})$) used to interrogate the atomic motion.

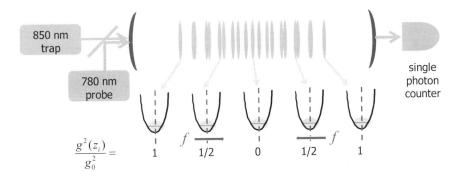

Fig. 1. Scheme for opto-mechanics with ultracold atoms in the Lamb-Dicke confinement regime. A high finesse cavity supports two longitudinal modes – one with wavelength of about 780 nm that is near the D2 resonance of ^{87}Rb atoms trapped within the resonator, and another with wavelength of about 850 nm. Light at the 850 nm resonance produces a one-dimensional optical lattice, with trap minima indicated in orange, in which atoms are confined within the lowest vibrational band. These atoms induce frequency shifts on the 780 nm cavity resonance. The strength of this shift, and of its dependence on the atomic position, varies between the different sites of the trapping optical lattice, as shown. Nevertheless, in the Lamb-Dicke confinement regime, the complex atoms-cavity interactions reduce to a simple opto-mechanical Hamiltonian wherein a single collective mode of harmonic motion, characterized by position and momentum operators $\hat{Z}$ and $\hat{P}$, respectively, is measured, actuated, and subjected to backaction by the cavity probe.

The opto-mechanics picture of atomic motion in cavity QED has also been considered recently by the Esslinger group in Zürich.[19] There, a continuous Bose-Einstein condensate of ^{87}Rb is trapped in a large-volume optical trap within the cavity volume. Yet, in spite of the stark differences in the external confinement and the motional response of the condensed gas, a similar opto-mechanical Hamiltonian emerges. In our prior description of the Lamb-Dicke regime, optical forces due to cavity probe light are found to excite and, conversely, to make the cavity sensitive to a specific collective motion in the gas. In the case of a continuous condensate, the cavity optical forces excite atoms into a specific superposition of the $\pm 2\hbar k_p$ momentum modes. Interference between these momentum-excited atoms and the underlying condensate creates a spatially (according to k_p) and temporally (according to the excitation energy) periodic density grating that is sensed via the cavity resonance frequency. Thus, by identifying operators $\hat{a}$ and $\hat{a}^\dagger$ with this momentum-space excitation and the operator $\hat{Z}$ with the density modulation, we arrive again at the Hamiltonian of Eq. 2.

We can attempt to bridge these two opto-mechanical treatments by tracking the response of an extended atomic gas to spatially periodic optical forces (due to probe light at wavelength 780 nm) as we gradually turn up the additional optical lattice potential (due to trapping light at wavelength 850 nm). In the absence of the lattice potential, a zero-temperature Bose gas forms a uniform Bose-Einstein condensate. The excitations of this system are characterized by their momentum and possess an energy determined by the Bogoliubov excitation spectrum; in Fig. 2(a), we present this spectrum as a free-particle dispersion relation, neglecting the effects of weak interatomic interactions. The spatially periodic optical force of the cavity probe excites a superposition of momentum excitations as described above.

Adding the lattice potential changes both the state of the Bose-Einstein condensate, which now occupies the lowest Bloch state, and also the state of excitations, which are now characterized by their quasi-momentum and by the band index. There are now many excitations of the fluid that may be excited at the quasi-momentum selected by the spatially periodic cavity probe. In the case that the lattice is very shallow, shown in Fig. 2(a), the cavity probe will still populate only one excited state nearly exclusively. Given the relation between the wavelengths of the trapping (850 nm) and cavity-probe light (780 nm), this excited state lies in the second excited band. As the lattice is deepened, however, matrix elements connecting to quasi-momentum states on other bands will grow (shown in Fig. 2(c)). Now our simple opto-mechanical picture is made substantially more complex, with multiple mechanical modes oscillating with differing mechanical frequencies all influencing the optical properties of the cavity.

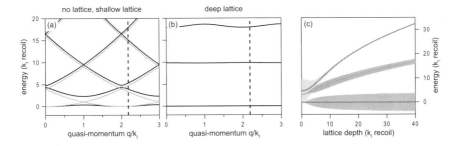

Fig. 2. Influence of band structure on the opto-mechanical response of an ultracold atomic gas confined within a high-finesse Fabry-Perot optical resonator. We consider the relevant macroscopic excitation produced by cavity probe light at wavevector $k_p = 2\pi/(780\,\text{nm})$ within a Bose gas confined within a one-dimensional optical lattice formed by light at wavevector $k_t = 2\pi/(850\,\text{nm})$ and with variable depth. The gas is cooled to zero temperature, non-interacting, and extended evenly across many lattice sites. Energies are scaled by the recoil energy $E_r = \hbar^2 k_t^2/2m$ and wavevectors by k_t. (a) With a weak lattice applied ($2E_r$), the band structure for atomic excitations (black lines) is slightly perturbed from the free-particle excitations in the absence of a lattice (gray). The cavity probe excites atoms primarily to states with quasi-momenta $\pm 2k_p$ within the second excited band, corresponding closely to momentum eigenstates in the lattice-free regime. (b) In a deep lattice ($15E_r$), energy bands show little dispersion and are spaced by energies scaling as the square root of the lattice depth. (c) Lines show the energies of the three lowest energy states at quasi-momentum $2k_p$ as a function of the lattice depth. The relative probability for excitation by cavity probe light to each of these states, taken as the square of the appropriate matrix element, is shown by the width of the shaded regions around each line. At zero lattice depth, cavity probe light excites the second excited band exclusively. At large lattice depth, the excitation probability to the first excited band grows while excitation to higher bands is suppressed. At intermediate lattice depths, several excited states are populated, indicating the onset of complex multi-mode behaviour.

Continuing to deepen the optical lattice, this complexity will be alleviated when we reach the Lamb-Dicke regime, i.e. as the Lamb-Dicke parameter $k_p \delta z$ becomes ever smaller, the probabilities of excitation from the ground state via the cavity probe zero in on the first excited band. We calculate such probabilities as $p_i \propto |< 2k_p; i| \cos(2k_p z)|g >|^2$ where the bra is the $2k_p$ quasi-momentum Bloch state in the ith band, and the ket is the ground state in the lattice considered. Here, we interpret excitations to higher bands as being controlled by terms of higher order in the Lamb-Dicke parameter, e.g. excitations to the second excited band result from couplings that are quadratic in the atomic positions.

Thus, we confirm that a simple opto-mechanics picture emerges for the collective atomic motion within a cavity both in the shallow- and deep-lattice limits. We note, however, that these limits differ in two important ways. First, we see that the mechanical oscillation frequency for the collective atomic motion is constrained to lie near the bulk Bragg excitation frequency in the shallow-lattice limit, whereas it may be tuned to arbitrarily high frequencies (scaling as the square root of the lattice depth) in the deep-lattice limit. The ready tunability of the mechanical frequency in the latter limit may allow for explorations of quantum opto-mechanical systems in various regimes, e.g. in the the resolved side-band regime where ground-state cavity cooling and also quantum-limited motional amplification are possible.[20,21] Second, we see that the mechanical excitation frequency has a significant quasi-momentum (Doppler) dependence in the shallow-lattice limit. This dependence makes it advantageous to use low-temperature Bose-Einstein condensates for experiments of opto-mechanics, as done in the Zürich experiments, so as to minimize the Doppler width of the Bragg excitation frequency. In contrast, the excitation bandwidth is dramatically reduced (exponentially with the lattice depth) in the deep-lattice limit. Thus, one can conduct opto-mechanics experiments with long-lived mechanical resonances in the deep-lattice limit without bothering to condense the atomic gas. Nevertheless, we note that variations in the mechanical frequency due to the presence of significant radial motion (not considered in this one-dimensional treatment) do indeed limit the mechanical quality factor in the Berkeley experiments.

4. Effects of the conservative optomechanical potential: optomechanical bistability

The observation of cavity nonlinearity and bistability arising from collective atomic motion is described in recent work.[12] Briefly, we find that the optical force due to cavity probe light will displace the equilibrium collective atomic position $\langle \hat{Z} \rangle$, leading to a probe-intensity-dependent shift of the cavity resonance frequency. By recording the cavity transmission as the cavity probe light was swept across the cavity resonance, we observed asymmetric and shifted cavity resonance lines, and also hallmarks of optical bistability.

Refractive optical bistability is well studied in a variety of experimental systems.[22] One unique aspect of our experiment is the observation of both branches of optical bistability at average cavity photon numbers as low as 0.02. The root of such strong optical nonlinearities is the presence within the cavity of a medium that responds significantly to the presence of infrequent cavity photons (owing to strong collective effects) and recalls the presence of such photons for long coherence times. Here, the coherence is stored within the long-lived collective motion of the gas. It is interesting to consider utilizing such long-lived motional coherence, rather than the shorter-lived internal state coherence typically considered, for the various applications of cavity QED and nonlinear optics in quantum information science, e.g. photon storage and generation, single-photon detection, quantum logic gates, etc.

Such motion-induced cavity bistability can also be understood in the context of the opto-mechanical forces described by Eq. (3). Neglecting the non-adiabatic following of the cavity field to the collective motion (essentially taking $\kappa/\omega_z \to \infty$ so that dynamical backaction effects are neglected) and also the quantum fluctuations of the cavity field, we may regard atomic motion in an optically driven cavity to be governed by an opto-mechanical potential of the form

$$U(Z) = \frac{1}{2} M \omega_z^2 Z^2 + n_{\max} \hbar \kappa \arctan \left(\frac{\Delta_{pc} - FZ/\hbar}{\kappa} \right). \qquad (4)$$

Here Δ_{pc} is the detuning of the constant frequency probe from the modified cavity resonance frequency ω_c', and $n_{\max}$ is the average number of cavity photons when the cavity is driven on resonance.

The form of this potential is sketched in Fig. 3 for different operating conditions of the atoms-cavity system. Cavity bistability[12] is now understood as reflecting an effective potential for the collective atomic variable Z that has two potential minima. Remarkably, these potential minima may be separated by just nanometer-scale displacements in Z. Even though the inherent quantum position uncertainty of each individual atom (10's of nm) is much larger than this separation, the reduced uncertainty in the collective variable Z allows for these small displacements to yield robust and distinct experimental signatures in the cavity transmission.

5. Quantum fluctuations of the optomechanical potential: measurement backaction

Aside from the conservative forces described above, the intracavity atomic medium is subject also to dipole force fluctuations arising from the quantum nature of the intracavity optical field. Indeed, should these force fluctuations be especially large, the picture of cavity optical non-linearity and bistability described in the previous section, in which we implicitly assume that the collective atomic motion may follow adiabatically into a local minimum of an opto-mechanical potential, must be dramatically modified. To assess the strength of such force fluctuations, let us consider the impact on

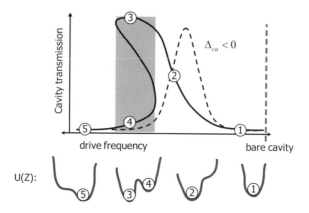

Fig. 3.   For different stable regimes of cavity operation, the cavity-relevant collective mode of the intracavity atomic ensemble is trapped in a particular minimum of the effective potential $U(Z)$ (bottom). In the regime of bistability, the two stable cavity states reflect the presence of two potential minima.

the atomic ensemble of a single photon traversing the optical cavity. During its residence time of $\sim 1/2\kappa$, such a photon would cause a dipole force that imparts an impulse of $\Delta P = f/(2\kappa)$ on the atomic medium, following which the collective mode is displaced by a distance $\Delta Z = \Delta P/(M\omega_z)$; in turn, this displacement will shift the cavity resonance frequency by $F\Delta Z/\hbar$. Comparing this single-photon-induced, transient frequency shift with the cavity half-linewidth leads us to define a dimensionless "granularity parameter" as

$$\epsilon = \sqrt{\frac{F\Delta Z}{\hbar\kappa}} = \frac{FZ_{\mathrm{ho}}}{\hbar\kappa}, \tag{5}$$

where $Z_{\mathrm{ho}} = \sqrt{\hbar/2M\omega_z}$ is the harmonic oscillator length for the atomic collective mode. The condition $\epsilon > 1$ marks the granular (or strong) opto-mechanical coupling regime in which the disturbance of the collective atomic mode by single photons is discernible both in direct quantum-limited measurements of the collective atomic motion and also in subsequent single-photon measurements of the cavity resonance frequency. In our experiments, the granularity parameter is readily tuned by adjusting frequency difference between the cavity and atomic resonance Δ_{ca}. Under conditions of our recent work, the granular regime is reached at $|\Delta_{ca}|/(2\pi) < 27$ GHz.

In recent work, we have focused on effects of fluctuations of the dipole force in the non-granular regime, attained at atom-cavity detunings in the 100 GHz range. As described in our work,[6] and also derived in earlier

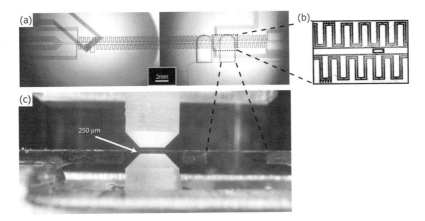

Fig. 4. An integrated cavity QED/atom chip. (a) A top view of the microfabricated silicon chip shows etched trenches, later electroplated with copper and used to tailor the magnetic field above the chip surface. The left portion of the image shows wire patterns used for producing the spherical-quadrupole field of a magneto-optical trap and also the Ioffe-Pritchard fields for producing stable magnetic traps. Serpentine wires spanning the entire chip form a magnetic conveyor system to translate atoms to the optical cavities that are located in the right half of the image. (b) A detailed view shows the serpentine wires and also a two-wire waveguide surrounding a central, rectangular hole that pierces the atom chip. (c) Fabry-Perot cavities are formed by mirrors straddling the atom chip. Between the mirrors, the chip is thinned to below 100 μm – outlines of the thinned areas are seen also in (a). The cavity mode light passes unhindered through the chip via the microfabricated holes shown in (b).

treatments,[20,23–25] these fluctuations will cause the motional energy of the collective atomic mode to vary according to the following relation:

$$\frac{d}{dt}\langle a^\dagger a \rangle = \kappa^2 \epsilon^2 \left[S_{nn}^{(-)} + \left(S_{nn}^{(-)} - S_{nn}^{(+)} \right) \langle a^\dagger a \rangle \right] \tag{6}$$

Here, the relevant dipole force fluctuations are derived from the spectral density of intracavity photon number fluctuations at the mechanical frequency ω_z, calculated for a coherent-state-driven cavity as $S_{nn}^{(\pm)} = 2\langle n \rangle \kappa / \left(\kappa^2 + (\Delta_{pc} \pm \omega_z)^2 \right)$ Eq. 6, which can be derived readily from a rate-equation approach,[20] reveals two manners in which the mechanical oscillator responds to a cavity optical probe: momentum diffusion, which raises the mechanical oscillator energy at a constant rate, and the dynamic back-action effects of cavity-based cooling or amplification of the mechanical motion, described by an exponential damping or gain.

The mechanical momentum diffusion in an opto-mechanical system plays the essential metrological role of providing the backaction necessary in a quantum measurement, as discussed, for example, by Caves in the context of optical interferometry.[26] For $\omega_z \ll \kappa$, we see that, at constant circulating power in the cavity, this diffusion is strongest for probe light at the cavity resonance and weaker away from resonance. This dependence on the intensity and detuning of the cavity probe light precisely matches the rate of information carried by a cavity optical probe on the state of the mechanical oscillator. To elucidate this point, we recall that, under constant drive by a monochromatic input field, the intracavity electric field oscillates at the input field frequency with complex amplitude $E_{cav} = \eta/(\kappa - i\Delta_{pc})$. A displacement by ΔZ of the mechanical oscillator varies the probe-cavity detuning by $F\Delta Z/\hbar$. In response, the electric field in the cavity varies as

$$E_{cav} \simeq E_0 \left(1 + \frac{i}{\kappa - i\Delta_{pc}} \frac{F\Delta Z}{\hbar}\right) = E_0 + E_{sig}, \qquad (7)$$

where E_0 is the cavity field with the cantilever at its equilibrium position and we expand to first order in ΔZ. The sensitivity of the cavity field to the cantilever displacement, at constant intracavity intensity (constant E_0), is determined by the magnitude of $|E_{sig}/E_0|^2 \propto 1/(1 + \Delta_{pc}^2/\kappa^2)$; this functional dependence matches that of the momentum diffusion term, supporting its representing measurement backaction.

To measure this backaction heating, we take advantage of several features of our experiment. First, by dint of the low temperature of our atomic ensemble, we ensure that the effects of dynamical backaction (cooling and amplification) are negligible. Second, the low quality-factor of our mechanical oscillator ensures that the momentum diffusion of the collective atomic motion leads to an overall heating of the atomic ensemble, allowing us to measure this diffusion bolometrically. Third, the large single-atom cooperativity in our cavity QED system implies that this backaction heating of the entire atomic ensemble dominates the single-atom heating due to atomic spontaneous emission. And, fourth, owing to the finite, measured depth of our intracavity optical trap, backaction heating can be measured via the light-induced loss rate of atoms from the trap. The measured light-induced heating rate was found to be in good agreement with our predictions, providing the first quantification of measurement backaction on a macroscopic object at a level consistent with quantum metrology limits.

6. Future developments: cavity QED/atom chips

While continuing explorations of quantum opto-mechanics in our existing apparatus, we are also developing an experimental platform that integrates the capabilities of single- and many-atom cavity QED onto microfabricated atom chips. Similar platforms have been developed recently by other groups.[27,28] Aside from enabling myriad applications in quantum atom optics and atom interferometry, we anticipate the cavity QED/atom chip to provide new capabilities in cold-atoms-based opto-mechanics. For instance, the tight confinement provided by microfabricated magnetic traps will allow atomic ensembles to be confined into single sites of the intracavity optical lattice potential, providing a means of tuning the opto-mechanical coupling between terms linear or quadratic in $\hat{Z}$. As emphasized by Harris and colleagues,[29] a purely quadratic coupling may allow for quantum non-demolition measurements of the energy of the macroscopic mechanical oscillator.

References

1. V. Braginskii and F. Y. Khalili, *Quantum Measurement* (Cambridge University Press, Cambridge, 1995).
2. D. M. Meekhof, C. Monroe, B. E. King, W. M. Itano and D. J. Wineland, *Phys. Rev. Lett.* **76**, 1796 (1996).
3. I. Bouchoule, H. Perrin, A. Kuhn, M. Morinaga and C. Salomon, *Phys. Rev. A* **59**, R8 (1999).
4. M. D. LaHaye, O. Buu, B. Camarota and K. C. Schwab, *Science* **304**, 74 (2004).
5. C. A. Regal, J. D. Teufel and K. W. Lehnert, *Nat. Phys.* **4**, 555 (2008).
6. K. W. Murch, K. L. Moore, S. Gupta and D. M. Stamper-Kurn, *Nat. Phys.* **4**, 561 (2008).
7. T. Kippenberg and K. Vahala, *Science* **321**, 1172 (2008).
8. B. S. Sheard, M. B. Gray, C. M. Mow-Lowry, D. E. McClelland and S. E. Whitcomb, *Phys. Rev. A* **69**, 051801 (2004).
9. T. Corbitt, D. Ottaway, E. Innerhofer, J. Pelc and N. Mavalvala, *Phys. Rev. A* **74**, 021802 (2006).
10. T. Corbitt, Y. B. Chen, E. Innerhofer, H. Muller-Ebhardt, D. Ottaway, H. Rehbein, D. Sigg, S. Whitcomb, C. Wipf and N. Mavalvala, *Phys. Rev. Lett.* **98**, 150802 (2007).
11. A. Dorsel, J. D. Mccullen, P. Meystre, E. Vignes and H. Walther, *Phys. Rev. Lett.* **51**, 1550 (1983).
12. S. Gupta, K. Moore, K. Murch and D. Stamper-Kurn, *Phys. Rev. Lett.* **99**, 213601 (2007).
13. V. B. Braginsky and A. B. Manukin, *Sov. Phys. JETP* **25**, 653 (1967).
14. T. J. Kippenberg, H. Rokhsari, T. Carmon, A. Scherer and K. J. Vahala, *Phys. Rev. Lett.* **95**, 033901 (2005).
15. O. Arcizet, P. F. Cohadon, T. Briant, M. Pinard and A. Heidmann, *Nature* **444**, 71 (2006).
16. S. Gigan, H. R. Bohm, M. Paternostro, F. Blaser, G. Langer, J. B. Hertzberg, K. C. Schwab, D. Bauerle, M. Aspelmeyer and A. Zeilinger, *Nature* **444**, 67 (2006).

17. P. Maunz, T. Puppe, I. Schuster, N. Syassen, P. W. H. Pinkse and G. Rempe, *Nature* **428**, 50 (2004).
18. H. W. Chan, A. T. Black and V. Vuletić, *Phys. Rev. Lett.* **90**, 063003 (2003).
19. F. Brennecke, S. Ritter, T. Donner and T. Esslinger, *Science* **322**, 235 (2008).
20. F. Marquardt, J. P. Chen, A. A. Clerk and S. M. Girvin, *Phys. Rev. Lett.* **99**, 093902 (2007).
21. I. Wilson-Rae, N. Nooshi, W. Zwerger and T. J. Kippenberg, *Phys. Rev. Lett.* **99**, 093901 (2007).
22. R. W. Boyd, *Nonlinear Optics*, 2nd edn. (Academic Press, Boston, 2003).
23. P. Horak, G. Hechenblaikner, K. M. Gheri, H. Stecher and H. Ritsch, *Phys. Rev. Lett.* **79**, 4974 (1997).
24. V. Vuletić and S. Chu, *Phys. Rev. Lett.* **84**, 3787 (2000).
25. K. Murr, P. Maunz, P. W. H. Pinkse, T. Puppe, I. Schuster, D. Vitali and G. Rempe, *Phys. Rev. A* **74**, 043412 (2006).
26. C. Caves, *Phys. Rev. D* **23**, 1693 (1981).
27. I. Teper, Y.-J. Lin and V. Vuletic, *Phys. Rev. Lett.* **97**, 023002 (2006).
28. Y. Colombe, T. Steinmetz, G. Dubois, F. Linke, D. Hunger and J. Reichel, *Nature* **450**, 272 (2007).
29. J. D. Thompson, B. M. Zwickl, A. M. Jayich, F. Marquardt, S. M. Girvin and J. G. E. Harris, *Nature* **452**, 72 (2008).

IMPROVED "POSITION SQUARED" READOUT USING DEGENERATE CAVITY MODES

J. C. SANKEY[1], A. M. JAYICH[2], B. M. ZWICKL[2], C. YANG[2], J. G. E. HARRIS[*,1,2]

[1]*Department of Applied Physics, Yale University*
[2]*Department of Physics, Yale University, New Haven, CT 06520, USA*
* E-mail: jack.harris@yale.edu
http://www.yale.edu/harrislab/

Optomechanical devices in which a flexible SiN membrane is placed inside an optical cavity allow for very high finesse and mechanical quality factor in a single device. They also provide fundamentally new functionality: the cavity detuning can be a quadratic function of membrane position. This enables a measurement of "position squared" (x^2) and in principle a QND phonon number readout of the membrane. However, the readout achieved using a single transverse cavity mode is not sensitive enough to observe quantum jumps between phonon Fock states.

Here we demonstrate an x^2-sensitivity that is orders of magnitude stronger using two transverse cavity modes that are nearly degenerate. We derive a first-order perturbation theory to describe the interactions between nearly-degenerate cavity modes and achieve good agreement with our measurements using realistic parameters. We also demonstrate theoretically that the x^2-coupling should be easily tunable over a wide range.

Keywords: Optomechanics; micromechanics; QND; cantilevers; radiation pressure; cavity QED; quantum jumps.

1. Introduction

In quantum mechanics a system's behavior is not independent of how it is measured. As a result, the readout used in an experiment must be tailored to the phenomena of interest. Likewise, for a given type of readout not all quantum effects are observable.

Experiments on mechanical oscillators have to date used readouts that couple directly to the oscillator's displacement. The most common example is an optical interferometer in which the oscillator serves as one of the interferometer's mirrors. In such a system the phase ϕ of the light reflected from the interferometer is proportional to the mirror's displacement x. An oscillator that is subject to continuous monitoring of x is predicted to show a number of striking quantum features, including the standard quantum limit of displacement detection.[1] Additionally, the linear coupling between x and ϕ can be used both to laser-cool the oscillator (perhaps eventually to its ground state)[2-4] and to squeeze the light leaving the cavity.[5,6] The connection between the readout of the mechanical oscillator and its manipulation highlights the fact that these are two aspects of the same optomechanical coupling.

In a recent paper[7] it was shown that a modest rearrangement of the usual optomechanical setup can realize a fundamentally different type of readout. When a nearly-transparent dielectric membrane is placed inside a cavity formed by two fixed, macroscopic mirrors, the phase of the light reflected from the cavity can be adjusted so that it is proportional either to x or to x^2. The quadratic coupling occurs when the membrane is placed at a node (or anti-node) of the intracavity standing wave. In such a situation the membrane is at a minimum (maximum) of the optical intensity, and so detunes the cavity resonance by a small (large) amount. As the membrane moves in either direction it encounters an optical intensity that is larger (smaller) by an amount quadratic in its displacement (to lowest order), and hence detunes the cavity by an amount which is also quadratic (to lowest order) in the displacement. If on the other hand the membrane is originally placed at a point which is neither a node nor an antinode, the cavity detuning is (to lowest order) linear in the displacement.   ·

Mechanical oscillators coupled to an x^2-readout have been discussed theoretically for some time. It has been shown that such a readout, coupled to a mechanical oscillator inside a sufficiently high-finesse optical cavity, can in principle provide a quantum nondemolition (QND) measurement of the energy (or equivalently the phonon number) of the mechanical oscillator.[8] With a sufficiently sensitive x^2-readout it should be possible to observe, in real time, the individual quantum jumps of the mechanical oscillator. This is in contrast to an oscillator coupled to an x-readout, in which the repeated measurements of the oscillator's position extract information which prevents the oscillator from remaining in an energy eigenstate. This is because the quantity x does not commute with the oscillator's energy, whereas the quantity x^2 does (at least in the rotating-wave approximation, whose validity is ensured by the cavity's high finesse).[8]

Although the x^2-readout demonstrated in Ref. 7 represented a major advance towards realizing the goal of QND measurements of a mechanical oscillator's energy, the strength of the x^2-coupling was insufficient to realize such a measurement in practice. This is because for a low-reflectivity membrane the scale of the x^2-coupling is $\sim 1/\lambda^2$, where λ is the wavelength of the light (the cavity detuning oscillates each time the membrane is displaced by $\lambda/2$). If the membrane's (field) reflectivity r approaches unity, the finesse of the "half-cavities" on either side of the membrane begins to increase, and the curvature of the cavity detuning (and hence the strength of the x^2-coupling) increases, diverging for $r \to 1$.[7] However the technical challenges involved in combining a high reflectivity mirror and a high-quality mechanical oscillator into a single element have proven considerable, so it would be highly advantageous to find a strong x^2-coupling which does not require a high-reflectivity membrane.

In this paper, we describe a new means for generating a strong x^2-coupling in this type of device. We show that the optical cavity's full spectrum of transverse modes contains many near-degeneracies, and that near these points the cavity's resonance frequencies display an avoided-crossing behavior as a function of the membrane displacement. This leads to a detuning proportional to x^2, but with a scale set by the symmetry-breaking aspects of the cavity/membrane geometry rather than the wavelength of light. We develop a perturbation theory that allows us to calculate the membrane-induced cavity detuning, and find that the x^2-coupling at these avoided crossings can be made orders of magnitude stronger than realized in earlier work. We compare these calculations to measurements and find quantitative agreement, indicating that the single-phonon QND measurements proposed in [7] may be feasible even with a low-r membrane.

2. Observed Effect of Membrane on Empty-Cavity Modes

Our experimental setup is shown in Fig. 1 and has also been described elsewhere.[7,10] A flexible silicon nitride membrane (1 mm $\times$ 1 mm $\times$ 50 nm thick) is situated near the waist of a high-finesse Fabry-Perot cavity so that its normal vector is roughly parallel to the cavity's long (x) axis. The membrane acts as the micromechanical resonator and its deflection is coupled to the cavity's optical modes via radiation pressure. The two macroscopic end mirrors are held fixed by an Invar cavity spacer. A motorized tilt stage holding the membrane is mounted to the spacer, and two piezoelectric actuators are used to displace the membrane along $\hat{x}$.

We can begin to characterize the optomechanical coupling in this system by measuring the transmission through the cavity as a function of membrane position and laser detuning, as shown in Fig. 2. Here the laser is aligned so that the dominant transmission peak corresponds to the $\text{TEM}_{0,0}$

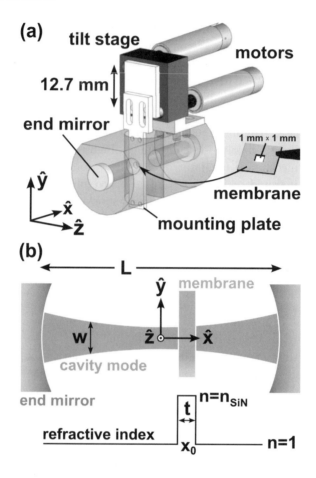

Fig. 1. (a) Schematic of our apparatus: A flexible SiN membrane mounted on a motorized tilt stage at the center of a Fabry-Perot cavity is coupled to the cavity's optical modes via radiation pressure. Piezoelectric actuators between the mounting plate and membrane enable displacements along the x-axis. (b) Simplified diagram of the cavity and membrane. The cavity length is $L = 6.7$ cm and the end mirror radius of curvature is 5 cm.

(singlet) mode, as confirmed by a camera monitoring transmission (inset). As the membrane moves along the longitudinal (x) axis, it perturbs the cavity resonance frequencies to lower values, producing a detuning that varies roughly sinusoidally with position.

When the membrane is located at an optical node, the perturbation is minimal, and the detuning is quadratic in position. As a result, light leaving the cavity contains only information about x^2. As discussed elsewhere,[7] this

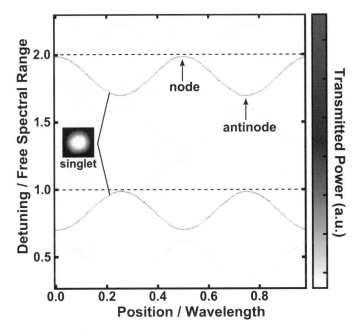

Fig. 2. Transmission through the cavity as a function of laser detuning and membrane position. The dominant signal corresponds to the $TEM_{0,0}$ (singlet) cavity mode. Dashed lines show the approximate position of the unperturbed singlet modes. We have labeled positions corresponding to a node and antinode of the upper singlet mode's electric field. At these points the detuning is proportional to x^2. (inset) An infrared camera image showing the transmitted beam profile.

may enable QND phonon number readout using the $TEM_{0,0}$ mode alone. Since the membrane is a thin (50 nm) dielectric ($n_{SiN} \approx 2$), it is a very poor reflector ($|r|^2 = 0.13$ where $|r|^2$ is the power reflectivity). As a result the curvature of the detuning is small and the x^2-sensitivity is weak. Practical estimates predict that in order to observe a phonon Fock state before it decays, the membrane reflectivity would need to be substantially higher, ~ 0.998.[7] This may represent the most difficult of the technical challenges to observing real-time quantum jumps of the membrane's mechanical energy.

A promising solution to this problem lies in the interactions between different transverse optical modes. We can couple to and identify many more of the cavity's transverse modes by intentionally misaligning the input laser, as shown in Fig. 3(a). We have identified all of the visible bands, such as the $\{TEM_{1,0}, TEM_{0,1}\}$ doublet, $\{TEM_{2,0}, TEM_{1,1}, TEM_{0,2}\}$ triplet, and so on up to the 13-fold degenerate (tridectet) modes.

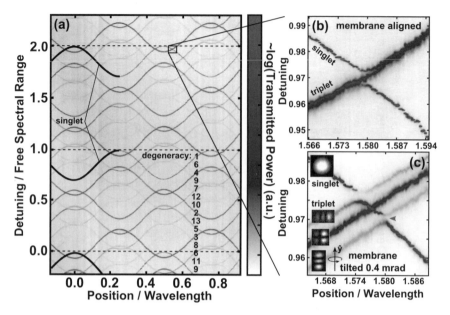

Fig. 3. (a) Transmission spectrum with the input laser misaligned, plotted on a log scale to enhance the faint features. The degeneracies of the different transverse modes are labeled and solid lines are drawn over the singlet mode for reference. (b) Close-up of the singlet-triplet crossing point for the membrane aligned with its normal vector parallel to the cavity axis ($\hat{x}$). (c) Singlet-triplet crossing with the membrane tilted about the y-axis by 0.4 mrad. The strength of the curvature at the marked gap corresponds to an effective membrane reflectivity of 0.994 ± 0.001 power.

The different transverse modes cross each other as a function of position at several places in Fig. 3. Figure 3 also shows a close-up of the crossing between the singlet and the triplet with the membrane's normal vector (b) aligned, and (c) with $\hat{x}$ tilted around the y-axis by 0.4 mrad. Tilting the membrane as in (c) lifts the degeneracy of the triplet in a predictable way: modes extended the furthest in the $\hat{z}$ direction shift the most. As is evident from Fig. 3(c), the crossing points between the singlet and the two *even* triplets (TEM$_{2,0}$ and TEM$_{0,2}$) are avoided, meaning that in addition to perturbing the individual modes, the membrane also couples them.

Most importantly, the quadratic detuning at the avoided crossing turning points is very strong. In Fig. 3(c), the curvature is 50 times stronger than at the single-mode turning points. This is the same curvature a membrane reflectivity of 0.994 ± 0.001 would generate using a single mode.[7] This is already very close to the QND target and we have not yet attempted to optimize the system.[a]

3. Model

While the membrane perturbs the empty cavity modes by up to a quarter of a free spectral range, its fractional effect relative to the laser frequency is minute ($\sim 10^{-6}$). We can therefore view the membrane perturbatively and develop a first-order theory to model the system, as discussed in the next section. We then outline a method by which to solve this problem analytically when the membrane is positioned near the cavity waist, and finally compare our results with measurements.

3.1. *First-Order Degenerate Perturbation Theory*

We start with the time-independent free-space electromagnetic wave equation

$$\nabla^2 \phi + \frac{\omega^2}{c^2} \phi = 0 \tag{1}$$

where ω is the angular frequency and c is the speed of light. As is drawn in Fig. 1(b), we define the origin to reside at the center of our cavity with the x-axis pointing toward one of the (spherical) end-mirrors. Under these boundary conditions, a convenient set of (Hermite-Gaussian) orthonormal solutions is given by[9]

$$\phi_j = \frac{H_m(\sqrt{2}y/w)H_n(\sqrt{2}z/w)}{w\sqrt{\pi L 2^{m+n-1}m!n!}} e^{-(y^2+z^2)/w^2}$$

$$\times\; e^{i(m+n+1)\Psi} e^{-ik(y^2+z^2)/2R} e^{-ikx-il\pi/2}, \tag{2}$$

Here H_m is the mth Hermite polynomial, $k = \omega/c$ is the wavenumber, $w(x) = \sqrt{2(x^2 + x_R^2)/kx_R}$ is the width of the cavity mode at x (where $x_R = 2.351$ cm is the Raleigh range and $w_0 = 89.2$ μm is the waist for our geometry), $L = 6.7$ cm is the cavity length, m and n are the transverse mode indices, l is the longitudinal mode index, $\Psi(x) = \tan^{-1}(w^2k/2R)$ is the Guoy phase shift, and $R(x) = (x^2 + x_R^2)/x$ is the wave fronts' radius of

---

[a]As discussed later, we have observed smaller gaps, but for that data the fit curvature is not very convincing due to vibrations limiting our displacement sensitivity. We can still use these smaller gaps to infer a lower bound on the curvature.

curvature. For a standing wave in our cavity, $\text{Re}(\phi_j)$ is proportional to the electric field amplitude, and the $l\pi/2$ term ensures that each longitudinal mode's electric field is zero at the end mirrors. The prefactors ensure that the inner product $\int dx \int dy \int dz \text{Re}(\phi_i)\text{Re}(\phi_j) = \delta_{ij}$.

As shown in Fig. 1(b), we represent the membrane as a block of refractive index $n_{\text{SiN}} \approx 2$ and thickness $t = 50$ nm, centered at position x_0. This modifies the speed of light in this short region, so that the wave equation in the cavity becomes

$$\nabla^2\psi + \frac{\omega^2}{c^2}(1 + V(x - x_c))\psi \tag{3}$$

where $V(x - x_c) = (n_{\text{SiN}}^2 - 1)\left(\Theta[x - (x_c - t/2)] - \Theta[(x_c + t/2) - x]\right)$ and Θ the Heaviside step function. For an "aligned" membrane (i.e. flat in the y-z plane) $x_c = x_0$ is constant. To incorporate tilt into the model, let $x_c = x_0 + \alpha_y y + \alpha_z z$ where α_y and α_z are the small rotations about the z and y axes, respectively.

The perturbed modes ψ can be expanded in terms of the empty-cavity modes:

$$\psi = c_1\phi_1 + c_2\phi_2 + c_3\phi_3 + \dots \tag{4}$$

where the c's are constants. We wish to study the region of near-degeneracy shown in Fig. 3(b-c), between the l-th longitudinal singlet mode ($m = n = 0$) and the three $(l - 1)$-th triplet modes ($m + n = 2$), so we make the assumption that ψ is composed mostly of these four empty-cavity modes

$$\psi = c_s\phi_s + c_y\phi_y + c_a\phi_a + c_z\phi_z + \sum \epsilon_j\phi_j \tag{5}$$

where the indices s, y, a, and z refer to the singlet, the triplet widest in the $\hat{y}$ direction ($m = 2, n = 0$), the antisymmetric triplet ($m = 1, n = 1$), and the triplet widest along $\hat{z}$ ($m = 0, n = 2$), respectively. The last term is a summation over all remaining modes, and its contribution is assumed to remain small ($\epsilon_j \ll 1$). We also assume ψ will have a new eigenvalue $\omega^2/c^2 \equiv \kappa$ that is not very different (i.e. within a fraction of a free spectral range) from any of the unperturbed eigenvalues of the four contributing ϕ's. Substituting this into Eq. 3,

$$(\nabla^2 + (1 + V)\kappa)(c_s\phi_s + c_y\phi_y + c_a\phi_a + c_z\phi_z + \sum \epsilon_j\phi_j) = 0. \tag{6}$$

If we now take an inner product of this equation with each of the four empty-cavity modes and divide through by κ, we obtain four new equations

$$(1 - \kappa_i/\kappa)c_i + V_{is}c_s + V_{iy}c_y + V_{ia}c_a + V_{iz}c_z + \sum V_{ij}\epsilon_j = 0 \tag{7}$$

where the index i is s, y, a, or z and V_{ij} is the inner product of the i-th and j-th mode with $V(x - x_c)$. The inner products V_{ij} involve an integral over thickness t between two modes that are normalized over length L and are small (of order $(n_{\text{SiN}}^2 - 1)t/L \sim 2 \times 10^{-5}$ or less), so the last term in Eq. 7 can be ignored. We can further simplify by writing $\kappa \equiv \kappa_s(1 + \delta)$ and $\kappa_{y,a,z} \equiv \kappa_s(1 + g)$ where δ is the fractional change due to the membrane and g is the (constant) fractional separation of the unperturbed singlet and triplet bands due to the Guoy phase. Both δ and g are of order 10^{-5}. To first order, the remaining equation can be written as a matrix

$$\begin{pmatrix} \delta + V_{ss} & V_{sy} & V_{sa} & V_{sz} \\ V_{sy} & \delta - g + V_{yy} & V_{ya} & V_{yz} \\ V_{sa} & V_{ya} & \delta - g + V_{aa} & V_{az} \\ V_{sz} & V_{yz} & V_{az} & \delta - g + V_{zz} \end{pmatrix} \begin{pmatrix} c_s \\ c_y \\ c_a \\ c_z \end{pmatrix} = 0. \quad (8)$$

Solving this eigenvalue problem for δ in terms of g and the V's is straightforward and, though time-consuming, it is also easy to numerically compute V_{ij}. The problem is in principle solved, and the result of such a calculation is shown in Fig. 4. Computation time can also be reduced by assuming the membrane is an infinitesimally thin sheet (also plotted), but even a small finite thickness of $t = 50$ nm produces a noticeable effect.

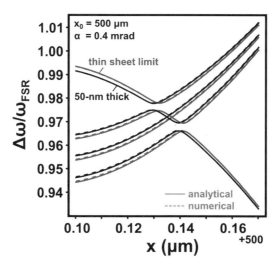

Fig. 4. Comparison of numerical (dashed) and analytical (solid) results near the singlet-triplet crossing points. The red curves correspond to the thin-membrane (delta function) limit, and the black lines include membrane thickness. For this plot, $x_0 = 500$ μm, $\alpha_z = 0.4$ mrad, and $\alpha_y = 0$.

3.2. *Analytical Solution Near the Cavity Waist*

We can also solve the inner products V_{ij} analytically by using an approximate form for the unperturbed modes near the cavity waist. We do this by expanding $1/R(x)$, $w(x)$, and $\Psi(x)$ in Eq. 2 in terms of the small parameter $\Delta = x/x_R$:

$$\frac{1}{R(x)} \approx \frac{\Delta}{x_R} + O^3 \tag{9}$$

$$w(x) \approx w_0\left(1 + \frac{1}{2}\Delta^2\right) + O^3 \tag{10}$$

$$\Psi(x) \approx \Delta + O^3 \tag{11}$$

Substituting this into Eq. 2 yields

$$\phi_j = \frac{H_m(\sqrt{2}y/w)H_n(\sqrt{2}z/w)}{w\sqrt{\pi L 2^{m+n-1}m!n!}}e^{-(y^2+z^2)/w^2}$$
$$\times\, e^{i\left[(n+m+1-\frac{y^2+z^2}{w^2}-kx_R)\Delta+l\pi/2\right]}. \tag{12}$$

We can now use this to estimate the inner products V_{ij}. We will not specify which modes are under consideration, so the results of this section may be applied to any set of nearly-degenerate cavity modes.

Before we attempt to solve these integrals, first note that

$$\iiint \mathrm{Re}(\phi_i)V\mathrm{Re}(\phi_j) = \mathrm{Re}\left[\frac{1}{2}\iiint \phi_i\phi_j V + \frac{1}{2}\iiint \phi_i\phi_j^* V\right] \tag{13}$$

which simplifies the calculation. The first integral is by far the most challenging and so we outline its solution here.

First, note that the singlet and triplet modes have slightly different unperturbed k's and w's. If $i = s$ and $j \in \{y, a, z\}$, then $k_j = k_i(1 + g)$ and $w_{0,j} \approx w_{0,i}(1 - g/2)$. By defining $A \equiv 1 + g/2 \sim 1$, we can easily keep track of this difference to first order (and $A = 1$ if the modes belong to the same degenerate manifold). If we plug Eq. 12 into the first integral of Eq. 13, make the substitutions $y \to wy/\sqrt{2A}$, $z \to wz\sqrt{2A}$ and $x \to x_R\Delta$, we have

$$\frac{1}{2}\iiint \phi_i\phi_j V = \frac{P_0}{\Delta_t}e^{-i(l_i+l_j)\pi/2}$$
$$\times \int dy \int dz \int_{\Delta_c-\Delta_t/2}^{\Delta_c+\Delta_t/2} d\Delta\, p(y,z)e^{-y^2-z^2}e^{-i(K+y^2+z^2)\Delta} \tag{14}$$

with

$$P_0 = \frac{(t/\cos\alpha)(n_{\text{SiN}}^2 - 1)}{\pi L \sqrt{2^{m_i+n_i+m_j+n_j} n_i! m_i! n_j! m_j!}} \tag{15}$$

$$\Delta_t = t/x_R \cos\alpha \tag{16}$$

$$\Delta_c = x_c/x_R = \Delta_0 + \beta_y y + \beta_z z \tag{17}$$

$$p(y,z) = H_{m_i}(y/\sqrt{A}) H_{m_j}(\sqrt{A}y) H_{n_i}(z/\sqrt{A}) H_{n_j}(\sqrt{A}z) \tag{18}$$

$$K = 2Ak_i x_R - (m_i + n_i + m_j + n_j + 2). \tag{19}$$

Here Δ_t is the dimensionless membrane thickness corrected for tilt $\alpha = \sqrt{\alpha_y^2 + \alpha_z^2}$, and we have allowed the position of the membrane center x_c to depend on y and z through small (rescaled) tilts in both directions, $\beta_{y,z} = \alpha_{y,z} w_i / x_R \sqrt{2A}$.

The membrane is 20 times thinner than the free-space wavelength, but it noticeably affects the cavity modes, as is evident in Fig. 4. We can approximate the integral over Δ by noting that for a smooth function $f(x)$,

$$\int_{x_0-\delta x/2}^{x_0+\delta x/2} f(x)dx = \delta x f(x_0) + \delta x^3 \frac{1}{24} f''(x_0) + O(\delta x^5) \tag{20}$$

Applying this to our integral,

$$\frac{1}{2} \iiint \phi_i \phi_j V \approx P_0 T e^{-i(l_i+l_j)\pi/2} \iint dy \, dz \, p(y,z) e^{-y^2} e^{-z^2}$$

$$\times \ e^{-i\left[(\Delta_0 + \beta_y y + \beta_z z)(y^2 + z^2)\right]}$$

$$\times \ e^{-i\left[(\Delta_0 + \beta_y y + \beta_z z)K\right]} \tag{21}$$

with $T = 1 - \Delta_t^2 K^2 / 24$. It should be noted that $K \sim 250,000$ is very large, and so when estimating the thickness correction T from Eq. 20, we ignored several terms smaller than $\Delta_t^2 K^2 / 24 \sim 10^{-3}$ by a factor of K or more.

The exponent in the second line of Eq. 21 contains only small quantities, so we can simplify this term by making the expansion $e^{i\epsilon} \approx 1 + i\epsilon$. Then we complete the square for y and z in the remaining exponential and make the variable change $y \to y - i\beta_y K/2$ and $z \to z - i\beta_z K/2$. If we then define the (analytically soluble) integral

$$\xi_{n_i n_j}^{q\beta K} = \int dx \, (x - i\beta K/2)^q e^{-x^2} H_{n_i}\left(\frac{x - i\beta K/2}{\sqrt{A}}\right) H_{n_i}((x - i\beta K/2)\sqrt{A}) \tag{22}$$

and the shorthand $\Gamma_{qp} \equiv \xi_{m_i m_j}^{q\beta_x K} \xi_{n_i n_j}^{p\beta_y K}$, Eq. 21 becomes

$$\frac{1}{2} \iiint \phi_i \phi_j V \approx P_0 T e^{-iK\Delta_0 - i(l_i+l_j)\pi/2} e^{-\frac{K^2(\beta_x^2 + \beta_y^2)}{4}}$$

$$\times [\Gamma_{00} - i\left((\Gamma_{20} + \Gamma_{02})\Delta_0 + (\Gamma_{30} + \Gamma_{12})\beta_y + (\Gamma_{03} + \Gamma_{21})\beta_z\right)]. \tag{23}$$

Applying a similar method to the second half of Eq. 13, and with definitions $K' \equiv m_j - m_i + n_j - n_i - gk_i x_R$ (much smaller than K) and $\Gamma'_{qp} \equiv \xi^{q\beta_x K'}_{m_i m_j} \xi^{p\beta_y K'}_{n_i n_j}$, it can be shown that

$$\frac{1}{2} \iiint \phi_i \phi_j^* V \approx P_0 e^{-iK\Delta_0 - i(l_i - l_j)\pi/2}$$

$$\times \left[\Gamma'_{00} - iK' \left(\Gamma'_0 \Delta_0 + \Gamma'_{10} \beta_y + \Gamma'_{01} \beta_z \right) \right]. \qquad (24)$$

Equations 23, 24 and 13 represent a very accurate analytical approximation of the inner products V_{ij} for small displacements (relative to x_R) from the cavity waist. These results are also plotted (with and without the thickness correction) in Fig. 4 of the previous section as solid lines. In practice, the agreement for our setup is excellent as long as $|x_c| < 1$ mm. As expected, the approximation is nearly perfect at the waist and breaks down as x_0 approaches x_R.

3.3. *Discussion and Comparison with Data*

We can gain insight into our system from the analytical results. By ignoring the off-diagonal terms in Eq. 8 (i.e. ignoring avoided crossings), the eigenvalues simplify substantially, and the detuning is given by

$$\frac{\Delta \omega_s}{\omega_0} \approx -\frac{t(n_{SiN}^2 - 1)}{2L}(1 - T\cos((2k_s x_R - 1)\Delta_0)) \qquad (25)$$

$$\frac{\Delta \omega_{y,a,z}}{\omega_0} \approx -\frac{t(n_{SiN}^2 - 1)}{2L}(1 + T\cos((2k_y x_R - 3)\Delta_0)) + g \qquad (26)$$

for the singlet and triplet modes respectively. These equations represent two sinusoidal bands oscillating (with opposite sign) below their unperturbed detunings, with peak-to-peak amplitudes $T(n_{SiN}^2 - 1)t/L \approx 27\%$ of the free spectral range, and separated from each other by the appropriate Guoy spacing. Applying this method to the other transverse modes, we can generate the entire band structure shown in Fig. 3(a), and for $n_{SiN} = 2.0$ the agreement is essentially perfect.

The spatial period of the singlet band is slightly smaller than the triplet, and so it should oscillate a little faster as a function of Δ_0. Though subtle, we do observe a phase difference between the bands (this is somewhat more visible in Fig. 3 when comparing the singlet and the nonet band), and we can use this phase difference to estimate the membrane's displacement x_0. In the time it takes to raster the band structure in Fig 3, drifts in the piezos and laser frequency cause noticeable distortions of the bands on the scale of this phase difference, but nonetheless by fitting the neighboring singlet and triplet sinusoids from Fig. 3(a) we estimate a positive (as defined by the x-axis) displacement of 300 ± 100 μm from the waist.

It is also worth noting that the effect of finite membrane thickness is to wash out the oscillations (i.e. T decreases from unity as t increases). This makes sense qualitatively because when the membrane is positioned at a node, the electric field is not zero everywhere inside it and so there will still be a small negative perturbation at the top of the band. Similarly, when the membrane is at an antinode, the field is not maximal everywhere inside it and so the perturbation is not as strong. Optical losses inside the membrane mean that even when positioned at a node the finite thickness will put an upper bound on the finesse this system can achieve. We have, however, already observed a finesse of 150,000 with the membrane inside the cavity.[10]

If we now look at the mode-coupling terms $V_{i \neq j}$, we can gain some insight into the avoided crossings of Fig. 3(b)–(c). First, all of the off-diagonal terms involving the antisymmetric mode ($\mathrm{TEM}_{1,1}$) are identically zero (even with $\alpha \neq 0$) in this approximation. This is a reflection of the fact that the $\mathrm{TEM}_{1,1}$ mode is an odd function in both the y and z directions, while the other three modes are even. The integrals across the membrane therefore all involve a function that is approximately odd and vanish. Hence there should be no avoided crossing between $\mathrm{TEM}_{0,0}$ and $\mathrm{TEM}_{1,1}$ to first order, which agrees with all of our observations (see Fig. 3(c), for example).

The other off-diagonal terms are not zero (thankfully), and the result is again relatively simple if we keep the membrane aligned (i.e. $\alpha = 0$).

$$V_{sy} = V_{sz} \approx -\Delta_0 \frac{t(n_{\mathrm{SiN}}^2 - 1)}{2L} T \cos\left[((k_s + k_z)x_R - 4)\Delta_0\right]. \quad (27)$$

When the membrane is aligned, the interaction is proportional to Δ_0 times a term that oscillates with a period close to that of the bands (though since the bands always cross each other at roughly the same phase, this term will modulate the coupling slowly as a function of Δ_0). Following this backward through the calculation in the previous section, we see that it arises from our expansion of the finite radius of curvature R. So (perhaps not surprisingly) the interaction between these modes arises from the mismatch between the curved wavefronts and the flat membrane.[b]

---

[b]Naturally, if the membrane distorted to follow a constant phase front as it moved from the waist, the different transverse modes would all remain orthogonal. Perhaps another way to think of the mismatch-induced coupling is to imagine a curved wave partially reflecting from a flat surface. It will certainly not scatter entirely back into the same mode.

This is an encouraging result because it implies a strong degree of tunability in the avoided gap and hence the x^2-sensitivity. Figure 5(a) shows our calculation of the gaps at each of the four avoided crossings as a function of displacement out to 500 μm from the waist. In this plot we show results for both the aligned case (dashed lines) and for a tilt of 0.25 mrad (solid lines). For the aligned case, the gaps collapse onto two similar curves (as they must by symmetry), and when the membrane is tilted, the top two gaps (TR and TL of the inset) move to larger initial values.[c] The top right (TR) gap is a more interesting function of position, as it is tunable through zero at finite offset. If instead we fix the position of the membrane, we should also be able to tune the TR gap over a wide range (including zero) with tilt, as shown in Fig. 5(b). Note the large quantitative and qualitative differences between the TR and TL gaps can help calibrate the magnitude as well as the direction of the membrane displacement if it is not already known.

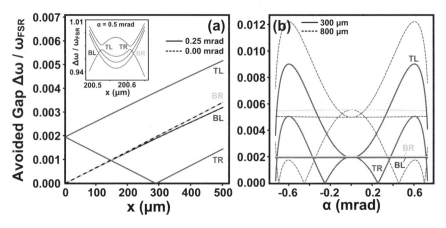

Fig. 5.   (a) Dependence of the four avoided crossing gaps on membrane position, fixing tilt at 0 (dashed lines) and 0.25 mrad. (inset) Plot of the mode detuning versus position near the singlet-triplet crossings. The four gaps in (a, inset) are labeled for reference. (b) Dependence of the four gaps on membrane tilt, fixing the position at 300 and 800 μm. The TR gap should be adjustable over a wide range.

As mentioned in Section 2, vibrations in this apparatus preclude reliable determinations of gap size and x^2-sensitivity for very small gaps. Nonetheless we have observed some smaller gaps as we tune the membrane tilt, two

---

[c]Similar intuition applies here. The flat membrane no longer encloses a constant phase front.

of which are shown in Fig. 6. In this data set, displacement noise due to ambient vibrations, coulomb forces on the membrane and/or piezo noise is quite evident. Further, during the left-to-right rastering of these data sets (acquired over $\sim 10-20$ minutes each) the laser temperature varied enough to cause a systematic detuning and sheer the data vertically. It is a large effect in this data set, and we are studying ways to compensate for it.

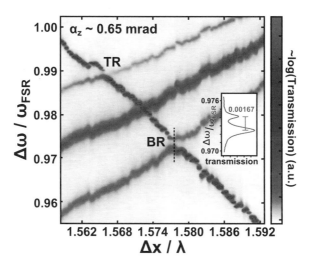

Fig. 6. Transmission data at large membrane tilt, $\alpha_z \approx 0.65$ mrad. The dashed line corresponds to the data shown in the inset. (inset) Single-shot measurement of transmission versus laser detuning.

In a given frequency sweep (vertical trace), however, the time it takes to traverse one of these gaps is roughly a millisecond; such a single-shot measurement of the mode spacing should therefore be much less susceptible to vibrations and drift. If we therefore record the smallest spacing in Fig. 6, we can put a lower bound on the detuning curvature using the form detuning takes near an avoided crossing, $\sqrt{(ax)^2 + (\Delta f/2)^2}$ where a is the asymptotic slope and Δf is the gap (both of which we estimate from Fig. 6). Doing so yields a lower bound on the effective membrane reflectivity of $|r|^2 > 0.992 \pm 0.004$ for the TR crossing and $|r|^2 > 0.9989 \pm 0.0005\%$ for the BR crossing (the sharper curvature of BR reflects the larger asymptotic slope a). This estimate is still subject to vibrations above a few kilohertz, which we have not characterized. On the other hand, when we fit the curvature explicitly as in Section 2, even lower-frequency vibrations (i.e. anything above about 0.1 Hz) can wash out sharp curvature, so that technique represents a very conservative estimate.

Figure 5 implies we can use the TR and BR gaps to estimate the position of our membrane relative to the waist. Figure 7 shows the aligned singlet-triplet crossing data from Fig. 3(b) along with curves generated by this model (for $\alpha = 0$) at several different membrane positions. We estimate the membrane's displacement from the waist to be about 550 μm here, which is a reasonable value for our apparatus and is in rough agreement with our previous estimate based on the horizontal offsets in the various bands shown in Fig. 3.

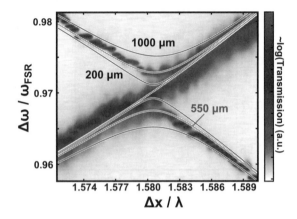

Fig. 7.   Analytical model plotted on top of transmission data for the aligned membrane. Here we show the analytical results for the membrane situated at 200, 550, and 1000 μm from the cavity waist.

We can further check the model for consistency by studying the interplay between tilt and the lifting of triplet degeneracy far from a crossing. It is relatively straightforward to show that this scales as α^2 for small α. The triplet splitting should also be quite insensitive to membrane position so we can use it to estimate the membrane's true tilt or even align the membrane.[d] Figure 8 shows the data from Fig. 3(c) along with the analytical result for $x_0 = 325$ μm and $\alpha_z = 0.395$ mrad. We obtain these parameters by first adjusting α_z until the triplet splitting is correct and then varying x_0 to match the avoided crossings. We have also plotted the result for displacement in the opposite direction, which essentially amounts to comparing our data with BL and TL in Fig. 5. The fit does not agree with the data here or

---

[d]This is in fact how we determined $\alpha \approx 0$.

at any other negative value of x_0. The smaller TR gap therefore confirms the sign of our membrane displacement (and it is the primary reason we chose to study the right side).

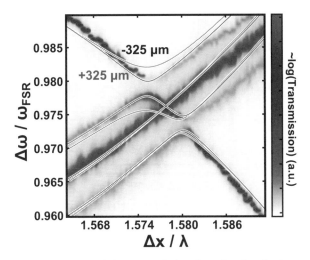

Fig. 8.   Model plotted on top of the transmission data for the tilted membrane. Here we show the analytical results for the membrane situated at +325 and -325 μm from the cavity waist.

We have performed similar analysis at several different values of the tilt stage's motor position, and these are summarized in Fig. 9. Assuming there is a small constant tilt α_y, we can fit this data with the form $\alpha = \sqrt{(aq_z)^2 + \alpha_y^2}$ where a is a mechanical conversion factor between motor position q_z and tilt. From the fit $\alpha_y = 0.16 \pm 0.01$ mrad and $a = 0.0756 \pm 0.0001$ mrad/μm. From the length of the tilt stage lever arm (12.7 mm) alone we estimate $a = 0.0787$ mrad/μm, implying a calibration error of $\sim 4\%$. We have plotted the expected result for the same α_y using a 12.7-mm lever arm for reference.

It is also important to note here that in the model we reproduce the ordering of the triplet modes: as we rotate the membrane about the y axis, the modes most extended in the z direction move the furthest.

The model is in reasonable agreement with the data thus far, and it implies that in future experiments we should be able to tune the x^2-sensitivity to essentially any desired value. This could be a very important tool in our attempt to perform QND measurements of a single phonon.

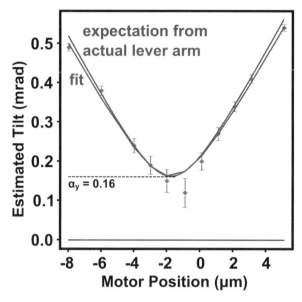

Fig. 9.   Plot of tilt estimated from triplet splitting versus our tilt stage motor position. The red curve is a fit allowing the motor's linearity constant and misalignment α_y to float and the blue curve is the expectation for the same α_y and determining the linearity constant from the tilt stage geometry.

4. Summary/Outlook

In this paper we have demonstrated that a SiN membrane can couple two nearly-degenerate transverse optical cavity modes, generating an avoided crossing and a cavity detuning that is strongly quadratic in membrane displacement x. Without optimizing the system, we have shown that this x^2-dependence (which is tunable over a wide range via membrane tilt) can be as strong as that generated using a single cavity mode and a membrane of reflectivity $|r|^2 \geq 0.9989 \pm 0.0005$. This means it might still be possible to perform QND measurements of phonon number in a membrane of modest reflectivity (i.e. $|r|^2 \sim 0.13$). We also derived a perturbative model of the system that quantitatively agrees with observations and further predicts the x^2-strength should be tunable to arbitrary strength through mm-scale membrane displacements.

These results should be taken with the caveat that the sharp avoided crossings described above occur when the membrane is not at a node of the intracavity field. As discussed previously,[7,10] this means that the optical loss in the membrane will limit the maximum cavity finesse. Whether or not the effect of this reduced finesse can be offset by the very strong quadratic coupling or reduced optical loss (e.g., via improved membrane materials or further engineering of the cavity modes) remains to be seen.

References

1. V. B. Braginsky, F. Ya. Khalili and Kip Thorne, *Quantum Measurement* (Cambridge University Press, 1995).
2. F. Marquardt, J. P. Chen, A. A. Clerk and S. M. Girvin, *Phys. Rev. Lett.* **99**, 093902 (2007).
3. I. Wilson-Rae, N. Nooshi, W. Zwerger and T. J. Kippenberg, *Phys. Rev. Lett.* **99**, 093901 (2007).
4. L. Diósi, *Phys. Rev. A* **78**, 021801 (2008).
5. C. Fabre *et al.*, *Phys. Rev. A* **49**, 1337 (1994).
6. S. Mancini and P. Tombesi, *Phys. Rev. A* **49**, 4055 (1994).
7. J. D. Thompson, B. M. Zwickl, A. M. Jayich, F. Marquardt, S. M. Girvin and J. G. E. Harris, *Nature* **452**, 72 (2008).
8. V. B. Braginsky, F. Ya. Khalili and K. Thorne, *Science* **209**, 547 (1980).
9. A. E. Siegman, *Lasers* (University Science Books, Sausalito, CA, 1986).
10. A. M. Jayich, J. C. Sankey, B. M. Zwickl, C. Yang, J. D. Thompson, S. M. Girvin, A. A. Clerk, F. Marquardt and J. G. E. Harris, *New J. Phys.* (in press, 2008).

TUNABLE INTERACTIONS IN A BOSE-EINSTEIN CONDENSATE OF LITHIUM: PHOTOASSOCIATION AND DISORDER-INDUCED LOCALIZATION

R. G. HULET*, D. DRIES, M. JUNKER, S. E. POLLACK, J. HITCHCOCK,

Y. P. CHEN[†], T. CORCOVILOS, and C. WELFORD

Department of Physics and Astronomy and Rice Quantum Institute, Rice University, Houston, Texas 77005, USA

*E-mail: randy@rice.edu, http://atomcool.rice.edu

[†]*Present Address: Dept. of Physics, Purdue University, West Lafayette, IN 47907*

Lithium-7 exhibits a broad Feshbach resonance that we exploit to tune the interactions in a Bose-Einstein condensate (BEC). We find that the rate of photoassociation can be enhanced by several orders of magnitude by tuning close to the resonance, and use this effect to observe saturation in the rate of association of a BEC for the first time. We have also used a lithium BEC to explore the effects of disorder on the transport and coherence properties of the condensate. We also show that the scattering length goes through a shallow zero-crossing far from the resonance, where it may be made positive or negative with a magnitude of less than 0.1 a_o, and have made preliminary transport measurements in the regime of weak repulsive and attractive interactions.

Keywords: BEC; photoassociation; association; disorder; soliton; Feshbach resonance.

1. Introduction

Our original BEC experiment with ^{7}Li used the $|F = 2, m_F = 2\rangle$ state,[1] for which the number of condensate atoms was limited by the negative scattering length a of the state. The $|1, 1\rangle$ state, on the other hand, has a broad Feshbach resonance located near 737 G (see Fig. 1), which enables large condensates to be formed when the interactions are repulsive. Moreover, the strength of the interactions may be adjusted from strongly interacting to essentially non-interacting. Three experiments are described in the following sections: (1) photoassociation in the strong coupling regime; (2) the effects of disorder on a BEC with tunable interactions; and (3) the measurement of a vs. magnetic field, and in particular, the achievement of $a < 0.1a_o$, where a_o is the Bohr radius.

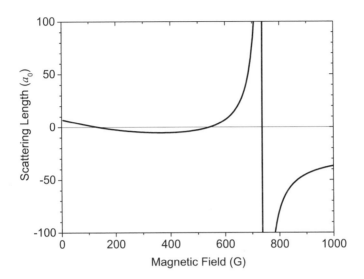

Fig. 1.   Coupled-channels calculation of the scattering length for the $|1, 1\rangle$ Feshbach resonance in ^{7}Li.

2. Photoassociation

A productive path for creating ultracold molecules is to associate ultracold atoms. Two methods have been employed for this purpose: magnetic field sweeps through Feshbach resonances, and photoassociation. Photoassociation is, in many ways, a more promising method because the strength of the atom-molecule coupling is adjustable, and the number of suitable systems is vastly greater than with Feshbach sweeps. A question of both fundamental and practical interest, for both methods, is what are the limitations on the rate of association?

Quantum mechanical unitarity limits the scattering amplitude for two-body scattering to the de Broglie wavelength. This mechanism usually sets the maximum rate of association for non-condensed atoms, as verified in several experiments, including our past work with ^{7}Li in the $|2, 2\rangle$ state cooled to the transition temperature T_c for BEC.[2] In the case of a condensate, the unitarity limit is extremely high and has been considered unreachable. Javanainen and his collaborators have suggested a process they term "rogue photodissociation" that should result in a more stringent rate limit than unitarity. In this process, atom pairs are stimulated back to the energetic continuum and are thereby lost from the condensate.[3] The maximum rate of association achieved in a previous condensate experiment was close to the rogue limit, but no saturation was observed.[4]

Motivated by this background, we designed an experiment in which the strength of the coupling between free atoms and an excited molecular state would be extremely large. The rates we achieved are unprecedented and sufficiently large to directly test the rogue model.[5] This was accomplished by varying the free-bound coupling via a Feshbach resonance. By tuning near the Feshbach resonance, the scattering wavefunction is enhanced at short internuclear distances where photoassociation occurs. The $v = 83$ vibrational level of the electronically excited $1^3\Sigma_g^+$ state was chosen as compromise between a large free-bound coupling strength, and a large detuning (60 GHz) from the $2P_{1/2}$ atomic resonance. Excited molecules created by the photoassociation laser pulse decay into pairs of energetic atoms that escape the trap and are detected as atom loss. The on-resonance rate coefficient K_p is defined by the time evolution of the density distribution:

$$\dot{n}(t, \mathbf{r}) = -K_p n^2(t, \mathbf{r})$$

2.1. *Results*

Figure 2 shows K_p for a thermal gas ($T > T_c$). The rate coefficient varies by more than 4 decades for fields near the Feshbach resonance. The enhancement of K_p at the resonance (737 G) is due to the large enhancement of the scattering wavefunction. The minimum at 710 G, on the other hand, is a result of a node in the scattering wave function that occurs when a is tuned to the Condon radius of the transition.[5]

The data of Fig. 2 demonstrate that K_p can be extraordinarily large near the Feshbach resonance, making it an ideal system for exploring saturation. Figure 3 shows K_p vs. the intensity I of the PA laser beam. The data in this plot correspond to a condensate with no visible thermal fraction. By achieving extremely large Feshbach-enhanced loss rates, saturation is observed in a condensate for the first time. The maximum K_p of 1.4×10^{-7} cm^3/s is nearly a factor of 10 larger than that of any previous photoassociation experiment.[2]

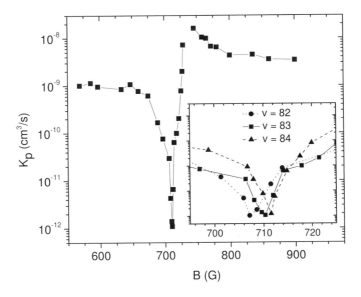

Fig. 2. K_p for a thermal gas ($\sim$10 μK). The inset shows K_p for three different excited state vibrational levels, using the same scale as the main figure. Reprinted from Ref. 5.

2.2. *Analysis*

A comparison of the data with theory is facilitated by defining K_p in terms of a characteristic length L, as $K_p = (\hbar/m)L$. The rogue photodissociation limit K_{pd} is obtained by taking L to be the average interatomic separation, $n_o^{-1/3}$, evaluated at the peak density n_o.[3] For the data of Fig. 3, $n_o = 1.6 \times 10^{12}$ cm^{-3}, giving $K_{pd} \sim 8 \times 10^{-9}$ cm^3/s. Surprisingly, the measured maximum K_p is nearly 20 times greater than K_{pd}. More recent calculations[6,7] show that while dissociation does impose a rate limit on condensate loss, it is not as stringent as K_{pd}. Our measured maximum K_p is, nonetheless, nearly 7 times greater then predicted from the equations given in Ref. 7.

An alternative explanation is provided by quantum mechanical unitarity. If we take $L \sim 2R_{TF}$, where $R_{TF} \simeq 10$ μm is the radial Thomas-Fermi radius, then $K_p \sim 1.8 \times 10^{-7}$ cm^3/s, in good agreement with the measured value of 1.4×10^{-7} cm^3/s. The observed saturation could also be explained by a higher than expected rogue limit, perhaps due to cross coupling between the photoassociation and Feshbach resonances, as discussed in Ref. 8.

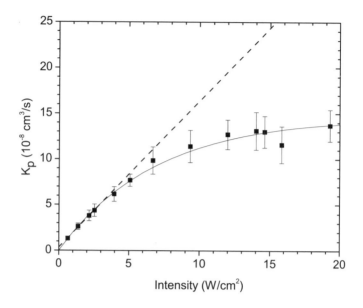

Fig. 3. K_p vs. I for a BEC at 732 G ($a \simeq 1000\,a_o$). Reprinted from Ref. 5.

3. Disorder

Materials, no matter how pure they may be or how carefully they are prepared, inevitably have some random disorder. This disorder can be caused
by crystal defects, impurities, or anything that changes the landscape of
how electrons move about in the material. Disorder can play an important role in the transport properties of real materials. Superconductors, for
example, can have zero resistance in the presence of material defects, but
with increasing disorder the electrons will localize, resulting in an insulating state. This effect has been explored in many systems experimentally,
including superfluid helium in porous media, and thin-film and granular
superconductors. Many fundamental questions, such as the nature of the
insulating state and the characterization of phase coherence, remain to be
resolved.

Gases of ultracold atoms have proven to be extremely useful stand-ins
for actual materials because of the ability to control many of the parameters,
including the characteristics of the disorder itself, as well as the particle
interactions via a Feshbach resonance. Following the pioneering work at
Florence,[9] Orsay,[10] and Hannover,[11] we use optical speckle to create a
highly-controllable disordered potential in a ^{7}Li $|1,1\rangle$ BEC.[12] The speckle

in our experiment is created by passing a 1030 nm laser beam through a microlens array. The resulting intensity autocorrelation function is Gaussian with a characteristic length $\sigma_d = 15$ μm. The strength of the disordered optical potential V_d is proportional to the intensity and is continuously controllable up to the chemical potential of the condensate ($\sim$1 kHz).

3.1. *Results*

We have performed two transport experiments and have also studied coherence as revealed by interference in time-of-flight (TOF) expansion imaging.[12] These experiments were performed near 720 G, where $a \simeq 200\,a_o$. Pinning of the condensate by disorder was studied by slowly dragging it through the disorder. This was accomplished by using a magnetic gradient to change the trap center. The data show that the condensate is pinned when $V_d \simeq \mu$,[12] where μ is the chemical potential of the condensate. In a related experiment, the trap center is suddenly displaced causing the condensate to undergo damped, dipole oscillations. The results, displayed in Fig. 4, show that even small disorder produces significant damping, and that the motion is overdamped for $V_d \gtrsim 0.4\mu$. The damping coefficient β is found to related to V_d by a power law, $\beta \propto (V_d/\mu)^{5/2}$. We do not have an explanation for the value of the exponent.

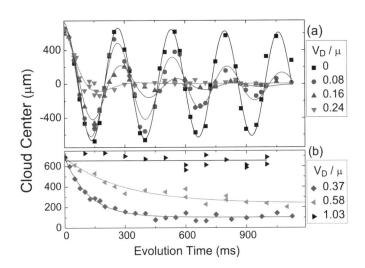

Fig. 4.   Damping of dipole oscillations for various V_d. Reprinted from Ref. 12.

Figure 5 shows both *in situ* and the corresponding TOF images for various values of V_d. The *in situ* images reveal how the density distribution is affected by the disorder, while the TOF images provide complementary information on condensate coherence. We find that the condensate density becomes increasingly modulated with increasing V_d. When $V_d \simeq \mu$, the condensate appears to fragment into disconnected pieces. Interference fringes are observed in the corresponding TOF images, but in the case of TOF, the maximum contrast occurs for intermediate disorder strength, $V_d \simeq 0.5\,\mu$.

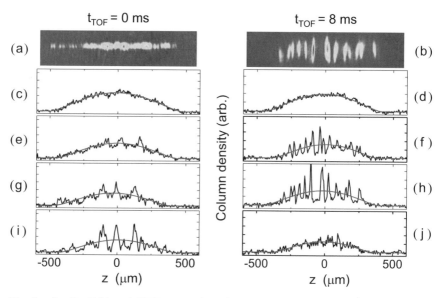

Fig. 5.   *In situ* (left) and TOF images (right) for various V_d. (c, d) $V_d = 0$; (e, f) $V_d \approx$ 0.3 μ; (a, b, g, h) $V_d \approx 0.5\ \mu$; (i, j) $V_d \approx 1.0\ \mu$. Reprinted from Ref. 12.

3.2. *Analysis*

The transport measurements indicate that global superfluidity is absent when $V_d \gtrsim 0.5\mu$, which is also where the condensate begins to fragment, as shown by the *in situ* density measurements. A key to interpreting the interference observed in the TOF images is that the interference patterns are completely *repeatable*. This observation rules out phase fluctuations in the initial condensate as the cause of the TOF interference. At intermediate V_d, sufficient local coherence remains in the initial condensate to produce these reproducible and high-contrast TOF interference patterns. We conclude that the condensate must remain connected under these conditions. For higher V_d ($\sim\mu$), the TOF contrast diminishes as the condensate fragments into multiple sites. Similar observations and conclusions were made by the authors of Ref. 13, although at lower V_d/μ.

4. Direct Measurement of the Scattering Length

The Feshbach resonance in ^{7}Li is unusually broad, and moreover, it has a very shallow zero-crossing near 544 G (see Fig. 1). These features make the Feshbach resonance particularly useful for experiments where fine control of a is desired. The zero-crossing was previously exploited to create bright matter-wave solitons,[14,15] and may also prove useful for achieving a non-interacting, or at least a very weakly interacting gas for disorder studies.

4.1. *Results*

We determine a by measuring the axial size and atom number for a BEC as a function of magnetic field, and compare with solutions of the Gross-Pitaevski equation. Figure 6 shows the extracted a for fields between the zero-crossing at 544 G and the resonance at 737 G, where 6 decades of dynamic range are resolved. The solid line in Fig. 6 is a fit to the standard Feshbach resonance form:

$$a(B) = a_{bg} \left(1 + \frac{\Delta B}{B - B_o} \right),$$

where $a_{bg} = -24.5\, a_o$, $\Delta B = 192$ G, and $B_o = 737$ G are the best fits to the data. The fit is remarkably good despite the large dynamic range.

The inset in Fig. 6 shows the extracted values of a near the zero-crossing in more detail. The two sets of points correspond to whether the magnetic dipolar interaction is accounted for in the Gross-Pitaevski equation, or is neglected. The dipolar interaction is generally quite small in lithium because its magnetic moment is only one Bohr magneton, yet its effect is significant when the a is smaller than $\sim 0.1\, a_o$. Since the slope of $a(B)$ near the zero-crossing is 0.1 a_o/G, only moderate field stability is needed to be in the regime where the dipolar interaction dominates. Similarly small scattering lengths have been measured in Cs (Ref. 16) and ^{39}K (Ref. 17), but since the slope of the zero-crossing in ^{7}Li is 6 times smaller than in ^{39}K and 600 smaller than for Cs, the ultimate resolution is significantly better for lithium.

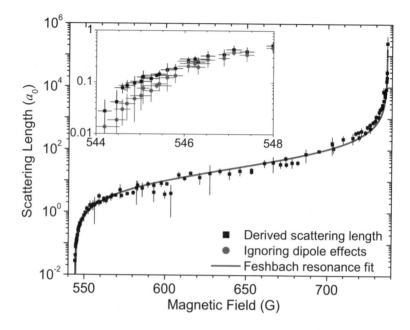

Fig. 6.   Scattering length a vs. B. The inset shows the zero-crossing in detail. The squares are the extracted values of a when the magnetic dipole interaction is accounted for, while the circles correspond to its neglect.

We have repeated the dipole oscillation experiment described in Section 3.1 with very small interaction strength. For small positive a the oscillations damp as before. Negative a can also be obtained by tuning B below the zero-crossing, where solitons are formed. Preliminary measurements indicate that damping is qualitatively different for solitons: the amplitude of the oscillation appears undamped while the number of atoms continuously decreases.

5. Conclusions

We have used a Feshbach resonance to tune the interactions in Bose condensates of ^{7}Li. The strong enhancement of the free-bound wavefunction overlap enables enormous photoassociation rates and the observation of saturation in a BEC for the first time. Rates far above the predicted "rogue photodissociation" limit are achieved. The transport and coherence properties of a BEC in a disordered potential have been explored with optical speckle. We have used the Feshbach resonance to make a as small as $0.1\,a_o$, and observe the damping of dipole oscillations in the presence of disorder for both weakly repulsive and attractive interactions. These preliminary experiments indicate that damping of dipole oscillations of solitons is manifested by the loss of atoms rather than damping of oscillation amplitude.

Acknowledgments

We gratefully acknowledge support of the NSF, ONR, the Welch Foundation (Grant No. C-1133), and the Keck Foundation.

References

1. C. C. Bradley, C. A. Sackett, J. J. Tollett and R. G. Hulet, *Phys. Rev. Lett.* **75**, 1687 (1995).
2. I. D. Prodan, M. Pichler, M. Junker, R. G. Hulet and J. L. Bohn, *Phys. Rev. Lett.* **91**, 080401 (2003).
3. J. Javanainen and M. Mackie, *Phys. Rev. Lett.* **88**, 090403 (2002).
4. C. McKenzie *et al.*, *Phys. Rev. Lett.* **88**, 120403 (2002).
5. M. Junker, D. Dries, C. Welford, J. Hitchcock, Y. P. Chen and R. G. Hulet, *Phys. Rev. Lett.* **101**, 060406 (2008).
6. T. Gasenzer, *Phys. Rev. A* **70**, 043618 (2004).
7. P. Naidon, E. Tiesinga and P. S. Julienne, *Phys. Rev. Lett.* **100**, 093001 (2008).
8. M. Mackie, M. Fenty, D. Savage and J. Kesselman, *Phys. Rev. Lett.* **101**, 040401 (2008).
9. J. E. Lye, L. Fallani, M. Modugno, D. S. Wiersma, C. Fort and M. Inguscio, *Phys. Rev. Lett.* **95**, 070401 (2005).
10. D. Clément, A. F. Varòn, M. Hugbart, J. A. Retter, P. Bouyer, L. Sanchez-Palencia, D. M. Gangardt, G. V. Shlyapnikov and A. Aspect, *Phys. Rev. Lett.* **95**, 170409 (2005).
11. T. Schulte, S. Drenkelforth, J. Kruse, W. Ertmer, J. Arlt, K. Sacha, J. Zakrzewski and M. Lewenstein, *Phys. Rev. Lett.* **95**, 170411 (2005).
12. Y. P. Chen, J. Hitchcock, D. Dries, M. Junker, C. Welford and R. G. Hulet, *Phys. Rev. A* **77**, 033632 (2008).
13. D. Clément, P. Bouyer, A. Aspect and L. Sanchez-Palencia, *Phys. Rev. A* **77**, 033631 (2008).
14. L. Khaykovich, F. Schreck, G. Ferrari, T. Bourdel, J. Cubizolles, L. D. Carr, Y. Castin and C. Salomon, *Science* **296**, 1290 (2002).
15. K. E. Strecker, G. B. Partridge, A. G. Truscott and R. G. Hulet, *Nature* **417**, 150 (2002).
16. M. Gustavsson, E. Haller, M. J. Mark, J. G. Danzl, G. Rojas-Kopeinig and H.-C. Nägerl, *Phys. Rev. Lett.* **100**, 080404 (2008).
17. M. Fattori, C. D'Errico, G. Roati, M. Zaccanti, M. Jona-Lasinio, M. Modugno, M. Inguscio and G. Modugno, *Phys. Rev. Lett.* **100**, 080405 (2008).

A PURELY DIPOLAR QUANTUM GAS

T. LAHAYE, J. METZ, T. KOCH, B. FRÖHLICH, A. GRIESMAIER and T. PFAU*

5. Physikalisches Institut, Universität Stuttgart, D-70550 Stuttgart, Germany
* E-mail: t.pfau@physik.uni-stuttgart.de

We report on experiments exploring the physics of dipolar quantum gases using a ^{52}Cr Bose-Einstein condensate (BEC). By means of a Feshbach resonance, it is possible to reduce the effects of short range interactions and reach a regime where the physics is governed by the long-range, anisotropic dipole-dipole interaction between the large $(6\,\mu_B)$ magnetic moments of Chromium atoms. Several dramatic effects of the dipolar interaction are observed: the usual inversion of ellipticity of the condensate during time-of flight is inhibited, the stability of the dipolar gas depends strongly on the trap geometry, and the explosion following the collapse of an unstable dipolar condensate displays d-wave like features.

Keywords: Bose-Einstein condensation, dipolar quantum gases, Feshbach resonances, condensate collapse, vortex rings.

1. Introduction

Although quantum gases are very dilute systems, most of their properties are governed by atomic interactions. This allows to use them, for example, as *quantum simulators* to study the many-body physics of systems usually encountered in condensed matter physics.[1] However, in all usual quantum gases, the interactions can be described extremely well by a short range, isotropic *contact* potential, whose magnitude is proportional to the *s*-wave scattering length a characterizing low energy collisions.

The dipole-dipole interaction taking place between particles having a permanent electric or magnetic dipole moment has radically different properties: it is long-range and anisotropic, as one readily sees from the expression

$$U_{\mathrm{dd}}(\boldsymbol{r}) = \frac{\mu_0 \mu^2}{4\pi} \frac{1 - 3\cos^2\theta}{r^3} \tag{1}$$

giving the interaction energy U_{dd} between two polarized dipoles $\boldsymbol{\mu}$ separated by $\boldsymbol{r}$ (θ is the angle between $\boldsymbol{r}$ and the direction along which the dipoles are pointing). These specific properties have attracted a lot of interest recently, and a large number of theoretical predictions have been made concerning dipolar quantum gases (see e.g. Ref. 2 for a review): for instance, the stability of a dipolar BEC depends crucially on the trap geometry (see section 3 below); in a quasi two-dimensional trap, the excitation spectrum can display a roton minimum instead of the usual Bogoliubov shape; finally, fascinating new quantum phases (including supersolids) are predicted to occur for dipolar bosons in an optical lattice.

In practice one always has a competition between contact and dipolar interactions; it is therefore useful to define the following (dimensionless) ratio of the dipolar and contact coupling constants:

$$\varepsilon_{dd} = \frac{\mu_0 \mu^2 m}{12\pi \hbar^2 a}. \tag{2}$$

The numerical factors are chosen in such a way that a *homogeneous* dipolar condensate is unstable against collapse for $\varepsilon_{dd} > 1$. For usual atomic magnetic moments μ (e.g. for the alkalis), ε_{dd} is very small (typically a few 10^{-3}) and dipolar effects are extremely small. Here, we report on experiments with ^{52}Cr, which has $\varepsilon_{dd} \simeq 0.16$ due to its large magnetic moment $\mu = 6\,\mu_B$, and which also allows, *via* Feshbach tuning of the scattering length a, to even enhance ε_{dd}.

This paper is organized as follows. We first describe briefly in Section 2 our experimental setup, with an emphasis on how we use a Feshbach resonance in order to enhance dipolar effects and create a 'quantum ferrofluid'. Section 3 is devoted to the study of the geometry dependence of the stability of a dipolar BEC. Finally, we describe in Section 4 the dynamics following the collapse of an unstable dipolar condensate.

2. Enhancing dipolar effects using a Feshbach resonance

A BEC of ^{52}Cr containing about 50,000 atoms was obtained in 2005 by evaporative cooling of optically trapped chromium atoms.[3] Shortly after the achievement of condensation, a first effect of the dipole-dipole interaction could be observed in time-of flight experiments:[4] The dipole-dipole interaction tends to *elongate* the BEC along the magnetization direction. However, due to the small value of $\varepsilon_{dd} \simeq 0.16$, the dipolar interaction was, in this experiment, only a small perturbation of the contact interaction, which essentially governed the expansion dynamics.

The existence of several Feshbach resonances[5] in ^{52}Cr opens the possibility to tune the scattering length a using an external magnetic field B,

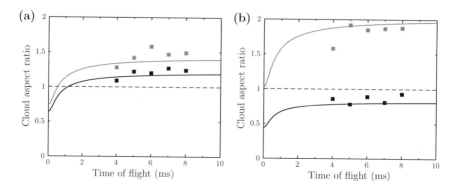

Fig. 1.   Free expansion of a dipolar condensate for two different orientations of the dipoles with respect to the trap axes. The black (resp. gray) squares and lines correspond to a situation where the dipoles point along the weak (resp. strong) axis of the trap. (a): Perturbative regime $\varepsilon_{dd} = 0.16$; (b): $\varepsilon_{dd} = 0.75$. In that case, the dipole-dipole interaction is strong enough to inhibit the usual inversion of ellipticity in time of flight.

according to

$$a = a_{bg} \left(1 - \frac{\Delta}{B - B_0} \right).$$

Here, $a_{bg} \simeq 100\, a_0$ is the s-wave scattering length, B_0 the resonance position, and Δ the resonance width. The broadest Feshbach resonance in ^{52}Cr is located at $B_0 = 589$ G and has a width Δ of only 1.5 G. This implies that the field control at the level of 3×10^{-5} r.m.s. that we implemented allows us to tune a close to 0 with a resolution of about one Bohr radius.

This 'knob' permits one to change a and allowed us to perform time of flight experiments for two different orientations of the dipoles with respect to the trap axes,[6] as in Ref. 4, but now for increasing values of ε_{dd}. Experimental results are shown in Fig. 1. One clearly sees the dramatic effect of an increase of ε_{dd} on the expansion dynamics. In particular, for $\varepsilon_{dd} \simeq 0.75$, the inversion of ellipticity of the condensate during time of flight (the usual 'smoking-gun' evidence for BEC) is inhibited by the strong dipole-dipole interaction.

The solid lines in Fig. 1 are theoretical predictions (without any adjustable parameters) based on the Gross-Pitaevskii equation (GPE) generalized to take into account the non-local dipole-dipole interaction in the description of the macroscopic wavefunction $\psi(\boldsymbol{r}, t)$ of the BEC:

$$i\hbar \frac{\partial \psi}{\partial t} = \left(-\frac{\hbar^2}{2m} \Delta + V_{ext} + g|\psi|^2 + \int |\psi(\boldsymbol{r}', t)|^2 U_{dd}(\boldsymbol{r} - \boldsymbol{r}')\, d\boldsymbol{r}' \right) \psi. \quad (3)$$

Here $g = 4\pi\hbar^2 a/m$ is the contact interaction coupling constant.

In this set of experiments, the trap geometry was not very far from spherical, which limited the study of dipolar condensates to values $\varepsilon_{\mathrm{dd}} \lesssim 1$. To go beyond this value and reach the purely dipolar regime $\varepsilon_{\mathrm{dd}} \gg 1$, we shall now see that one needs to tailor the confining potential, so that the attractive part of the dipole-dipole interaction does not destabilize the condensate.

3. Geometrical stabilization of a purely dipolar condensate

3.1. *Experimental study*

It is well known that a BEC with attractive contact interactions is unstable against long-wavelength fluctuations (this *phonon instability* leads to a collapse of the BEC having $a < 0$). As the dipole-dipole interaction has an attractive part for dipoles in a 'head-to-tail' configuration (see equation (1) for $\theta \simeq 0$), it is intuitively clear that in a prolate trap with the dipoles pointing along the weak direction of the trap [see Fig. 2(a)], the net effect of the dipolar interaction is attractive. Thus, in this configuration, one expects that when a is reduced, the condensate becomes unstable, at a critical value a_{crit} which should be positive (the small repulsive contact interaction being unable, at this point, to counteract the dipolar attraction). Conversely, in an oblate trap with the dipoles pointing along the strong confinement direction, the critical scattering length should be negative, and a *purely dipolar quantum gas* can be stabilized.

In Ref. 7, this geometry-dependent stability of a dipolar condensate was studied experimentally. A long period ($\simeq 8$ μm) optical lattice, obtained by interfering two laser beams at 1064 nm under a small angle of 8°, was superimposed onto the optical dipole trap, allowing us to realize traps with cylindrical symmetry around the z-axis (polarization direction) and having an aspect ratio $\lambda \equiv \omega_z/\omega_\rho$ that could be varied over two orders of magnitude (from ~ 0.1 to ~ 10) while keeping the average trapping frequency $\bar{\omega} = (\omega_z\omega_\rho^2)^{1/3}$ constant. The experiment consists of creating a BEC in a trap with a given aspect ratio λ, then ramping a to a final value a_{f}, and finally measuring the atom number N in the condensate. One observes that when a_{f} is decreased below a critical value a_{crit}, N suddenly drops to zero. We stress that for all the traps we used, the condensate density was roughly the same.

Figure 2(b) shows the measured a_{crit} as a function of λ and clearly displays the expected behavior: for small λ (prolate traps), a_{crit} is positive,

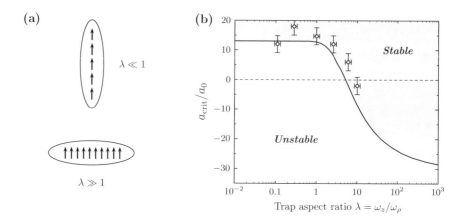

(a)

$\lambda \ll 1$

$\lambda \gg 1$

(b)

Stable

Unstable

$a_{\rm crit}/a_0$

Trap aspect ratio $\lambda = \omega_z/\omega_\rho$

Fig. 2.   (a): Intuitive picture of the geometry-dependent stability of the dipolar Bose gas. For a prolate trap (aspect ratio $\lambda = \omega_z/\omega_\rho$ smaller than one) with the dipoles pointing along the weak axis of the trap, the dipole-dipole interaction is essentially attractive; such a condensate is thus unstable. For an oblate trap, the dipole-dipole interaction is essentially repulsive and the BEC is stable. (b): Stability diagram of a dipolar BEC in the plane (λ, a). The points with error bars are the experimental results for the critical scattering length $a_{\rm crit}$ below which no condensate is observed; the solid line is the stability threshold obtained with a simple gaussian ansatz (see text).

and starts to decrease when the trap becomes more oblate. For $\lambda \simeq 10$, one has $a_{\rm crit} \simeq 0$, meaning that a purely dipolar quantum gas ($\varepsilon_{\rm dd} \to \infty$) can be stabilized by an appropriate trap geometry.

3.2. *A simple theoretical model*

A simple way to go beyond the qualitative picture above and obtain an estimate for the instability threshold $a_{\rm crit}(\lambda)$ is to use a variational method. Inserting a Gaussian ansatz (with the axial and radial sizes σ_z and σ_ρ as variational parameters) into the Gross-Pitaevskii energy functional whose minimization gives the GPE (3), one obtains the following energy to minimize:

$$E(\sigma_\rho, \sigma_z) = \frac{N\hbar\bar{\omega}}{4}\left(\frac{2}{\sigma_r^2} + \frac{1}{\sigma_z^2}\right) + \frac{N\hbar\bar{\omega}}{4\lambda^{2/3}}\left(2\sigma_r^2 + \lambda^2\sigma_z^2\right)$$
$$+ \frac{N^2\hbar\bar{\omega}a}{\sqrt{2\pi}\ell}\frac{1}{\sigma_\rho^2\sigma_z}\left[1 - \varepsilon_{\rm dd}f\left(\frac{\sigma_\rho}{\sigma_z}\right)\right], \qquad (4)$$

where $\ell = \sqrt{\hbar/(m\bar{\omega})}$. The first two terms are the kinetic and potential energies, while the third arises from contact and dipolar interactions. The function f is monotonically decreasing from 1 to -2 as a result of the anisotropy of the dipolar interaction. For a given λ, one can find a (possibly local) minimum of E at finite values of (σ_ρ, σ_z) if and only if a is larger than a critical value: this defines the stability threshold $a_{\text{crit}}(\lambda)$ within this model.

The solid line in Fig. 2(b) is the result obtained with this simple procedure, for the experimental parameters $\bar{\omega} = 2\pi \times 800$ Hz and $N \simeq 2 \times 10^4$. One obtains a relatively good agreement with experimental data. A numerical solution of the GPE (3) gives even better agreement with measurements.[8]

Equation (4) allows one to understand easily, in the $N \to \infty$ limit, the behavior of $a_{\text{crit}}(\lambda)$ for $\lambda \to 0$ and $\lambda \to \infty$. Indeed, in this limit, it is the sign of the interaction term which determines the stability; therefore one has

$$\begin{cases} \lambda \to 0 : \text{BEC unstable if } a < a_{\text{dd}} \\ \lambda \to \infty : \text{BEC unstable if } a < -2a_{\text{dd}}, \end{cases} \tag{5}$$

where a_{dd} is the length defined in such a way that $\varepsilon_{\text{dd}} = a_{\text{dd}}/a$ (for ^{52}Cr, one readily calculates, with the help of Eq. (2), that $a_{\text{dd}} \simeq 15a_0$). It is apparent on Fig. 2(b) that for $N = 2 \times 10^4$, the results are already close to the $N \to \infty$ limit (5).

4. *d*-wave collapse of a dipolar condensate

It is natural to ask what happens if one drives the condensate into the unstable regime, e.g. by decreasing the scattering length below a_{crit}. In the case of pure contact interactions, a collapse of the condensate, followed by an explosion of a 'remnant' BEC (*Bose-Nova*), has been observed in several systems.[9-12] More recently, the formation of soliton trains has also been reported.[13,14]

We have studied the collapse dynamics of a dipolar condensate[15] (in a roughly spherical trap) by ramping down rapidly the scattering length to a final value of $\sim 5a_0 < a_{\text{crit}}$, then waiting an adjustable holding time, and performing a time of flight of 8 ms before imaging the cloud. Figure 3(a) presents the evolution of the condensate when the holding time is varied. One observes that the cloud, initially elongated along the magnetization direction z (horizontal axis on the figure) acquires rapidly a complicated structure with a four-fold symmetry, corresponding to a density distribution having a torus-like component close to the plane $z = 0$ and two 'blobs' close to the z-axis. Interestingly, this angular symmetry of the cloud is very close to that of a d-wave $\propto (1 - 3\cos^2\theta)$, i.e. precisely the symmetry of the dipole-dipole interaction (1). During the same time period, the atom number in

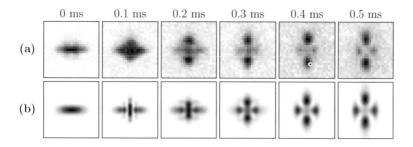

Fig. 3.   (a): Experimental images of the 'exploding' remnant condensate after the collapse, as a function of the holding time. (b): Result of a numerical simulation of the experiment, without any adjustable parameter. The field of view is $130\,\mu m \times 130\,\mu m$.

the condensate strongly decreases due to the three-body losses occurring because of the high densities transiently reached during the collapse.

The group of M. Ueda in Tokyo performed a three dimensional numerical simulation of the GPE (3), in which all input parameters were given their experimentally measured value.[15] Three-body losses were accounted for by adding the imaginary term

$$i\hbar \left.\frac{\partial\psi}{\partial t}\right|_{3\,body} = -\frac{i\hbar L_3}{2}|\psi|^4\psi$$

to Eq. (3), where $L_3 \simeq 2 \times 10^{-40}$ m^6/s is the measured three-body loss coefficient. Figure 3(b) represents the results of the simulation. The agreement is excellent, all the more if one keeps in mind that no adjustable parameter is introduced. Let's mention that the inclusion of a small delay (also measured independently) in the time variation $a(t)$ of the scattering length, due to eddy currents in the vacuum chamber, had to be included to achieve a quantitative agreement. The simulation also reproduces quantitatively the time dependence of the condensate atom number.

A fascinating prediction of the numerical simulation is the spontaneous formation, during the collapse, of two quantized vortex rings with opposite circulation (and charge ± 1), as a result of the strongly anisotropic collapse: the collapse in the radial directions is fast and quickly followed by an outward flow, while axially the flow is still inward, thus giving rise to the circulation. Detecting experimentally the presence of vortex rings is very challenging, but might be done by using interferometric techniques (e.g. matter wave "heterodyning") to reveal the winding of the phase of the BEC wavefunction around the topological defects.

5. Outlook

The results presented in this paper are the first dramatic manifestations of dipolar effects in quantum gases, and pave the way for future studies involving even more strongly interacting dipolar systems, especially the ones that may be obtained using the permanent electric dipole moments of heteronuclear molecules in their ground state. Due to the large value of such dipole moments (on the order of one Debye), the long-range character of the dipolar interaction could then be used to achieve novel quantum phases in optical lattices,[2] as well as to implement promising quantum information processing schemes.[16]

However, already in the case of the comparatively weaker magnetic dipoles, extremely interesting theoretical proposals deserve experimental study; to mention only one example, the generation of two-dimensional solitons[17,18] (whose stability arises from the long-range character of the dipole-dipole interaction) is a very appealing experiment.

Acknowledgments

We acknowledge support by the German Science Foundation (SFB/TRR 21, SPP 1116), the Landesstiftung Baden-Württenberg and the EU (Marie-Curie Grant MEIF-CT-2006-038959 to T. L.).

References

1. I. Bloch *et al.*, *Rev. Mod. Phys.* **80**, 885 (2008).
2. M. A. Baranov, *Phys. Rep.* **464**, 71 (2008).
3. A. Griesmaier *et al.*, *Phys. Rev. Lett.* **95**, 160401 (2005).
4. J. Stuhler *et al.*, *Phys. Rev. Lett.* **95**, 150406 (2005).
5. J. Werner *et al.*, *Phys. Rev. Lett.* **94**, 183201 (2005).
6. T. Lahaye *et al.*, *Nature* **448**, 712 (2007).
7. T. Koch *et al.*, *Nat. Phys.* **4**, 218 (2008).
8. J. L. Bohn, private communication.
9. C. A. Sackett *et al.*, *Phys. Rev. Lett.* **82**, 876 (1999).
10. J. M. Gerton *et al.*, *Nature* **408**, 692 (2000).
11. J. L. Roberts *et al.*, *Phys. Rev. Lett.* **86**, 4211 (2001).
12. E. A. Donley *et al.*, *Nature* **412**, 295 (2001).
13. K. S. Strecker *et al.*, *Nature* **417**, 150 (2002).
14. S. L. Cornish *et al.*, *Phys. Rev. Lett.* **96**, 170401 (2006).
15. T. Lahaye *et al.*, *Phys. Rev. Lett.* **101**, 080401 (2008).
16. D. DeMille, *Phys. Rev. Lett.* **88**, 067901 (2002).
17. P. Pedri and L. Santos, *Phys. Rev. Lett.* **95**, 200404 (2005).
18. I. Tikhonenkov *et al.*, *Phys. Rev. Lett.* **100**, 090406 (2008).

BOSE-EINSTEIN CONDENSATION OF EXCITON-POLARITONS

Y. YAMAMOTO

E.L. Ginzton Laboratory, Stanford University, Stanford, CA94305, USA
National Institute of Informatics, Tokyo, Japan
E-mail: yyamamoto@stanford.edu, www.stanford.edu/group/yamamotogroup

This article summarizes recent work on the exciton-polariton BEC at Stanford, which was presented at ICAP 2008. The covered topics include cooperative cooling of exciton-polariton spin mixtures, quantum degeneracy at thermal equilibrium condition, Bogoliubov excitation spectrum, first and second order coherence, and dynamical condensation at excited Bloch bands in a one-dimensional lattice.

Keywords: Exciton-polariton; BEC; spin dynamics; Bogoliubov spectrum.

1. Exciton-Polaritons and BEC

An experimental technique of controlling spontaneous emission of an atom by use of a cavity has been applied to Wannier-Mott excitons in a semiconductor quantum well.[1] Due to a strong collective dipole coupling between microcavity photon fields and QW excitons, a semiconductor planer microcavity features a normal mode splitting of the order of $1\sim10$ THz,[2] which corresponds to the reversible spontaneous emission of an oscillation period of 100 fs $\sim$ 1 ps.[3] Figure 1(a) shows the energy-momentum dispersion characteristics of such normal modes, often referred to exciton-polaritons. An observed absorption spectrum and time-dependent emission intensity feature the normal mode splitting and reversible spontaneous emission, as shown in Fig. 1(b) and (c).[3]

A ground state of lower exciton-polariton (LP) at zero in-plane momentum ($k = 0$) has been identified as a promising candidate for observing BEC is solids.[4] A LP at $k = 0$ has an effective mass of four orders of magnitude lighter than a bare exciton mass, so the critical temperature for polariton BEC is four orders of magnitude higher than that required for bare exciton BEC at the same particle density. The extremely light effective mass of a LP at $k = 0$ also solves a notorious enemy to exciton BEC, localization due to crystal defects, disorders and inhomogeneous potential fluctuations. The critical particle density for 2D polariton BEC is also four orders of magnitude smaller than that required for 2D exciton BEC at the same temperature, which resolves another serious problem of Auger recombination and dissociation of excitons at high densities. By inserting 12 to 24 multiple quantum wells into a microcavity, an exciton density per QW per polariton is further diluted.

2. Bosonic Final State Stimulation and Polariton Condensation

In the first cold collision experiment of LPs,[5] the counterpropagating two LPs, piled up at bottleneck momentum $\pm k$, collide and scatter into LP and UP at $k = 0$ with conservation of energy and momentum. When the LP population n_{LP} at $k = 0$ is injected with coherent laser excitation, an enhanced scattering rate proportioned to $(1 + n_{LP})$ is observed as increased population of the UP at $k = 0$. In the subsequent experiment of polariton amplifier,[6] the weak probe light is injected into the bottleneck LPs at $k \neq 0$. A linear gain of 10-20 dB due to LP-LP stimulated scattering was observed during a time interval of the bottleneck polariton lifetime of ~ 100 ps.

If a pump rate is further increased to above the condensation threshold, n_{LP} increases nonlinearly as shown in Fig. 2.[7] A threshold is identified as a quantum degeneracy point where $n_{LP} = 1$. This sample incorporates a tapered cavity resonance energy across the wafer. If we excite a particular spot where the cavity resonance energy is above the bandgap, we can observe a normal photon leasing behavior based on inverted electron-hole pairs as shown in Fig. 2. This result clearly demonstrates polariton condensation is based on the stimulated scattering of LPs so that electronic population inversion is not required. Another distinct difference between a polariton condensation and a photon laser is the pump dependence of lasing spot sizes. A rapid increase in the photon laser spot size above its threshold stems from the fact that the critical density is independent of the system size, while a higher density is required to maintain a quasi-BEC in larger 2D system.[8] The measured increase in the polariton condensation spot size is quantitatively described by this theory.[9]

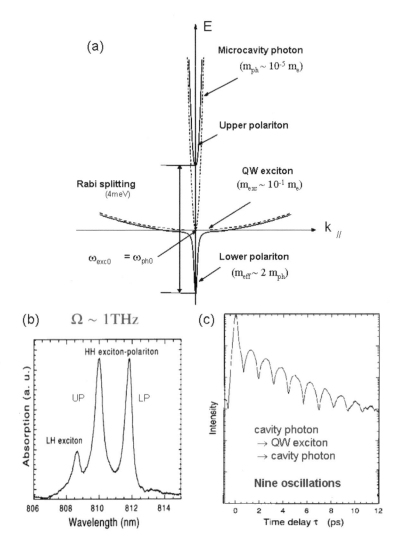

Fig. 1.   The energy-momentum dispersion relation (a), absorption spectrum (b), and time-dependent emission intensity of exciton-polaritons in a GaAs planar microcavity (c).[3]

3. Spin Dynamics and Cooperative Cooling of Exciton-Polariton Mixture

The above polariton condensation experiment satisfies a quantum degeneracy condition ($n_{LP} \gg 1$) above threshold. In order to satisfy a thermal

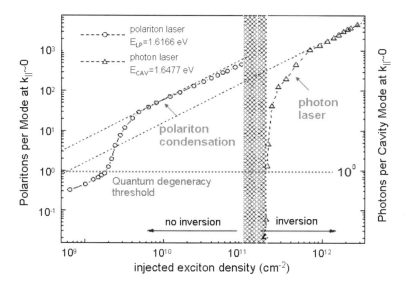

Fig. 2.   Emission intensity vs. injected electron-hole density for polariton condensation and photon laser.[7]

equilibrium condition simultaneously, the polariton cooling time must be much shorter than a polariton lifetime. Two tricks are useful to achieve this criterion: one is a blue detuning (cavity resonance energy is higher than exciton resonance energy at $k = 0$)[10] and the other is a cooperative cooling by spin mixtures.[11]

Figure 3 shows the energy-momentum $(E - k)$ and energy-position $(E - x)$ dispersion characteristics for varying pump rates when the pump polarization is linear and circular polarization.[11] At a pump rate below thereshold, a standard parabolic $(E - k)$ dispersion and constant $(E - x)$ dispersion are observed for two cases. At a pump rate above threshold, a linear pumping scheme realizes the smooth polariton condensation at $k = 0$, while a circular pumping scheme features a so-called bottleneck polariton condensation at $k \neq 0$. This result suggests that co-existence of two spin components, created by linear polarization pumping, accelerates a cooling process, compared to the case of single spin injection by circular polarization pumping. Under circular polarization pumping, the polarization of the emission at below threshold is almost completely random but the same circular polarization is maintained in the emission at above the threshold.[7] This indicates a spin relaxation time is shorter than a spontaneous cooling time (~ 100 ps) below threshold, but becomes much longer than a stimulated cooling time ($\lesssim$ 10 ps) above threshold. Under linear polarization pumping, the linear polarization is maintained in the emission above threshold[7] but the polarization direction is rotated by $\sim 90°$.[11] This unexpected result suggests a striking quantum interference effect, that is, the two scattering amplitudes between iso-spins due to their repulsive interaction and between hetero-spins due to their attractive interaction constructively interfere by rotating the polarization direction by 90°.

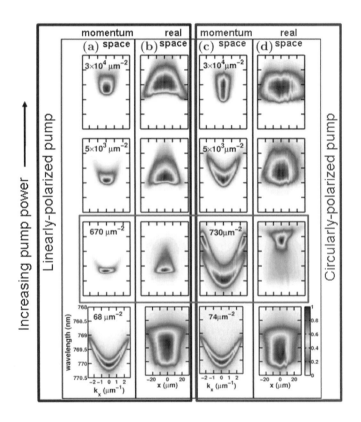

Fig. 3.   Energy-momentum and energy-position dispersion relations for linear and circular polarization pumping.[10]

4. Quantum Degeneracy at Thermal Equilibrium Condition

The first experimental result of satisfying the quantum degeneracy condition ($n_{LP} \gg 1$ or $\varepsilon_0 - \mu \ll k_B T$) and the thermal equilibrium condition

$(T_{LP} \sim T_{\text{phonon}})$ is summarized in Fig. 4.[10] A linearly polarized pump laser injects two spin components and takes advantage of a cooperative cooling mechanism. Fig. 4(a) shows a representative instantaneous polariton population vs. excitation energy relation, for which a Bose-Einstein distribution is fitted with a temperature $T_{LP} \sim 4.4$ K and a quantum degeneracy parameter $(\varepsilon_0 - \mu)/k_B T \sim 0.1$. As shown in Fig. 4(b) and (c), T_{LP} reaches a phonon reservoir temperature $T_{\text{phonon}} \sim 4$ K at 30 psec after the injection of hot LPs, while the quantum degeneracy condition is sustained up until 50 psec after the initial excitation.

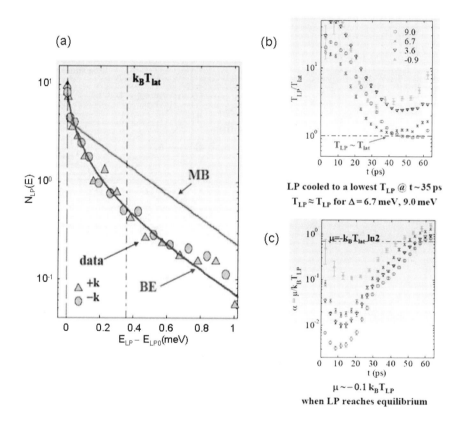

Fig. 4. Instantaneous polariton population vs. excitation energy (a), time-dependent polariton temperature (b) and time-dependent quantum degeneracy parameter (c).

5. Bogoliubov Excitation Spectrum

Observation of a Bogoliubov excitation spectrum, which is a unique signature of BEC with weakly interacting particles, is summarized in Fig. 5.[12] In the Bogoliubov spectrum, an exciton energy normalized by a mean-field energy E/U is a universal function of a momentum normalized by a healing length $k\xi$. This is experimentally confirmed by four traps with varying detuning parameters Δ as shown in Fig. 5. From the slope of linear dispersion at low momentum regime, the effective sound velocity is measured to be $C \sim 10^8$ cm/s which is eight orders of magnitude larger than that of atomic BEC.

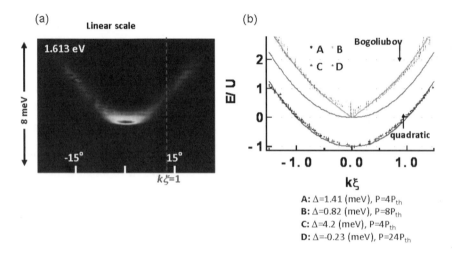

Fig. 5.   Excitation energy E/U vs. momentum $k\xi$ dispersion below and above condensation threshold.

6. First and Second Order Coherence

A first-order spatial coherence function $g^{(1)}(r)$ can be measured with a Young's double slit interferometer. The result is shown in Fig. 6(a), in which the first-order coherence abruptly builds up at BEC threshold and monotonically decreases with slit separation.[9] The solid line is the theoretical prediction based on the Fourier transform of experimental momentum distribution and explains well the experimental result.

A second-order temporal coherence function $g^{(2)}(0)$ can be measured with a Hanbury-Brown and Twiss interferometer. The result is shown in Fig. 6(b), in which the photon bunching effect $\left(g^{(2)}(0) > 1\right)$ due to bosonic final state stimulation is observed above threshold. An excess intensity noise, manifested by $g^{(2)}(0) > 1$ well above threshold, is explained by co-existance of thermal and quantum depletion.[13]

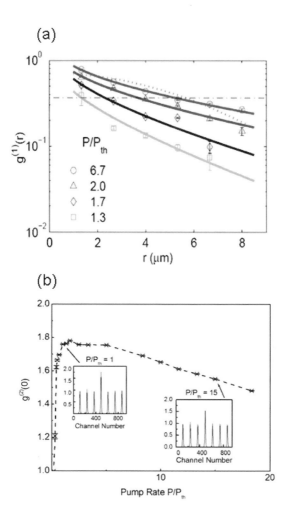

Fig. 6. (a) A first-order coherence function $g^{(1)}(r)$ for varying pump levels. (b) A second-order coherence function $g^{(2)}(0)$ vs. pump level P/P_{th}.

7. Dynamical Condensation at Excited Bloch Bands in One-Dimensional Lattice

When many exciton-polariton condensates couple with each other in a one-dimensional lattice structure, Bloch bands are formed due to periodic structure as shown in Fig. 7(a). In such a system, exciton-polariton condensation is observed at an excited Bloch band consisting of anti-phased p-waves, as well as at a ground state band consisting of in-phase coupled s-waves, as shown in Fig. 7(b).[14] A similar phenomenon was also observed in atomic BEC systems.[15]

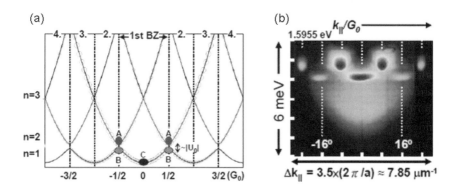

Fig. 7.   (a) A Bloch band structure of exciton-polaritons in a one -dimensional lattice. (b) Exciton-polariton condensation at ground (s-wave) and excited (p-wave) Bloch band states.

Acknowledgments

This work was supported by JST/SORST and Special Coordination Funds for Promoting Science and Technology.

References

1. Y. Yamamoto, S. Machida, K. Igeta and Y. Horikoshi, *Coherence and Quantum Optics* (Plenum Press, New York, 1989).
2. C. Weisbuch, M. Nishioka, A. Ishikawa and Y. Arakawa, *Phys. Rev. Lett.* **69**, 3314 (1992).
3. S. Jiang, S. Machida, Y. Takiguchi, Y. Yamamoto and H. Cao, *Appl. Phys. Lett.* **73**, 3031 (1998).
4. A. Imamoglu, R. J. Ram, S. Pau and Y. Yamamoto *Phys. Rev. A* **53**, 4250 (1996).

5. R. Huang, F. Tassone and Y. Yamamoto, *Phys. Rev. B* **61**, R7854 (2000).
6. R. Huang, Y. Yamamoto, R. André, J. Bleuse, M. Muller and H. Ulmer-Tuffigo, *Phys. Rev. B* **65**, 165314 (2002).
7. H. Deng, G. Weihs, D. Snoke, J. Bloch and Y. Yamamoto, *Proc. Nat. Acad. Sci.* **100**, 15318 (2003).
8. W. Ketterle and N. J. van Druten, *Phys. Rev. A* **54**, 656 (1996).
9. H. Deng, G. S. Solomon, R. Hey, K. H. Ploog and Y. Yamamoto, *Phys. Rev. Lett.* **99**, 126403 (2007).
10. H. Deng, D. Press, S. Götzinger, G. S. Solomon, R. Hey, K. H. Ploog and Y. Yamamoto, *Phys. Rev. Lett.* **97**, 146402 (2006).
11. G. Roumpos, C. W. Lai, T. C. H. Liew, Y. G. Rubo, A. V. Kavokin and Y. Yamamoto, submitted for publication.
12. S. Utsunomiya, L. Tian, G. Roumpos, C. W. Lai, N. Kumada, T. Fujisawa, M. Kuwata-Gonokami, A. Loefler, S. Hoeling, A. Forchel and Y. Yamamoto, *Nat. Phys.*, published online, doi:10.1038/nphys1034 (2008).
13. P. Schwendimann and A. Quattropani, *Phys. Rev. B*, **77**, 085317 (2008).
14. C. W. Lai, N. Y. Kim, S. Utsunomiya, G. Roumpos, H. Deng, M. D. Fraser, T. Byrnes, P. Recher, N. Kumada, T. Fujisawa and Y. Yamamoto, *Nature*, **450**, 529 (2007).
15. T. Müller, S. Fölling, A. Widera and I. Bloch, *Phys. Rev. Lett.* **99**, 200405 (2007).

ANDERSON LOCALIZATION OF MATTER WAVES

P. BOUYER, J. BILLY, V. JOSSE, Z. ZUO, A. BERNARD, B. HAMBRECHT,

P. LUGAN, D. CLÉMENT, L. SANCHEZ-PALENCIA, and A. ASPECT

Laboratoire Charles Fabry de l'Institut d'Optique,
Campus Polytechnique, rd 128, 91127 PALAISEAU CEDEX, France
E-mail: philippe.bouyer@institutoptique.fr; URL: www.atomoptic.fr

In 1958, P.W. Anderson predicted the exponential localization[1] of electronic wave functions in disordered crystals and the resulting absence of diffusion. It was realized later that Anderson localization (AL) is ubiquitous in wave physics[2] as it originates from the interference between multiple scattering paths, and this has prompted an intense activity. Experimentally, localization has been reported in light waves[3] microwaves,[4] sound waves,[5] and electron gases[6] but to our knowledge there was no direct observation of exponential spatial localization of matter-waves (electrons or others). We present in this proceeding the experiment that lead to the observation of Anderson Localization (AL)[7] of a Bose-Einstein Condensate (BEC) released into a one-dimensional waveguide in the presence of a controlled disorder created by laser speckle. Direct imaging allows for unambiguous observation of an exponential decay of the wavefunction when the conditions for AL are fulfilled. The disorder is created with a one-dimensional speckle potential whose noise spectrum has a high spatial frequency cut-off, hindering the observation of exponential localization if the expanding BEC contains atomic de Broglie wavelengths that are smaller than an effective mobility edge corresponding to that cut-off. In this case, we observe the density profiles that decay algebraically.[9]

Keywords: Bose-Einstein condensate; Anderson localization.

1. Introduction

Ultracold atomic systems are now widely considered to revisit standard problems of condensed matter physics (CM) with unique control possibilities. Dilute atomic Bose-Einstein condensates (BEC)[10] and degenerate Fermi gases (DFG)[11] are currently produced, taking advantage of the recent progress in cooling and trapping of neutral atoms.[12] For example, periodic potentials (optical lattices) with no defects can be designed in a wide variety of geometries.[13] In these lattices, transport has been investigated at length, showing lattice-induced reduction of mobility[14] and interaction-induced self-trapping.[15]

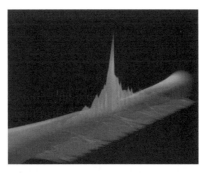

Fig. 1.   Observation of Anderson localization in 1D with an expanding Bose-Einstein condensate in the presence of a 1D speckle disorder.

Introducing disorder in a quantum system can dramatically change its properties and result in a variety of non-intuitive phenomena, many of which are not yet fully understood. Quantum systems in disorder can show intriguing phenomena such as Anderson localization,[1,16] percolation,[17] disorder-driven quantum phase transitions and the corresponding Bose-glass[18] or spin-glass[20] phases. The transport of quantum particles in non ideal material media (eg the conduction of electrons in an imperfect crystal) is strongly affected by scattering from impurities of the medium. Even for weak disorder, semi-classical theories, such as those based on the Boltzmann equation for matter-waves scattering from the impurities, often fail to describe transport properties, and fully quantum approaches are necessary. The basic idea is that contrary to Bloch's theory which predicts a frictionless transport of non-interacting particles[21] or to the Drude theory of transport in the presence of impurities which predicts ohmic conduction, localization effects in disordered potentials may result in a strong suppression of the electronic transport in solids in the presence of defects or impurities.[1]

In the case of degenerate atomic quantum gases, disordered potentials can be produced optically as demonstrated in several recent experiments.[22,23] In addition to the possibility to design perfectly controlled and phonon-free disordered potentials, these systems offer the possibility to implement systems in any dimensions, to control the inter-atomic interactions, either by density control or by Feshbach resonances and to measure *in situ* atomic density profiles via direct imaging.[7,8]

Our experiment allows us to study the 1D Anderson Localization of an expanding BEC in the presence of a weak disorder potential created by laser speckle (Fig. 1). A ^{87}Rb BEC is created in a hybrid optomagnetic trap[24] where the transverse confinement ($\omega_\perp/2\pi = 70$ Hz) is given by an optical wave guide (Nd:YAG laser at 1064 nm) whereas a weak inhomogeneous magnetic field ensures the longitudinal trapping ($\omega_z/2\pi = 5.4$ Hz) as shown in Fig. 2. Typically we produced small BECs with a few 10^4 atoms, corresponding to transverse and longitudinal radii around 3 μm and 35 μm. When the magnetic field is switch off, the BEC starts to expands along the optical guide under the effect of the initial interaction energy, which is rapidly converted into kinetic energy (the initial chemical potential μ_{in} is around a few 100 Hz, corresponding to expansion velocities below 2 mm/s). In the early time of the expansion, the interactions decrease rapidly, and thus the residual interactions play no role during the subsequent expansion at longer time. They are furthermore completely negligible in the wings of the atomic wavefunction. To observe AL, we use a weak optical disorder, i.e. much weaker that the typical kinetic energy of the expanding atoms. This avoids classical reflections from large peaks, so that localization results from the destructive interference between multiple quantum reflections of small amplitude.[1,2,9]

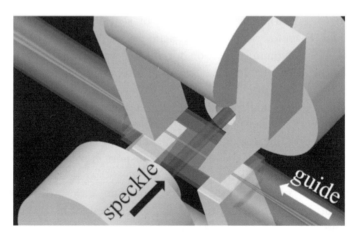

Fig. 2.   Schematic representation of the experimental set-up. The BEC is made in a hybrid magneto-optical trap. The optical waveguide (transverse confinement) is made by a far off resonance red-detuned Nd:YAG. The magnetic field (created by a ferromagnet) is used for the longitudinal confinement. The speckle is shone perpendicularly to the propagation axis.

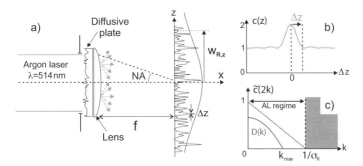

Fig. 3. (a) Experimental realization of laser speckle by shining an argon laser ($\lambda =$ 514 nm) on a diffusive plate. The speckle is focused on the BEC propagation axis with a lens of focal length $f = 14$ mm. The numerical aperture corresponds to $N.A. = 0.30\pm0.03$ (b) Autocorrelation function $c(\delta z) = \langle V(z)V(z+\delta z)\rangle$ of the disorder potential in case of a uniform illumination. The speckle grain size, given by the width of the autocorrelation function, is $\Delta z = \pi\sigma_R = \lambda/2N.A. = 0.82 \mu$m. (c) Normalized spatial frequency distribution $\tilde{c}(2k)$ (Fourier transform of $c(\Delta z)$) of the speckle potential with the high frequency cut off at $1/\sigma_R = 3.85 \mu$m^{-1}. The k-momentum distribution $\mathcal{D}(k)$ of the expanding BEC is also plotted (in arbitrary units) to illustrate the condition for exponential localization $k_{\max} < 1/\sigma_R$ ($k_{\max}\sigma_R < 1$).

2. Random potential created by laser speckle

The random potential is created by focusing an argon laser (wavelength $\lambda = 514$ nm) passing through a diffusing plate[25] onto the atoms, so that it results in an optical speckle pattern (see Fig. 3(a)). Since the speckle grain size (defined by the radius of the autocorrelation function $c(\delta z) = \langle V(z)V(z+\delta z)\rangle$) is directly related to the numerical aperture NA, we create an anisotropic speckle pattern by anisotropically illuminating the diffusive plate. With our numerical aperture (NA= 0.3), the speckle pattern has a very thin grain size of $\Delta z = \lambda/(2\text{ NA}) = 0.82 \mu$m along the BEC propagation direction (see Fig. 3b). It is convenient to introduce the correlation length $\sigma_R = \Delta z/\pi = 0.26 \pm 0.03 \mu$m, such that the spatial frequency spectrum $\tilde{c}(2k)$ of $V(z)$ (i.e. the rescaled Fourier transform of $c(\delta z)$) has a cut-off at $1/\sigma_R$. In the transverse directions, the typical speckle grain sizes (97 and 10 μm) are larger than the BEC dimension and the atoms feel a homogeneous potential. The disorder potential can then be considered as 1D.

The disorder amplitude, referred to as V_R in the following, is characterized by the standard deviation of the speckle potential σ_V. Since the random intensity is exponentially distributed, it is simply given by the mean value $V_R = \sigma_V = \langle V\rangle$, which is the quantity measured experimentally.

Finally, the spatial extension of the speckle field is independently controlled by the diffraction angles of the diffusers. To match the BEC expansion, we choose anisotropic diffusing angles so that the illuminating area has an elongated gaussian shape characterized by the longitudinal and transverse radii $w_{R,z}$ and $w_{R,y}$ with $w_{R,z} \gg w_{R,y}$. We use two diffusing plates with different diffusing angles along the z axis, corresponding respectively to $w_{R,z} = 1.8$ and $w_{R,z} = 5.3$ mm (with $w_{R,y} = 0.6$ mm for both realizations). The first one, which allows for higher disorder amplitude ($V_R|_{max} \simeq 100$ Hz) is the one used for the results presented in Ref. 7.

3. How to observe Anderson Localization with an expanding BEC

Our experiment starts with a small BEC released from a loose hybrid opto-magnetic trap ($\omega_\perp/2\pi = 70$ Hz and $\omega_z/2\pi = 5.4$ Hz) and expanding in the wave guide (see Fig. 4). Once interactions are negligible, the BEC is well described by the sum of non-interacting k-momentum waves, for which the momentum distribution ($\mathcal{D}(k)$) has a maximum momentum k_{max} (directly related to μ_{in}) that we measure directly (see Fig. 4(c)) by monitoring the evolution of the BEC in a flat 1D-potential [a].

Naively, k_{max} must be seen as one of the key parameters that will set the localisation length. A low k_{max} will allow us to observe localized profiles in our experimental field of observation and a large k_{max} will set a localisation length so large that the profiles are too broad to be observed. Experimentally, we controlled k_{max} by varying the number of atoms and we achieved its smallest value by decreasing this number down to $N = 1.7 \times 10^4$, which corresponds to a chemical potential $\mu_{in}/h = 219$ Hz. There the BEC expands at a velocity $v_{max} = \dot{R}_z(t) = 1.7$ mm/s that gives $k_{max} = m_{Rb} v_{max}/\hbar = 2.47 \pm .25 \ \mu m^{-1}$.

We have theoretically investigated the scenario of the experiment in presence of weak disorder ($V_R \ll \mu_{in}$), in Ref. 9. At short times, the atom-atom interactions drive the expansion of the BEC and the disorder does not play a significant role. Then, when the density has significantly decreased, the expansion is governed by the scattering of an almost non-interacting cloud and can be described by the same momentum distribution $\mathcal{D}(k)$ as

---

[a]The optical trap induces a loose longitudinal trapping frequency ($\omega_z/2\pi \simeq 0.5$ Hz), that we cancel by adding a weakly anti-trapping magnetic potential, created by a pair of coils along one transverse direction. We choose the current carried by the two coils in order to compensate carefully the residual trapping frequency ($\omega_z/2\pi < 0.05$ Hz). We thus achieve a quasi free ballistic expansion over a few millimeters.

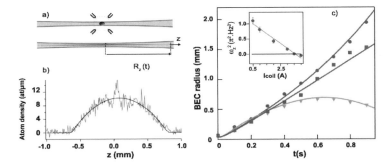

Fig. 4. Expansion of the BEC in the absence of a disorder potential. (a) Experimental scheme : the BEC is formed in a hybrid trap that is the combination of an horizontal waveguide ensuring a strong transverse confinement and a loose magnetic longitudinal trap. The expansion is set along the optical waveguide by switching off the magnetic field. (b) Determination of the expanding BEC radius by fitting the profile with an inverted parabola $(1 - z^2/R_z^2)$ (straight line). The profile shown corresponds to an expansion time of 400 ms for a BEC of N=1.7 10^4 atoms. (c) BEC radius versus time for different residual potentials controlled by an external magnetic field. The trapping frequencies $\omega_z/2\pi$, obtained by comparing the radius evolution $R_z(t)$ with numerical simulations (straight lines), correspond respectively to 0.40±0.05 Hz (green triangles), 0.0±0.05 Hz (red squares) and -0.23±0.05 Hz (blue dots) for the anti-trapping case. The dependence of the residual trapping frequency on the current in the coils creating the expelling magnetic field is shown in the inset. In the free ballistic case (red squares) where the residual trapping potential has been suppressed, the radius evolves with a constant slope from which we determine the maximum k-momentum $k_{\max}$.

in the absence of disorder. For a weak disorder potential, the Born approximation holds and implies that each k-momentum wave will be scattered only if there is a corresponding momentum (the Bragg condition) in the diffuser spatial spectrum $\tilde{c}(2k)$, i.e. the Fourier transform of the disorder correlation function $c(\delta z)$. For a speckle potential, the diffraction imposes a high frequency cut-off on the spatial spectrum: it vanishes for $k > 1/\sigma_R$. This imposes an *effective mobility edge* to the observation of an exponential profile: when $k_{\max}\sigma_R < 1$, each k-wave of the expanding BEC is scattered in first order by the disorder potential and localizes exponentially with a k-dependent localization length $L(k)$. In the stationary regime, the BEC localizes exponentially, with a localization length given by $L(k_{\max})$. On the contrary, when $k_{\max}\sigma_R > 1$, the k-waves with $1/\sigma_R < k < k_{\max}$ are not scattered in first order. The *localization length* is significantly increased and localization is observed through an algebraic profile, with a power law decay $n_{1D} \propto 1/|z|^\beta$ (with $\beta = 2$).

4. Direct observation of localized atomic profiles

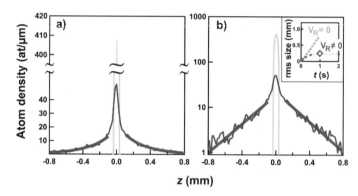

Fig. 5. Stationary profile in (a) linear and (b) semi-logarithmic scale of the BEC one second after release in disorder potential. The initial BEC is made of $1.7 \; 10^4$ atoms ($\mu_{\mathrm{in}} = 219$ Hz) which corresponds to a measured value of $k_{\max} = 2.47 \pm 0.25 \; \mu\mathrm{m}^{-1}$ ($k_{\max}\sigma_{\mathrm{R}} = 0.65 \pm 0.09$). The disorder potential amplitude is weak compared to the typical kinetic energy of the expanding BEC ($V_{\mathrm{R}}/\mu_{\mathrm{in}} = 0.12$). In the inset of (b), we display the rms width of the profiles versus time in the presence or absence of the disorder. This shows that a stationary regime is reached after 0.5 s. Straight lines in (a) are exponential fits ($\exp(-2|z|/L_{\mathrm{loc}})$) to the wings and correspond to the straight lines in (b). The narrow central peak (pink) represents the trapped condensate before release ($t = 0$ s). Note that the profiles are obtained by averaging over five runs of the experiment with the same disorder realization.

With $k_{\max} = 2.47 \; \mu\mathrm{m}^{-1}$ and $\sigma_{\mathrm{R}} = 0.26 \; \mu\mathrm{m}$ as experimental conditions, we observe the localization in the *exponential regime* $k_{\max}\sigma_{\mathrm{R}} = 0.65 \pm 0.09 < 1$. As soon as we switch off the longitudinal trapping, in the presence of weak disorder, the BEC starts expanding, but the expansion rapidly stops, in stark contrast with the free expansion case (see inset of Fig. 5(b) showing the evolution of the rms width of the observed profiles). A plot of the density profile, in linear and semi logarithmic coordinates (Fig. 5a,b), then shows clear exponential wings, a signature of Anderson Localization. In addition, we verified that we rapidly reach (in less than one second) a stationary situation when the exponential profile no longer evolves, as seen on Fig. 6(a). An exponential fit to the wings of the density profiles yields the localization length L_{loc}, which also no longer evolves when the stationary situation is reached. We can then compare the measured localization length with the theoretical value.[9]

Figure 7 shows the variation of L_{loc} with the amplitude of the disorder, V_{R}, for the same number of atoms, i.e. the same $k_{\max}$. In addition

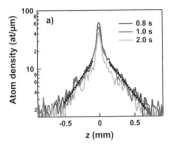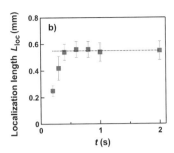

Fig. 6. (a) Three successive density profiles, from which the localization length is extracted by fitting an exponential to the atomic density in the wings. (b) Localization length L_{loc}, vs expansion time t. The error bars indicate 95% confidence intervals on the fitted values

to the results published in,[7] we present here the measurement obtained with a larger speckle field extension ($w_{R,z} = 5.3$ mm compared to $w_{R,z} = 1.8$ mm). The dash-dotted line is a plot of the localization length calculated for the values of k_{max} and σ_R determined experimentally. The shaded area reflects the variations of the dash-dotted line when one takes into account the uncertainties in k_{max} and σ_R. The uncertainty in the calibration of V_R does not appear in Fig. 7. We estimate this to be not larger than 30 %, which does not affect the agreement between theory and experiment. At low disorder amplitude, the profile extension can become comparable to the size of the speckle with small extension ($w_{R,z} = 1.8$ mm) and the measurement can be affected. The larger speckle field, which is much larger than the atomic wavefunction, indeed shows a much better agreement than the one presented in Ref. 7.

We have investigated the regime where the initial interaction energy is large enough that a fraction of the atoms have a k-vector larger than $1/\sigma_R$ by repeating the experiment with a BEC containing a larger number of atoms. In this regime, the localization of the BEC becomes *algebraic* for the scales accessible experimentally. Indeed, the part of the BEC wavefunction corresponding to the waves with momenta in the range $1/\sigma_R < k$ will expand further before it eventually localizes at much longer scales : this plays the role of an *effective mobility edge* where a significant change in the wavefunction behavior is expected. Figure 8a shows the observed density profile in such a situation ($k_{max}\sigma_R = 1.16 \pm 0.14$). The log-log plot suggests a power law decrease in the wings, with an exponent of 1.97 ± 0.05, in agreement with the theoretical prediction of wings decreasing as $1/|z|^2$. In this regime, where no localization length can be extracted, we verified that the *algebraic decay* does not depend on the amplitude of the disorder (Fig. 8(b)).

5. Conclusion

Coherent transport of waves within a large variety of media has been attracting a considerable amount of interest. The subject is of primary importance within the context of condensed matter physics, for example to understand normal metallic conduction, superconductivity, superfluid flows in low temperature quantum liquids, but also in optics, acoustics and atomic physics with special interest in coherent diffusion in inhomogeneous systems. The main difficulty in understanding quantum transport results from the subtle interplay of interferences, scattering onto the potential landscape, and possibly interatomic interactions. Direct imaging of atomic quantum gases in controlled optical disordered potentials reveals here again that it is a promising technique to investigate this variety of open questions. Firstly, as in other problems of condensed matter simulated with ultra-cold atoms,

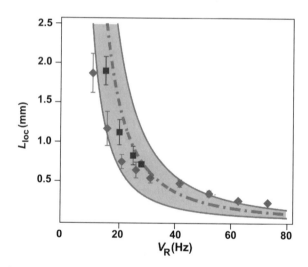

Fig. 7.   Localization length L_{loc} versus the disorder amplitude V_R for $k_{max}\sigma_R = 0.65 \pm 0.09$. The data are obtained with two different diffusing plates, inducing two different extensions for the disordered potential: $w_{R,z} = 1.8$ mm (diamond light blue) and $w_{R,z} = 5.3$ mm (square dark blue). The dash-dot red curve shows the theoretical predictions for L_{loc} and the two straight lines represent the uncertainty associated with the evaluations of k_{max} and σ_R. For low amplitude values of V_R (typically below 25 Hz), the smaller speckle realization gives much more reliable values as its extension remains much larger than the measured localization lengths. For intensity reasons, larger disorder amplitude were not accessible in the experiment.

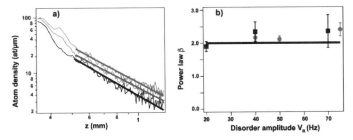

Fig. 8. (a) Stationary profiles in the algebraic localization regime for different disorder amplitudes V_R. The initial BEC is made of 1.7×10^5 atoms ($\mu_{in} = 520$ Hz) which corresponds to a measured value of $k_{max} = 4.47 \pm 0.30 \ \mu m^{-1}$ ($k_{max}\sigma_R = 1.16 \pm 0.14$). The straight lines are the fits of the wings to a power law decay ($n_{1D} \propto 1/|z|^\beta$). The different values obtained for β are shown in (b) (blue squares). In addition the red circles correspond to values found with higher number of atoms (2.8×10^5) with $k_{max}\sigma_R = 1.30 \pm 0.2$.

direct imaging of atomic matter-waves offers unprecedented possibilities to measure important properties, such as the localization length in this problem. Secondly, our experiment can be extended to quantum gases with controlled interactions where localization of quasi-particles, Bose glass or Lifshits glass are expected, as well as to Fermi gases and to Bose-Fermi mixtures where rich phase diagrams have been predicted. The reasonable quantitative agreement between our measurements and the theory of 1D Anderson localization in a speckle potential demonstrates the high degree of control in our set-up. It opens the path to the realization of "real" quantum simulators for investigating Anderson localization in a wider variety of models. Extending the technique to two and three dimensions, and better controlling interactions, it might be possible to better understand the behavior of real materials. We could experience situations where current theory can not provide precise predictions. In the long run, a better understanding of these phenomena will allow the improvement of semi-conductor devices, such as amorphous silicon-based electronic devices, for example.

References

1. P. W. Anderson, *Phys. Rev.* **109**, 1492 (1958).
2. B. van Tiggelen, in *Wave Diffusion in Complex Media*, Lectures Notes at Les Houches 1998, ed. J. P. Fouque, NATO Science (Kluwer, Dordrecht, 1999).
3. D. S. Wiersma, *et al.*, *Nature* **390**, 671 (1997); F. Scheffold *et al.*, *Nature* **398**, 206 (1999); M. Störzer, *et al.*, *Phys. Rev. Lett.* **96**, 063904 (2006); T. Schwartz, *et al.*, *Nature* **446**, 52 (2007); Y. Lahini, *et al.*, *Phys. Rev. Lett.* **100**, 013906 (2008).

4. R. Dalichaouch, *et al.*, *Nature* **354**, 53 (1991); A. A. Chabanov, M. Stoytchev and A. Z. Genack, *Nature* **404**, 850 (2000).

5. R. L. Weaver, *Wave Motion* **12**, 129 (1990).

6. E. Akkermans and G. Montambaux, *Mesoscopic Physics of Electrons and Photons* (Cambridge U. Press, 2006).

7. J. Billy, *et al.*, *Nature* **453**, 891 (2008).

8. G. Roati, *et al.*, *Nature*, **453**, 895 (2008)

9. L. Sanchez-Palencia, *et al.*, *Phys. Rev. Lett.* **98**, 210401 (2007).

10. E. A. Cornell and C. E. Wieman, *Rev. Mod. Phys.* **74**, 875 (2002); W. Ketterle, *ibid.* **74**, 1131 (2002).

11. A. G. Truscott, K. E. Strecker, W. I. McAlexander, G. B. Partridge and R. G. Hulet, *Science* **291**, 2570 (2001); F. Schreck, L. Khaykovich, K. L. Corwin, G. Ferrari, T. Bourdel, J. Cubizolles and C. Salomon, *Phys. Rev. Lett.* **87**, 080403 (2001); Z. Hadzibabic, C. A. Stan, K. Dieckmann, S. Gupta, M. W. Zwierlein, A. Görlitz and W. Ketterle, *Phys. Rev. Lett.* **88**, 160401 (2002).

12. S. Chu, *Rev. Mod. Phys.* **70**, 685 (1998); C. Cohen-Tannoudji, *ibid.* **70**, 707 (1998); W. D. Phillips, *ibid.* **70**, 721 (1998).

13. G. Grynberg and C. Robilliard, *Phys. Rep.* **355**, 335 (2000).

14. S. Burger, F. S. Cataliotti, C. Fort, F. Minardi, M. Inguscio, M. L. Chiofalo and M. P. Tosi, *Phys. Rev. Lett.* **86**, 4447 (2001); M. Krämer, L. P. Pitaevskii and S. Stringari, *Phys. Rev. Lett.* **88**, 180404 (2002); C. D. Fertig, K. M. O'Hara, J. H. Huckans, S. L. Rolston, W. D. Phillips and J. V. Porto, *Phys. Rev. Lett.* **94**, 120403 (2005).

15. A. Trombettoni and A. Smerzi, *Phys. Rev. Lett.* **86**, 2353 (2001); Th. Anker, M. Albiez, R. Gati, S. Hunsmann, B. Eiermann, A. Trombettoni and M. K. Oberthaler, *Phys. Rev. Lett.* **94**, 020403 (2005).

16. Y. Nagaoka and H. Fukuyama (eds.), *Anderson Localization*, Springer Series in Solid State Sciences, Vol. 39 (Springer, Berlin, 1982); T. Ando and H. Fukuyama (eds.), *Anderson Localization*, Springer Proceedings in Physics, Vol. 28 (Springer, Berlin, 1988).

17. A. Aharony and D. Stauffer, *Introduction to Percolation Theory* (Taylor & Francis, London, 1994).

18. T. Giamarchi and H. J. Schulz, *Phys. Rev. B* **37**, 325 (1988).

19. M. P. A. Fisher, P. B. Weichman, G. Grinstein and D. S. Fisher, *Phys. Rev. B* **40**, 546 (1989).

20. M. Mézard, G. Parisi and M. A. Virasoro, *Spin Glass and Beyond* (World Scientific, Singapore, 1987).

21. N. W. Ashcroft and N. D. Mermin, *Solid State Physics* (Saunders College Publishing, New York, 1976).

22. J. E. Lye, L. Fallani, M. Modugno, D. Wiersma, C. Fort and M. Inguscio, *Phys. Rev. Lett.* **95**, 070401 (2005); D. Clément, A. F. Varón, M. Hugbart, J. A. Retter, P. Bouyer, L. Sanchez-Palencia, D. Gangardt, G. V. Shlyapnikov and A. Aspect, *Phys. Rev. Lett.* **95**, 170409 (2005); C. Fort, L. Fallani, V. Guarrera, J. Lye, M. Modugno, D. S. Wiersma and M. Inguscio, *Phys.*

Rev. Lett. **95**, 170410 (2005); T. Schulte, S. Drenkelforth, J. Kruse, W. Ertmer, J. Arlt, K. Sacha, J. Zakrzewski and M. Lewenstein, *Phys. Rev. Lett.* **95**, 170411 (2005).

23. L. Fallani, J. Lye, V. Guarrera, C. Fort and M. Inguscio, *Phys. Rev. Lett.* **98**, 130404 (2007).

24. W. Guerin, J.-F. Riou, J. Gaebler, V. Josse, P. Bouyer and A. Aspect, *Phys. Rev. Lett.* **97**, 200402 (2006).

25. D. Clément, A. F. Varon, J. Retter, L. Sanchez-Palencia, A. Aspect and P. Bouyer, *New J. Phys.* **8**, 165 (2006).

ANDERSON LOCALIZATION OF A NON-INTERACTING BOSE-EINSTEIN CONDENSATE

G. ROATI[1], C. D'ERRICO[1], L. FALLANI[1], M. FATTORI[1,2], C. FORT[1],

M. ZACCANTI[1], G. MODUGNO[1], M. MODUGNO[1,3], and M. INGUSCIO[1]

[1] *LENS and Dipartimento di Fisica, Università di Firenze, and INFM-CNR, Sesto Fiorentino, 50019, Italy*
[2] *Museo Storico della Fisica e Centro Studi e Ricerche E. Fermi, 00184 Roma, Italy*
[3] *BEC-INFM Center, Università di Trento, 38050 Povo, Italy*

One of the most intriguing phenomena in physics is the localization of waves in disordered media. This phenomenon was originally predicted by P. W. Anderson, fifty years ago, in the context of transport of electrons in crystals, but it has never been directly observed for matter waves. Ultracold atoms open a new scenario for the study of disorder-induced localization, due to the high degree of control of most of the system parameters, including interactions. For the first time we have employed a noninteracting Bose-Einstein condensate to study Anderson localization. The experiment is performed in a 1D lattice with quasi-periodic disorder, a system which features a crossover between extended and exponentially localized states as in the case of purely random disorder in higher dimensions. We clearly demonstrate localization by investigating transport properties, spatial and momentum distributions. Since the interaction in the condensate can be controlled, this system represents a novel tool to solve fundamental questions on the interplay of disorder and interactions and to explore exotic quantum phases.

Keywords: Anderson localization; Bose-Einstein condensate.

1. Introduction

Localization of particles and waves in disordered media is a ubiquitous problem in modern physics, originally studied by P. W. Anderson in his famous paper "Absence of diffusion in certain random lattices",[1] that appeared exactly fifty years ago. In that paper Anderson considered the case of electrons in a crystal lattice, described by a single particle tight binding model with random on-site energies, showing that the transport (diffusion of an initially localized wavepacket) is suppressed when the amplitude Δ of disorder exceeds a critical value of the order of the tunneling amplitude J

between adjacent sites. For this discovery, in 1977 Anderson was awarded the Nobel Prize in Physics.

Since then, the issue of localization in the presence of disorder has been investigated in many other systems, and nowadays the term Anderson localization, also known as strong localization, refers to the general phenomenon of the localization of waves in a random medium, and applies e.g. to the transport of electromagnetic waves, acoustic waves, quantum waves, spin waves, etc. This phenomenon has been experimentally observed in a variety of systems.[2] For example, evidence of the Anderson localization for light waves in disordered media has been provided by an observed modification of the classical diffusive regime, featuring a conductor-insulator transition.[3,4] Recently, the first effects of weak nonlinearities have also been shown in experiments with light waves in photonic lattices.[5,6] Despite this, the transition between extended and localized states originally studied by Anderson for non-interacting electrons has not been directly observed in crystals, owing to the high electron-electron and electron-phonon interactions.[7] Indeed, a clear understanding of the interplay between disorder and nonlinearity is still lacking and is considered a crucial issue to be addressed in contemporary condensed matter physics.[8]

The combination of ultracold atoms and optical potentials offers a novel platform for the study of disorder-related phenomena where most of the relevant physical parameters, including those governing interactions, can be controlled.[9,10] The introduction of laser speckles[11] and quasi-periodic optical lattices[12] has made possible the investigation of the physics of disorder. After preliminary investigations in regimes where localization was precluded either by the size of the disorder or by delocalizing effects of nonlinearity,[11,13–16] the first observation of Anderson localization of a matter wave has been recently reported.[17,18]

In Ref. 17 we have studied the disorder induced localization of a quantum wavefunction in a lattice system, following the original idea of Anderson,[1] by using a Bose-Einstein condensate in which the atom-atom interactions can be tuned independently of the other parameters.[19] The lattice is realized by means of a one-dimensional periodic optical potential, and disorder is introduced by using a weaker incommensurate secondary lattice. The resulting quasi-periodic lattice constitutes an experimental realization of the so called Harper[20] or Aubry-André model.[21] This system displays a transition from extended to localized states analogous to the Anderson transition, but already in one dimension,[22,23] whereas in the case of pure random disorder, more than two dimensions would be needed.[24] We have clearly observed this transition by studying transport and both spatial and momentum distributions, also verifying the scaling behaviour of the critical disorder strength.

2. The quasi-periodic lattice

The quasi-periodic bichromatic potential, obtained by superimposing two one-dimensional optical lattices, has the form

$$V_b(x) = s_1 E_{R1} \sin^2(k_1 x) + s_2 E_{R2} \sin^2(k_2 x + \phi) \qquad (1)$$

where $k_i = 2\pi/\lambda_i$ ($i = 1, 2$) are the lattice wavenumbers, s_i are the heights of the two lattices in units of their recoil energies $E_{Ri} = h^2/(2m\lambda_i^2)$, and ϕ is an arbitrary phase. The potential of wavelength λ_1 creates the primary lattice, of period $d = \lambda_1/2$, that is weakly perturbed by the secondary lattice of wavelength λ_2.

In the case of a non-interacting condensate, the axial and transverse degrees of freedom are separable. Along the direction of the bichromatic lattice, the system is described by the single-particle Hamiltonian $H_{1D} = -(\hbar^2/2m)\nabla_x^2 + V_b(x)$. In the tight-binding limit this system can be mapped to the Aubry-André model[21] by expanding the particle wavefunction $\psi(x)$ over a set of maximally localized Wannier states $|w_j\rangle$ at the lattice site j. The resulting Hamiltonian is[23]

$$H = -J \sum_j (|w_j\rangle\langle w_{j+1}| + |w_{j+1}\rangle\langle w_j|) + \Delta \sum_j \cos(2\pi\beta j + \phi)|w_j\rangle\langle w_j| \ (2)$$

where J is the tunneling amplitude between adjacent sites, that depends on the height of the main lattice according to $J \simeq 1.43 s_1^{0.98} \exp(-2.07\sqrt{s_1})$,[25] $\beta = \lambda_1/\lambda_2$ is the ratio of the two lattice wave numbers, and Δ the strength of disorder that can be written as $\Delta \simeq s_2 E_{R2}/(2E_{R1}) = s_2\beta^2/2$.[26] In the experiment, the two relevant energies J and Δ (see Fig. 1a) can be controlled independently by changing the heights of the primary and secondary lattice potentials, respectively.

The bichromatic potential can display features of a perfectly ordered system, when the two wavelengths are commensurate, but also of quasidisorder when β is irrational.[21,22] In the latter case, a common choice in the study of the Aubry-André model is the inverse of the golden ratio, $\beta = (\sqrt{5} - 1)/2$, for which the model displays a "metal-insulator" phase transition from extended to localized states at $\Delta/J = 2$.[23] The localization properties of incommesurate potentials have been extensively investigated in the literature and it was soon recognized that the underlying mechanism

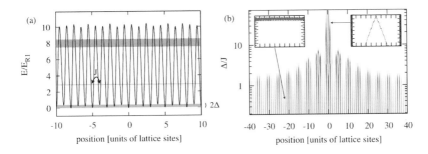

Fig. 1.   (a) Plot of the bichromatic potential for $s_1 = 10$, $\Delta/J = 10$, together with the first and second Bloch bands of the primary lattice (gray stripes around 3 and 8 E_r respectively). The lowest gray stripe represents the amplitude 2Δ of the disorder. (b) Typical density plot of a low-lying eigenstate of the bichromatic potential, as a function of Δ/J (vertical axis), obtained by direct diagonalization of the full hamiltonian H_{1D}. For small values of Δ/J the state is delocalized over many lattice sites (left inset). For $\Delta/J \approx 7$ the state becomes exponentially localized on lengths smaller than the lattice constant (right inset).

is essentially the same as for Anderson localization for pure disordered systems.[22,23,27,28]

In the experiment the value of β has been fixed to $\beta = 1.1972\ldots$ and the transition is broadened and shifted towards larger values of Δ/J (see Fig. 1(b)). Owing to the quasi-periodic nature of the potential, the localized states appear approximately every five sites of the main lattice (2.6 μm). Figure 1(b) represents a typical density plot of a lowest lying eigenstate of the bichromatic potential, obtained by direct diagonalization of the full hamiltonian H_{1D}.

3. The non-interacting condensate

The non-interacting Bose-Einstein condensate is prepared by sympathetically cooling a cloud of interacting ^{39}K atoms in an optical trap, and then tuning the s-wave scattering length almost to zero by means of a Feshbach resonance.[19,29]

The condensate is initially prepared in a homogeneous magnetic field of about 396 G, where a broad Feshbach resonance raises the value of the s-wave scattering length from the background value of $-29a_0$ to about $180a_0$ ($a_0 = 0.529 \times 10^{-10}$ m).[19,30] This allows the efficient formation of a stable condensate. The condensate is trapped in a crossed dipole trap with an average harmonic frequency of 100 Hz, and contains about 10^5 atoms. The scattering length is then reduced by shifting the magnetic field to about 350 G, a the zero-crossing position. This magnetic field is adiabatically changed with a combined linear and exponential ramp lasting 110 ms, to avoid shape excitations of the cloud. We estimate a residual scattering length of the order of $0.1a_0$, limited by magnetic field instability (100 mG) and by the contribution to the scattering of higher-order partial waves, corresponding to an atom-atom interaction energy of $U < 10^{-5}J$.[9]

The spatial size of the condensate can be controlled by changing the harmonic confinement provided by the trap. For most of the measurements the size along the direction of the lattice is $\sigma \approx 5$ mm. The quasi-periodic potential is imposed by using two lasers in a standing-wave configuration.[16] The Gaussian shape of the laser beams forming the primary lattice also provides radial confinement of the condensate in the absence of the harmonic trap.

4. Absence of diffusion

In a first experiment we have investigated transport, by abruptly switching off the main harmonic confinement and letting the atoms expand along the one-dimensional bichromatic lattice. Fig. 2(a) shows the spatial distribution of the atoms at increasing evolution times using detection by absorption imaging. In the regular lattice ($\Delta = 0$) the eigenstates of the potential are extended Bloch states, and the system expands ballistically. For large disorder ($\Delta/J > 7$) we observe no diffusion, because in this regime the condensate can be described as the superposition of several localized eigenstates whose individual extensions are less than the initial size of the condensate. In the crossover between these two regimes we observe a ballistic expansion with reduced speed. This crossover is summarized in Fig. 2(b), which shows the width of the atomic distribution versus the rescaled disorder strength Δ/J for a fixed evolution time of 750 ms, for three different values of J. In all three cases, the system enters the localized regime at the same disorder strength, providing compelling evidence for the scaling behaviour of the model in Eq. (1).

In this regime, the eigenstates of the Hamiltonian in Eq. (1) are exponentially localized, and the tails of diffusing wave packets are expected to behave like stretched exponentials.[31] We have therefore analysed the tails of the spatial distributions with an exponential function of the form $f_\alpha(x) = A\exp(-|(x - x_0)/l|^\alpha)$, the exponent α being a fitting parameter. Two examples of this analysis, one for weak disorder and one for strong disorder, are shown in Figs. 3(a), 3(b). The exponent α exhibits a smooth

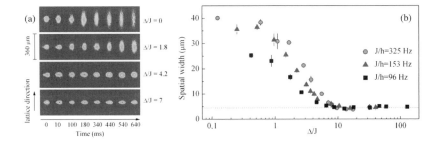

Fig. 2.   (a) In situ absorption images of the Bose-Einstein condensate diffusing along the quasi-periodic lattice for different values of Δ and $J/h = 153$ Hz (h is Planck's constant). For $\Delta/J \geq 7$ the size of the condensate remains at its original value, reflecting the onset of localization. (b) Root-mean-squared size of the condensate for three different values of J, at a fixed evolution time of 750 ms, versus the rescaled disorder strength Δ/J. The dashed line indicates the initial size of the condensate. The onset of localization appears in the same range of values of Δ/J in all three cases.

crossover from a value of two to a value of one as Δ/J increases (Fig. 3(c)), signalling the onset of an exponential localization. The value $\alpha = 2$ that we obtain for small Δ/J corresponds to the expected ballistic evolution of the initial Gaussian momentum distribution of the non-interacting condensate. We note that in the radial direction, where the system is only harmonically trapped, the spatial distribution is always well fitted by a Gaussian function ($\alpha = 2$).

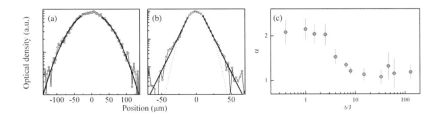

Fig. 3.   (a), (b) Experimental profiles and fitting function $f_\alpha(x)$ (thick line) for $\Delta/J = 1$ (a) and $\Delta/J = 15$ (b) (note the vertical log scale). The dotted line in (b) represents a Gaussian fit, $\alpha = 2$. (c) Dependence of the fitting parameter α on Δ/J, indicating a transition from a Gaussian to an exponential distribution.

5. The momentum distribution

Information on the eigenstates of the system has also been extracted from the analysis of the momentum distribution of the stationary atomic states in the presence of a shallow harmonic confinement. The width of the axial momentum distribution $P(k)$ is inversely proportional to the spatial extent of the condensate in the lattice. We have measured it by releasing the atoms from the lattice and imaging them after a ballistic expansion.

In Fig. 4, we show examples of the experimental momentum distributions that are in agreement with the model predictions for the low-lying eigenstates. Without disorder, we observe the typical grating interference pattern with three peaks at $k = 0, \pm 2k_1$, reflecting the periodicity of the primary lattice. The tiny width of the peak at $k = 0$ indicates that the wavefunction is spread over many lattice sites.[32] For weak disorder, the eigenstates of the Hamiltonian in Eq. (1) are still extended, and additional momentum peaks appear at momentum space distances $\pm 2(k_1 - k_2)$ from the main peaks, corresponding to the beating of the two lattices. As we further increase Δ/J, $P(k)$ broadens and its width eventually becomes comparable with that of the Brillouin zone, k_1, indicating that the extension of the localized states becomes comparable with the lattice spacing. From the theoretical analysis of the Aubry-André model, we have a clear indication that in this regime the eigenstates are exponentially localized on individual lattice sites. We note that the side peaks in the two bottom profiles of Figs. 4(a), 3(b) indicate that the localization is non-trivial, that is, the tails of the eigenstates extend over several lattice sites even for large disorder. The small modulation on top of the profiles is due to the interference between the several localized states over which the condensate is distributed. In Fig. 4(c), we present the root-mean-squared width of the central peak of $P(k)$ as a function of Δ/J, for three different values of J. The three data sets lie on the same line, confirming the scaling behaviour of the system. A visibility of the interference pattern, $V = (P(2k_1) - P(k_1))/(P(2k_1) + P(k_1))$, can be defined to highlight the appearance of a finite population in the momentum states $\pm k_1$ and, therefore, the onset of exponential localization with an extension comparable with the lattice spacing. In Fig. 4(d), we show the visibility extracted from the same data as Fig. 4. Experiment and theory are again in good agreement, and feature a sudden decrease in the visibility for $\Delta/J \approx 6$.

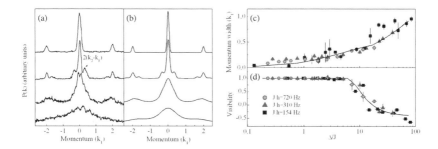

Fig. 4. (a), (b) Experimental and theoretical momentum distributions $P(k)$ for increasing Δ/J (0, 1.1, 7.2 and 25, from top to bottom). The interference pattern of a regular lattice observed at $\Delta = 0$ is at first modified by the appearance of peaks at the beating between the two lattices, and then increasingly broadened. Momentum is measured along the horizontal axes in units of k_1. (c) Root-mean-squared size of the central peak of $P(k)$ versus Δ/J, for three different values of J. The experimental data follow a unique scaling behaviour, as expected from theory (continuous line). The width of the peak is measured in units of k_1. (d) Visibility V of the interference pattern versus Δ/J. In both experiment and theory (continuous line), V decreases abruptly for $\Delta/J \approx 6$, indicating localization on distances comparable to the lattice period.

6. Interference between multiple localized states and effects of weak interactions

Further information on the localized states can be extracted from the interference of a small number of them. This can be obtained by simply reducing the spatial extent of the condensate through an increase of the harmonic confinement. Typical profiles of $P(k)$ are displayed in Figs. 5(a)–5(c). Depending on the degree of confinement, we observe one, two or three states, featuring a smooth distribution or a clear multiple-slit interference pattern. The spacing of the fringes yields a spatial separation between the localized states of about five sites, as expected.

We have also observed first effects of the interactions on the interference pattern of multiple localized states. In the non interacting regime the states are independent, owing to the large separation with respect to their axial extent, and the phase of the interference pattern varies randomly in the range $[0, \pi]$, from shot to shot. When a weak interaction is turned on, the eigenstates of the system become a superposition of an increasing number of noninteracting eigenstates, and the effective tunneling between them increases. This produces a decrease of the phase variance, and eventually the states lock in phase.

The independent localized states have a quasi-two-dimensional geome-

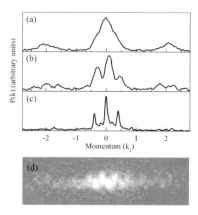

Fig. 5.   (a)-(c) Momentum distribution of the condensate prepared in a disordered lattice with $\Delta/J \approx 10$ for different values of the harmonic confinement; (a) Profile of a single localized state (initial spatial size of the condensate, $\sigma = 1.2$ μm); (b) Interference of two localized states ($\sigma = 1.2$ μm); (c) Three states ($\sigma = 2.1$ μm). (d) Dislocated interference pattern.

try, because their axial extents are much smaller than their radial extents. This feature makes our system an excellent testing ground for studying the physics of quasi-two-dimensional systems,[33] which were recently investigated using widely spaced optical lattices.[34] Actually, we have also observed interference patterns which present a dislocation (Fig. 5d), possibly produced by thermal activation of a vortex in one of the two localized states as in Ref. 34, but in our case for non-interacting atoms.

7. Conclusions and perspectives

We have observed Anderson localization of coherent non-interacting matter waves in a disorderd bichromatic lattice. This system offers a high degree of theoretical and experimental control, making it a novel platform for the study of the interplay between interaction and disorder, paving the way for possible new exotic quantum phases.[9,10,26,35] Preliminary studies already reveal how a weak, controllable interaction affects the observed localization transition.

Acknowledgments

We thank J. Dalibard for discussions, S. Machluf for contributions, and all the colleagues of the Quantum Gases group at LENS. This work has been supported by MIUR, EU (IP SCALA), ESF (DQSEuroQUAM), INFN and Ente CRF.

References

1. P. W. Anderson, *Phys. Rev.* **109**, 1492 (1958).
2. B. Kramer and A. MacKinnon, *Rep. Prog. Phys.* **56**, 1469 (1993).
3. M. P. Van Albada and A. Lagendijk, *Phys. Rev. Lett.* **55**, 2692 (1985).
4. D. S. Wiersma, P. Bartolini, A. Lagendijk and R. Righini, *Nature* **390**, 671 (1997).
5. T. Schwartz, G. Bartal, S. Fishman and M. Segev, *Nature* **446**, 52 (2007).
6. Y. Lahini *et al.*, *Phys. Rev. Lett.* **100**, 013906 (2008).
7. P. A. Lee and T. V. Ramakrishnan, *Rev. Mod. Phys.* **57**, 287 (1985).
8. Y. Dubi, Y. Meir and Y. Avishai, *Nature* **449**, 876 (2007).
9. B. Damski, J. Zakrzewski, L. Santos, P. Zoller and M. Lewenstein, *Phys. Rev. Lett.* **91**, 080403 (2003).
10. M. Lewenstein *et al.*, *Adv. Phys.* **56**, 243 (2007).
11. J. E. Lye *et al.*, *Phys. Rev. Lett.* **95**, 070401 (2005).
12. L. Fallani, J. E. Lye, V. Guarrera, C. Fort and M. Inguscio, *Phys. Rev. Lett.* **98**, 130404 (2007).
13. D. Clément *et al.*, *Phys. Rev. Lett.* **95**, 170409 (2005).
14. C. Fort *et al.*, *Phys. Rev. Lett.* **95**, 170410 (2005).
15. T. Schulte *et al.*, *Phys. Rev. Lett.* **95**, 170411 (2005).
16. J. E. Lye *et al.*, *Phys. Rev. A* **75**, 061603 (2007).
17. G. Roati, C. D'Errico, L. Fallani, M. Fattori, C. Fort, M. Zaccanti, G. Modugno, M. Modugno and M. Inguscio, *Nature* **453**, 895 (2008).
18. J. Billy *et al.*, *Nature* **453**, 891 (2008).
19. G. Roati *et al.*, *Phys. Rev. Lett.* **99**, 010403 (2007).
20. P. G. Harper, *Proc. Phys. Soc. A* **68**, 874878 (1955).
21. S. Aubry and G. André, *Ann. Israel Phys. Soc.* **3**, 133 (1980).
22. D. R. Grempel, S. Fishman and R. E. Prange, *Phys. Rev. Lett.* **49**, 833 (1982); R. E. Prange, D. R. Grempel and S. Fishman, *Phys. Rev. B* **29**, 6500 (1984).
23. C. Aulbach, A. Wobst, G.-L. Ingold, P. Hänggi and I. Varga, *New J. Phys.* **6** (2004).
24. E. Abrahams, P. W. Anderson, D. C. Licciardello and T. V. Ramakrishnan, *Phys. Rev. Lett.* **42**, 673 (1979).
25. F. Gerbier *et al.*, *Phys. Rev. A* **72**, 053606 (2005).
26. G. Roux *et al.*, preprint arxiv:0802.3774 (2008).
27. G.-L. Ingold *et al.*, *Eur. Phys. J. B* **30**, 175 (2002).
28. D. J. Thouless, *Phys. Rev. B* **28**, 4272 (1983).
29. M. Fattori *et al.*, *Phys. Rev. Lett.* **100**, 080405 (2008).
30. C. D'Errico *et al.*, *New J. Phys.* **9** (2007).
31. J. Zhong *et al.*, *Phys. Rev. Lett.* **86**, 2485 (2001).
32. P. Pedri *et al.*, Phys. Rev. Lett. **87**, 220401 (2001).
33. S. Burger *et al.*, *Europhys. Lett.* **57**, 1 (2002).

34. Z. Hadzibabic, P. Krüger, M. Cheneau, B. Battelier and J. Dalibard, *Nature* **441**, 1118 (2006).
35. P. Lugan *et al.*, *Phys. Rev. Lett.* **98**, 170403 (2007).
36. Yu. B. Ovchinnikov *et al.*, *Phys. Rev. Lett.* **83**, 284 (1999).

FERMI GASES WITH TUNABLE INTERACTIONS

J. E. THOMAS, L. LUO, B. CLANCY, J. JOSEPH, Y. ZHANG,

C. CAO, X. DU, and J. PETRICKA

Physics Department, Duke University, Durham, NC 27708-0305, USA
E-mail: jet@phy.duke.edu

Fermi gases with magnetically tunable interactions provide a clean and controllable laboratory system for modelling interparticle interactions between fermions. Near a Feshbach resonance, the s-wave scattering length diverges and Fermi gases are strongly interacting, enabling tests of nonperturbative many-body theories in a variety of disciplines, from high temperature superconductors to neutron matter and quark-gluon plasmas. We measure the entropy and energy of this model system, enabling model-independent comparison with thermodynamic predictions. Our experiments on the expansion dynamics of rotating strongly interacting Fermi gases reveal extremely low viscosity hydrodynamics. Combining the thermodynamic and hydrodynamic measurements enables an estimate of the ratio of the shear viscosity to the entropy density. A strongly interacting Fermi gas in the normal fluid regime is found to be a nearly perfect fluid, where the ratio of the viscosity to the entropy density is close to a universal minimum that has been conjectured by string theory methods. In the weakly interacting regime near a zero crossing in the s-wave scattering length, we observe coherently prepared Fermi gases that slowly evolve into long-lived spin-segregated states that are far from equilibrium and weakly damped.

Keywords: Fermi gas, hydrodynamics, entropy, viscosity, spin-state segregation.

1. Introduction

Interacting fermionic particles play a central role in the structure of matter. For this reason, cold Fermi gases with magnetically tunable interactions serve as a paradigm for testing the predictive capability of theories in a variety of disciplines, from high temperature superconductivity[1] to minimum viscosity hydrodynamics in quark-gluon plasmas.[2] To understand these systems in the regime of very strong interactions between fermionic particles, such as the strong coupling between electrons in high-T_c superconductors and the strong interactions between neutrons in neutron matter, nonperturbative many-body theories are required.

In recent years, based on progress in optical cooling and trapping of fermionic atoms, a clean and controllable strongly interacting Fermi system, comprising a degenerate, strongly interacting Fermi gas,[3] is now of interest to the whole physics community. Strongly interacting Fermi gases are produced near a Feshbach resonance,[3-5] where the zero energy s-wave scattering length a_S is large compared to the interparticle spacing, while the interparticle spacing is large compared to the range of the two-body interaction. In this regime, the system is known a unitary Fermi gas, where the unitarity limit determines the size of the two-particle scattering cross section and the properties are universal and independent of the details of the two-body scattering interaction.[6,7] In contrast to other strongly interacting Fermi systems, in atomic gases, the interactions, energy, and spin population can be precisely adjusted, enabling a variety of experiments for exploring this model system.

Many studies of strongly interacting Fermi gases have been implemented over the past several years. Some of the first experiments observed the expansion hydrodynamics of the strongly interacting cloud.[3,8] Evidence for superfluid hydrodynamics was first observed in collective modes.[9,10] Collective modes were later used to study the $T = 0$ equation of state throughout the crossover regime.[11-13] Recently, measurements of sound velocity have also been used to explore the $T = 0$ equation of state.[14] Below a Feshbach resonance fermionic atoms join to form stable molecules and molecular Bose-Einstein condensates.[15-19] Fermionic pair condensation has been observed by projection experiments using fast magnetic field sweeps.[18,19] Above resonance, strongly bound pairs have been probed by radio frequency and optical spectroscopy.[20-23] Phase separation has been observed in spin polarized samples.[24,25] Rotating Fermi gases have revealed vortex lattices in the superfluid regime[26,27] as well as irrotational flow in both the superfluid and normal fluid regimes.[28] Measurement of the thermodynamic properties of a strongly interacting Fermi gas was first accomplished by adding a known energy to the gas, and then determining an empirical temperature that was calibrated using a pseudogap theory.[29] Recent model-independent measurements of the energy and entropy[30] provide very important information, because they enable direct, precision tests that distinguish predictions from recent many-body theories, without invoking any specific theoretical model.[31,32]

2. Measuring the Energy and Entropy of a Strongly Interacting Fermi Gas

In a strongly interacting Fermi gas, both the energy per particle E and entropy per particle S can be measured in a model-independent way, without invoking any specific theoretical predictions. We describe our recent measurements and compare the results to recent nonperturbative many-body calculations.

2.1. *Model-independent energy measurement*

Model-independent energy measurement is a consequence of the virial theorem, which holds for a unitary Fermi gas near a broad Feshbach resonance, where the zero-energy s-wave scattering length is very large compared to the interparticle spacing, while the range of the potential is very small. The virial theorem has been demonstrated both theoretically,[33–37] and experimentally,[33] and is a consequence of universal thermodynamics.[7] While the local density approximation (LDA) was assumed in our proof of the virial theorem,[33] several other proofs show that the result holds even when the LDA breaks down.[34–37] For a strongly interacting Fermi gas in a harmonic trap, $E = 2\langle U \rangle$, where $\langle U \rangle$ is the average single particle trapping potential. For a scalar pressure in the local density approximation, the potential energy is the same in each direction, and one obtains[33]

$$E = 3m\omega_z^2 \langle z^2 \rangle. \tag{1}$$

The mean square size of the cloud is most easily measured in the long (axial) z-direction of the trap. The harmonic oscillation frequency ω_z is precisely measured by parametric resonance. Corrections arising from trap anharmonicity are readily incorporated.[30]

Equation (1) is a remarkable result: The energy of the strongly interacting gas obeys the same virial theorem as an ideal gas, despite the fact that the strongly interacting gas generally contains condensed superfluid fermion pairs, noncondensed pairs, and unpaired atoms, all in the nonperturbative regime. A simple measurement of the axial mean square cloud size enables a model-independent determination of the energy.

2.2. *Entropy measurement*

The entropy is measured by means of an adiabatic sweep of the bias magnetic field from the strongly interacting regime near the Feshbach resonance to a weakly interacting regime. In ^{6}Li, the bias field is swept from 840 G, just above the Feshbach resonance at 834 G, to a weakly interacting regime at 1200 G.

The entropy S_W of the weakly interacting gas is essentially the entropy S_I of an ideal Fermi gas in a harmonic trap, which can be calculated in terms of the mean square axial cloud size $\langle z^2 \rangle_{1200}$ measured after the sweep, as shown in Fig. 1. Since the sweep is adiabatic, we have

$$S = S_W. \tag{2}$$

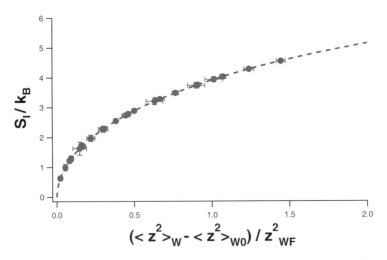

Fig. 1.   The conversion of the mean square size at 1200 G to the entropy. The dashed line is the calculated entropy for a noninteracting Fermi gas in the gaussian trap with a trap depth $U_0/E_F = 10$. $\langle z^2 \rangle_0 = 0.71\, z_F^2$ is the measured ground state size for a weakly interacting Fermi gas, where z_F^2 is the mean square size for an energy equal to the Fermi energy of an ideal gas at the trap center. The calculated error bars of the entropy are determined from the measured error bars of the cloud size at 1200 G.

The adiabaticity of the magnetic field sweep is verified by employing a round-trip-sweep: The mean square size of the cloud at 840 G after a round-trip-sweep lasting 2 s is found to be within 3% of mean square size of a cloud that remains at 840 G for a hold time of 2 s. The nearly unchanged atom number and mean square size proves the sweep does not cause any significant atom loss or heating, which ensures entropy conservation for the sweep. The background heating rate is the same with and without the sweep and increases the mean square size by about 2% over 2 s. We correct the mean square size data by subtracting the increase arising from the background heating rate.

2.3. *Energy versus Entropy*

To measure the energy versus entropy for the strongly interacting gas at 840 G, we prepare the gas with a selected total energy and measure the mean square cloud size at 840 G. Then we prepare the gas again at 840 G with the same energy and adiabatically sweep to 1200 G before the cloud size is measured. We generate the energy-entropy curve for a strongly interacting Fermi gas, as shown in Fig. 2. Here, the energy measured from the mean square axial cloud size at 840 G is plotted versus the entropy measured at 1200 G after an adiabatic sweep of the magnetic field.

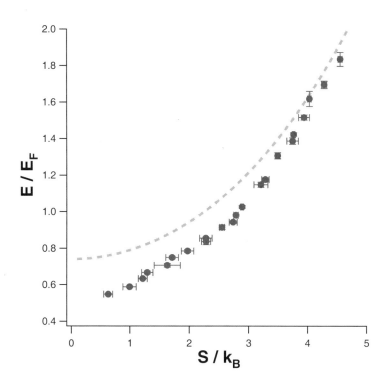

Fig. 2. Measured total energy per particle in units of E_F of a strongly interacting Fermi gas at 840 G versus its entropy per particle in units of k_B. For comparison, the dot-dashed green curve shows $E(S)$ for an ideal Fermi gas. E_F is the Fermi energy of an ideal Fermi gas at the trap center.

2.4. *Testing predictions from many-body theories*

Perhaps the most important application of the energy-entropy measurements is to test strong coupling many-body theories and simulations. Since the energy and entropy are obtained in absolute units without invoking any specific theoretical model, the data can be used to distinguish recent predictions for a trapped strongly interacting Fermi gas.

Figure 3 shows how four different predictions compare to the measured energy and entropy data. These include a pseudogap theory,[38,39] a T-matrix calculation using a modified Nozières and Schmitt-Rink (NSR) approximation,[31,32] a quantum Monte Carlo simulation,[40,41] and a combined Luttinger-Ward and De Dominicis-Martin (LW-DDM) variational formalism.[42] The most significant deviations appear to occur near the ground state, where the precise determination of the energy seems most difficult. The pseudogap theory predicts a ground state energy that is well above the measured value while the LW-DDM prediction of is slightly low compared to the measurement. All of the different theories appear to converge at the higher energies.

3. Viscosity Measurement in a Rotating Strongly Interacting Fermi Gas

We have measured the expansion dynamics of a rotating Fermi gas.[28] The gas is cooled by evaporation to near the ground state and a controlled amount of energy is added. The trap is rotated abruptly to excite a scissors mode. Then the gas is released and imaged after a selected expansion time. Figure 4 shows typical data, where the initial angular velocity Ω_0 is given in terms of the axial trap frequency ω_z. As the gas expands, the angular velocity increases, which is a consequence of irrotational hydrodynamics: the moment of inertia decreases as the aspect ratio approaches unity.

Remarkably, the cloud for the normal fluid at $E/E_F = 2.1$ behaves almost identically to the superfluid cloud for $E/E_F = 0.56$. Indeed, the moment of inertia is quenched well below the rigid body value in both cases, and is in very good agreement with expectations for irrotational flow, Fig. 5.

While irrotational flow is expected for the superfluid, since the velocity field is the gradient of the phase of a macroscopic wavefunction, irrotational flow in the normal fluid requires very low shear viscosity. To estimate the shear viscosity, we have included in the hydrodynamic equations the divergence of the pressure tensor arising from shear viscosity. Figure 6 shows

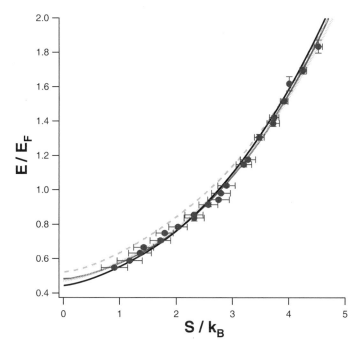

Fig. 3. Comparison of the experimental energy versus entropy data with the calculations from strong coupling many-body theories. The dashed grean line is a pseudogap theory.[38,39] The blue dotted line is an NSR calculation.[31,32] The red solid line is a quantum Monte Carlo simulation.[40,41] The solid black line is a LW-DDM variational calculation.[42]

how the estimated shear viscosity depends on the energy of the cloud, The shear viscosity is given in units of the quantum viscosity, i.e. in units of $\hbar n$, where n is the density.

By combining the entropy and viscosity measurements, we are able to estimate the ratio of the shear viscosity to entropy density. The results are compared to the string theory conjecture[43] for the minimum ratio in Fig. 7. We find that a strongly interacting Fermi gas in the normal fluid regime (above $0.8\,E_F$) is a nearly perfect fluid.

4. Spin Segregation in Weakly Interacting Fermi Gases

Near the zero crossing of the Feshbach resonance, the s-wave scattering length is smoothly tunable from small and negative to small and positive. In this weakly interacting regime, we have studied the behavior of coherently prepared samples, and observe anomalous spin segregation.[44]

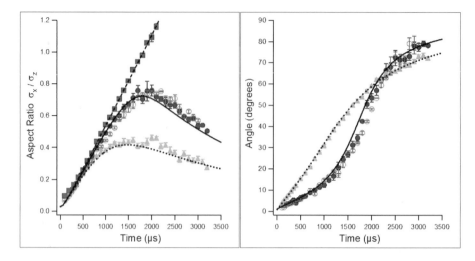

Fig. 4.   Aspect ratio and angle of the principal axes versus time. Purple squares (no angular velocity); Blue solid circles ($\Omega_0/\omega_z = 0.40$, $E/E_F = 0.56$); Red open circles ($\Omega_0/\omega_z = 0.40$, $E/E_F = 2.1$); Green triangles ($\Omega_0/\omega_z = 1.12$, $E/E_F = 0.56$). The solid, dashed, and dotted lines are the theoretical calculations using the measured initial conditions. E_F is the Fermi energy of an ideal gas at the trap center.

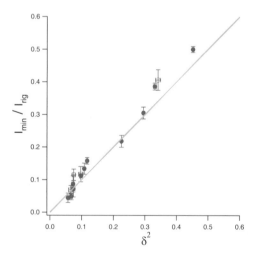

Fig. 5.   Quenching of the moment of inertia versus the square of the measured cloud deformation factor δ. Blue solid circles: Initial energy before rotation below the superfluid transition energy $E_c = 0.83\,E_F$. Red open circles: Initial energy before rotation above the superfluid transition energy. Green solid line: Prediction for irrotational flow.

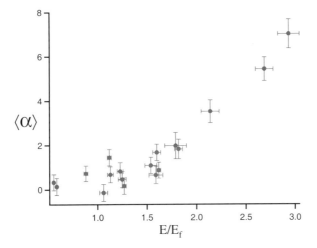

Fig. 6.   Estimated shear viscosity in units of $\hbar n$ versus energy in units of E_F.

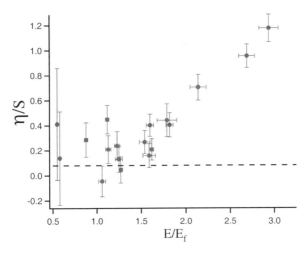

Fig. 7.   Estimated ratio of the shear viscosity to the entropy density. Dotted line shows the string theory conjecture for the minimum ratio.[43]

In the experiments, a cold sample of [6]Li fermions is prepared in one spin state. Then, a radiofrequency pulse is used to create a coherent 50-50 superposition of the two lowest hyperfine states $|1\rangle$ and $|2\rangle$. Initially, the two states have identical axial density profiles in the trap. After 200 ms, we observe the behavior shown in Fig. 8: For slightly negative scattering length of a few bohr, spin state $|1\rangle$ moves outward creating a two-peaked density distribution and state $|2\rangle$ moves inward, creating a narrow density distribution. Reversing the sign of the scattering length reverses the roles of the two spin states.

The highly nonequilibrium spatial distributions created by the segregation relax slowly, over several seconds, back to identical spatial profiles. No segregation is observed unless the sample is coherently prepared.

In contrast to the spin segregation observed in Bose gases,[45] which is explained by an overdamped spin wave, the spin segregation observed in a very weakly interacting Fermi gas is not explained by existing theory: We find the theory predicts that the difference in the spin densities oscillates with the axial period. For our trap, this period is 7 ms, while the observed spatial profiles evolve slowly over several hundred ms. Further, the theory predicts an amplitude which is a factor of 200 too small. We believe that the long correlation time for opposite spins in the weakly interacting regime may require a completely different treatment than previous work.

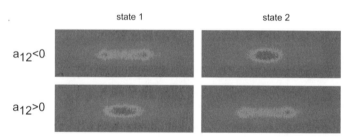

Fig. 8.   Absorption images (state 1 and state 2) taken at 200 ms after the RF pulse for 526.2 G (scattering length $a_{12} < 0$) and for 528.8 G ($a_{12} > 0$). Each image is 1.2 mm in the horizontal direction.

5. Conclusions

Studies of Fermi gas mixtures near a Feshbach resonance enable tests of recent many-body theories in the strongly interacting unitary regime, where the properties of the system are independent of the details of the two-body scattering interactions and hence universal. Model-independent measurements of the entropy and energy are directly compared to predictions without invoking any specific model for the data analysis. Agreement is found to be reasonably good. However, the precise behavior of the superfluid-normal fluid transition has not yet been determined. Measurements of the hydrodynamic expansion of a rotating strongly interacting Fermi gas reveal very low viscosity and a ratio of shear viscosity to entropy density close to the minimum conjectured by string theory methods. At the opposite extreme, measurement of coherently prepared Fermi gases near the zero crossing in the s-wave scattering length reveal very strong, long-lived spin segregation that is not predicted by previous theories based on overdamped spin waves, which successfully explained spin segregation in Bose gases.

Acknowledgments

This research is supported by the Physics Divisions of the Army Research Office and the National Science Foundation and the Chemical Sciences, Geosciences and Biosciences Division of the Office of Basic Energy Sciences, Office of Science, U.S. Department of Energy.

References

1. Q. Chen, J. Stajic, S. Tan and K. Levin, *Phys. Rep.* **412**, 1 (2005).
2. U. Heinz, *Nucl. Phys. A* **721**, 30 (2003).
3. K. M. O'Hara, S. L. Hemmer, M. E. Gehm, S. R. Granade and J. E. Thomas, *Science* **298**, 2179 (2002).
4. M. Houbiers, H. T. C. Stoof, W. I. McAlexander and R. G. Hulet, *Phys. Rev. A* **57**, R1497 (1998).
5. L. Luo, *et al.*, *New J. Phys.* **8**, 213 (2006).
6. H. Heiselberg, *Phys. Rev. A* **63**, 043606 (2001).
7. T.-L. Ho, *Phys. Rev. Lett.* **92**, 090402 (2004).
8. T. Bourdel, *et al.*, *Phys. Rev. Lett.* **91**, 020402 (2003).
9. J. Kinast, S. L. Hemmer, M. Gehm, A. Turlapov and J. E. Thomas, *Phys. Rev. Lett.* **92**, 150402 (2004).
10. M. Bartenstein, *et al.*, *Phys. Rev. Lett.* **92**, 203201 (2004).
11. J. Kinast, A. Turlapov and J. E. Thomas, *Phys. Rev. Lett.* **94**, 170404 (2005).
12. A. Altmeyer, *et al.*, *Phys. Rev. Lett.* **98**, 040401 (2007).
13. M. J. Wright, *et al.*, *Phys. Rev. Lett.* **99**, 150403 (2007).
14. J. Joseph, *et al.*, *Phys. Rev. Lett.* **98**, 170401 (2007).
15. M. Greiner, C. A. Regal and D. S. Jin, *Nature* **426**, 537 (2003).
16. S. Jochim, *et al.*, *Science* **302**, 2101 (2003).
17. M. W. Zweirlein, *et al.*, *Phys. Rev. Lett.* **91**, 250401 (2003).
18. C. A. Regal, M. Greiner and D. S. Jin, *Phys. Rev. Lett.* **92**, 040403 (2004).
19. M. W. Zwierlein, *et al.*, *Phys. Rev. Lett.* **92**, 120403 (2004).
20. C. Chin, *et al.*, *Science* **305**, 1128 (2004).
21. G. B. Partridge, K. E. Strecker, R. I. Kamar, M. W. Jack and R. G. Hulet, *Phys. Rev. Lett.* **95**, 020404 (2005).
22. C. H. Schunck, Y. il Shin, A. Schirotzek and W. Ketterle, *Nature* **454**, 739 (2008).

23. J. T. Stewart, J. P. Gaebler and D. S. Jin, *Nature* **454**, 744 (2008).
24. M. W. Zwierlein, A. Schirotzek, C. H. Schunck and W. Ketterle, *Science* **311**, 492 (2005).
25. G. B. Partridge, W. Li, R. I. Kamar, Y. Liao and R. G. Hulet, *Science* **311**, 503 (2006).
26. M. W. Zwierlein, J. R. Abo-Shaeer, A. Schirotzek, C. H. Schunck and W. Ketterle, *Nature* **435**, 1047 (2005).
27. C. H. Schunck, M. W. Zwierlein, A. Schirotzek and W. Ketterle, *Phys. Rev. Lett.* **98**, 050404 (2007).
28. B. Clancy, L. Luo and J. E. Thomas, *Phys. Rev. Lett.* **99**, 140401 (2007).
29. J. Kinast, *et al.*, *Science* **307**, 1296 (2005).
30. L. Luo, B. Clancy, J. Joseph, J. Kinast and J. E. Thomas, *Phys. Rev. Lett.* **98**, 080402 (2007).
31. H. Hu, P. D. Drummond and X.-J. Liu, *Nature Physics* **3**, 469 (2007).
32. H. Hu, X.-J. Liu and P. D. Drummond, *Phys. Rev. A* **77**, 061605(R) (2008).
33. J. E. Thomas, J. Kinast and A. Turlapov, *Phys. Rev. Lett.* **95**, 120402 (2005).
34. F. Werner and Y. Castin, *Phys. Rev. A* **74**, 053604 (2006).
35. F. Werner, *Phys. Rev. A* **78**, 025601 (2008).
36. D. T. Son, *Phys. Rev. Lett.* **98**, 020604 (2007).
37. J. E. Thomas, *Phys. Rev. A* **78**, 013630 (2008).
38. Q. Chen, Pseudogap theory of a trapped Fermi gas, private communication.
39. Q. Chen, J. Stajic and K. Levin, *Phys. Rev. Lett.* **95**, 260405 (2005).
40. A. Bulgac, J. E. Drut and P. Magierski, *Phys. Rev. Lett.* **96**, 090404 (2006).
41. A. Bulgac, J. E. Drut and P. Magierski, *Phys. Rev. Lett.* **99**, 120401 (2007).
42. R. Haussmann and W. Zwerger, *Phys. Rev. A* **78**, 063602 (2008).
43. P. K. Kovtun, D. T. Son and A. O. Starinets, *Phys. Rev. Lett.* **94**, 111601 (2005).
44. X. Du, L. Luo, B. Clancy and J. E. Thomas, Observation of anomalous spin segregation in a trapped fermi gas, to appear in *Phys. Rev. Lett.* (September 2008).
45. H. J. Lewandowski, D. M. Harber, D. L. Whitaker and E. A. Cornell, *Phys. Rev. Lett.* **88**, 070403 (2002).

PHOTOEMISSION SPECTROSCOPY FOR
ULTRACOLD ATOMS

D. S. JIN,* J. T. STEWART and J. P. GAEBLER

*JILA, Quantum Physics Division, National Institute of Standards and Technology
and Department of Physics, University of Colorado, Boulder, CO 80309, USA*
* E-mail: jin@jilau1.colorado.edu, http://jilawww.colorado.edu/~jin/

We perform momentum-resolved rf spectroscopy on a Fermi gas of ^{40}K atoms in the region of the BCS-BEC crossover. This measurement is analogous to photoemission spectroscopy, which has proven to be a powerful probe of excitation gaps in superconductors. We measure the single-particle spectral function, which is a fundamental property of a strongly interacting system and is directly predicted by many-body theories. For a strongly interacting Fermi gas near the transition temperature for the superfluid state, we find evidence for a large pairing gap.

Keywords: ARPES; BCS-BEC crossover; fermions; superfluidity.

We realize a powerful new technique to probe ultracold atoms and use this technique to probe the BCS-BEC crossover.[1] The phase diagram of the BCS-BEC crossover was first mapped out at JILA using observations of pair condensation.[2] Since then, there have been many experiments exploring this crossover. Many of these experiments have examined macroscopic quantities such as thermodynamics, collective excitations, and superfluidity. In this paper, we focus instead on probing microscopic behavior in the crossover. This allows direct access to the excitation gap, which is an essential feature of fermionic superfluidity.

In the BCS limit, the excitation gap Δ is the order parameter which characterizes the onset of the new order at the superfluid (or superconductor) phase transition. The gap is zero above the transition temperature T_c and non-zero below T_c. This excitation gap arises because of the pairing of fermions, which results in a minimum energy, 2Δ, that must be added to create excitations with fermionic character. In other words, 2Δ is the minimum energy required to break a pair. In the BCS-BEC crossover, things get even more interesting because an excitation gap is expected to exist above the superfluid phase transition temperature.[3-8] Here, 2Δ is still the minimum energy to break a pair, but Δ is now referred to as the pseudogap and does not necessarily signal the onset of the superfluid phase transition. This concept of preformed pairs (pairing that occurs at temperatures above T_c) is perhaps easiest to see when thinking about the limit of a BEC of diatomic molecules. Here, clearly the pairing (molecule formation) can happen at temperatures well above the BEC transition temperature.

Clearly, one would like to probe the excitation gap in the BCS-BEC crossover and see the predicted pseudogap as well as the pairing that occurs in the superfluid phase. Previous experiments have used photoassociation[9] or rf spectroscopy[10-14] to probe microscopic behavior in the BCS-BEC crossover. Our new photoemission spectroscopy technique presented here is based on rf spectroscopy.

In 2003, our group used rf spectroscopy to measure the binding energy of potassium Feshbach molecules.[10] The minimum energy to break a pair, 2Δ, is simply the binding energy in the BEC limit. In 2004, the Innsbruck group reported rf spectroscopy of the BCS-BEC crossover and the observation of a double-peak structure in the spectrum.[12] This was interpreted as observation of the gap.[12,15] In 2007, the MIT group used this same interpretation of rf spectra for a strongly imbalanced Fermi gas mixture and came to the incorrect conclusion that this system had pairs without superfluidity in the $T = 0$ limit.[14] This conclusion was difficult to believe because it meant that there exist bosons (the pairs) that do not Bose condense at $T = 0$. The problem lies with the rf spectroscopy and a number of theorists have pointed out problems with the simple interpretation of double-peaked rf spectra in terms of a pairing gap.[16]

Two issues can affect the interpretation of the rf spectroscopy results. The first issue is the fact that the density of the gas varies spatially because the atoms are confined in a harmonic potential. The gas density is highest in the center of the cloud and falls to zero at the edges of the cloud. This inhomogeneous density is important if one is probing many-body behavior, which of course depends on density. In particular, the size of the excitation gap in the BCS-BEC crossover depends on density (except in the BEC limit where the pairing becomes a two-body effect, namely molecules). A number of theory papers have pointed out that the two features in the double-peaked rf spectra come from different parts of the cloud.[15,17-21] One peak, which occurs near the single-atom Zeeman frequency, is due to atoms at low density at the edges of the cloud. The second peak, which is shifted in frequency with respect to the single-atom Zeeman frequency, comes from the center of the cloud where the density is high and many-body effects are important. Ignoring the unshifted feature due to atoms at the edges of the cloud, one is left with a single, frequency-shifted peak. This feature shows that there is a many-body (density-dependent) shift; this could be due to a pairing gap or could instead be simply a mean-field, or cold-collision, shift.

A second issue is final-state effects, which is relevant for the situation where the rf spectroscopy involves transfer of the strongly interacting atoms into a spin state that is also strongly interacting. This was the case for the ^{6}Li experiments.[12–14] Even if one assumes that the shifted peak is due to pairing, extracting a value for the gap from the measured frequency shift is a difficult theoretical problem when there are final-state effects.[22–27] These final-state effects can be avoided if the strongly interacting atoms are transferred into a spin-state where the atoms are only weakly interacting with the remaining cloud.

Our recent work[1] and also recent results from the MIT group[28] report rf spectroscopy of a strongly interacting Fermi gas without strong final-state effects. The MIT paper[28] directly contrasts data for ^{6}Li atoms with and without strong final-state effects. The rf spectra in the two cases are dramatically different, both qualitatively and quantitatively. These recent results show that density inhomogeneity and final-state effects are likely causes of the double-peak structure in previous rf spectra, which therefore can not be simply interpreted as evidence for pairing. In ^{40}K rf spectroscopy, final-state effects are easily avoided and in fact the final spin-state in the rf transfer is so weakly interacting that atoms transferred into this state can pass through the gas with a low probability of experiencing even one collision. Beyond just permitting rf spectroscopy without complicated final-state effects, this circumstance allows us to extract additional information and obtain a clear signature of a pairing gap.

We can now obtain momentum-resolved rf spectra, since the momenta of the atoms transferred into the new spin-state are not scrambled by collisions. This turns out to provide a very close analogue to photoemission spectroscopy (PES) of solids[29] (see Fig. 1). For electronic systems, PES has proven to be a powerful probe of the excitation spectra.[30] In a typical photoemission experiment, a beam of photons ejects electrons from the sample via the photoelectric effect. These photons are counted as a function of their energy and momentum. Using conservation of energy, the energy of the single-particle states in the solid can be determined. Thus, PES reveals the density of states and the dispersion, E_S vs. k, for the single-particle states

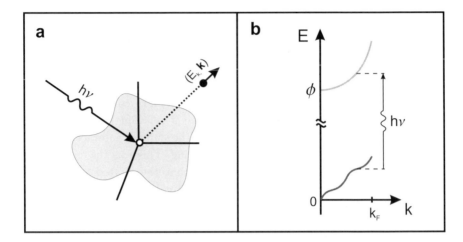

Fig. 1. **Photoemission spectroscopy for ultracold atom gases. a** In electron PES, one measures the energy of electrons emitted from solids, liquids, or gases by the photoelectric effect. Using energy conservation, the original energy of the electrons in the substance can be determined. Similarly, in photoemission spectroscopy for atoms, rf photon with energy, $h\nu$, transfers atoms into a weakly interacting spin state. **b** The rf photon drives a vertical transition where the momentum $\hbar k$ is essentially unchanged. By measuring the energy and momentum of the out-coupled atoms (upper curve) we can determine the quasiparticle excitations and their dispersion relation (lower curve). Here ϕ is the Zeeman energy difference between the two different spin states of the atom.

in a strongly interacting electron system. This technique has been used to probe the excitation gap in high-T_c superconductors and other strongly correlated materials. Note that in PES, as well as in our experiment, the photon momentum is negligible compared to the typical momentum of the strongly interacting particles.

The basic steps of our technique for atoms are (1) apply a short rf pulse, (2) turn off the trap, (3) selectively image the transferred atoms after a period of expansion from the trap, (4) obtain the three-dimensional number of atoms as a function of momentum $N(k)$ using an inverse Abel transform, and (5) repeat for different rf frequencies.[1] If we simply count the number of atoms vs. rf frequency, we obtain an rf spectra without final-state effects. If we use conservation of energy, we can extract the occupation of single-particle states as a function of energy E_s and momentum k in the strongly interacting atom gas.

Figure 2 shows the intensity map (proportional to number of atoms as a function of E_s and k) measured for the strongly interacting Fermi gas. In the intensity map, the observed energy width of the data is larger than the

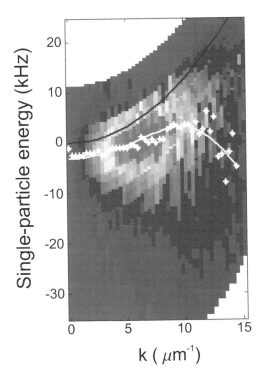

Fig. 2. **Single-particle excitation spectrum obtained using photoemission spectroscopy for ultracold atoms.** Data is for a strongly interacting Fermi gas where $1/k_F^0 a = 0$ and $T \approx T_c$. Plotted is an intensity map of the number of atoms out coupled to a weakly-interacting spin state as a function of the single-particle energy E_s and wave vector k. The black line is the expected dispersion curve for an ideal Fermi gas. The white points (*) mark the center of each fixed energy distribution curve. The Fermi wave vector k_F^0 is $8.6 \pm 0.3 \mu m^{-1}$. The white line is a fit of the centers to a BCS-like dispersion.

measurement resolution and can be caused by a finite lifetime of the single-particle excitations. The black line shows the dispersion for free particles, $E = \hbar^2 k^2 / 2m$. The white points show the measured dispersion curve, which was obtained by fitting a Gaussian to the intensity vs E_s for each value of k. This dispersion curve shows a back-bending behavior that is characteristic of a pairing gap.

One of the aspects of PES that makes it a useful probe of microscopic behavior is that it measures the spectral function, which is a quantity that is directly predicted by many-body theories.[30] The spectral function for the BCS-BEC crossover has been discussed in many theory papers. The peaks of the spectral function are predicted to follow a "BCS-like" dispersion curve where the BCS gap is replaced by the pseudogap.[3-8] This dispersion has two branches, with one corresponding to the occupied part of the excitation spectrum while the other corresponds to the Bogoliubov excitations. With our technique, as in PES of solids, we only measure the states that are occupied and so would not expect to see the excited branch when probing a low temperature gas.

The white points in Fig. 2 fit well to the BCS-type dispersion with the chemical potential and the gap Δ as fit parameters. It should be noted that these best fit values can be influenced by the density inhomogeneity of the trapped gas and by the fact that we use Gaussian fits to extract the peak (white points) for each value of k. Our photoemission spectroscopy for ultracold atoms reveals the dispersion curve $E_s(k)$ and we are able to see the back-bending behavior characteristic of an excitation gap due to pairing of fermions. (Note also that one can easily discriminate against atoms at very low density at the edges of the cloud, which would give a signal that follows the simple quadratic dispersion shown by the black line.)

In conclusion, we have used photoemission spectroscopy, accomplished by momentum resolving the out-coupled atoms in rf spectroscopy, to probe the occupied single-particle density of states and energy dispersion through the BCS-BEC crossover. In the future, it may be possible to use spatially resolved photoemission spectroscopy to probe the local pairing gap. Another extension of this work will be to study the BCS-BEC crossover as a function of temperature and/or unbalanced spin population. Photoemission spectroscopy for ultracold atoms is a powerful and conceptually simple probe of strongly correlated atom gases that could be applied to many other atom gas systems. In the studies presented here, the atoms are interacting via isotropic s-wave interactions and therefore considering different directions of the out-coupled atoms' momenta was not necessary. However, like angle-resolved photoemission spectroscopy (ARPES) for solids, this technique could also be applied to non-isotropic systems such as atoms in an optical lattice, low dimensional systems, or higher partial wave pairing of atoms.[31]

Acknowledgments

We thank E. Cornell, D. Dessau and the JILA BEC group for stimulating discussions. We acknowledge funding from the US NSF. A more complete account of this work has appeared in Ref. 1.

References

1. J. T. Stewart, J. P. Gaebler and D. S. Jin, *Nature* **454**, 744 (2008).
2. C. A. Regal, M. Greiner and D. S. Jin, *Phys. Rev. Lett.* **92**, 040403 (2004).
3. B. Janko, J. Maly and K. Levin, *Phys. Rev. B* **56**, R11407 (1997).
4. Y. Yanase and K. Yamada, *J. Phys. Soc. Japan* **68**, 2999 (1999).
5. A. Perali, P. Pieri, G. C. Strinati and C. Castellani, *Phys. Rev. B* **66**, 024510 (2002).
6. G. M. Bruun and G. Baym, *Phys. Rev. A* **74**, 033623 (2006).
7. A. Bulgac, J. E. Drut, P. Magierski and G. Wlazlowski (2008), arXiv:0801.1504v1.
8. N. Barnea, arXiv:0803.2293v1 (2008).
9. G. B. Partridge, K. E. Strecker, R. I. Kamar, M. W. Jack and R. G. Hulet, *Phys. Rev. Lett.* **95**, 020404 (2005).
10. C. A. Regal and D. S. Jin, *Phys. Rev. Lett.* **90**, 230404 (2003).
11. C. A. Regal, C. Ticknor, J. L. Bohn and D. S. Jin, *Nature* **424**, 47 (2003).
12. C. Chin, M. Bartenstein, A. Altmeyer, S. Riedl, S. Jochim, J. H. Denschlag and R. Grimm, *Science* **305**, 1128 (2004).
13. Y. Shin, C. H. Schunck, A. Schirotzek and W. Ketterle, *Phys. Rev. Lett.* **99**, 090403 (2007).
14. C. H. Schunck, Y. Shin, A. Schirotzek, M. W. Zwierlein and W. Ketterle, *Science* **316**, 867 (2007).
15. J. Kinnunen, M. Rodriguez and P. Torma, *Science* **305**, 1131 (2004).
16. S. Giorgini, L. P. Pitaevskii and S. Stringari, *Rev. Mod. Phys.* (2008).
17. Y. He, Q. Chen and K. Levin, *Phys Rev. A* **72**, 011602 (2005).
18. Y. Ohashi and A. Griffin, *Phys Rev. A* **72**, 063606 (2005).
19. Y. He, C. C. Chien, Q. Chen and K. Levin, *Phys Rev. A* **77**, 011601(R) (2008).
20. P. Massignan, G. M. Bruun and H. T. C. Stoof, arXiv:0709.3158v2 (2007).
21. E. J. Mueller (2008), arXiv:0711:0182v2.
22. Z. Yu and G. Baym, *Phys. Rev. A* **73**, 063601 (2006).
23. M. Punk and W. Zwerger, *Phys. Rev. Lett.* **99**, 170404 (2007).
24. A. Perali and G. C. Strinati, *Phys. Rev. Lett.* **100**, 010402 (2008).
25. S. Basu and E. J. Mueller, arXiv:0712.1007v1 (2007).
26. M. Veillette, E. G. Moon, A. Lamacraft, L. Radzihovsky, S. Sachdev and D. E. Sheehy, arXiv:0803.2517v1 (2008).
27. Y. He, C. C. Chien, Q. Chen and K. Levin, arXiv:0804.1429v1 (2008).
28. C. H. Schunck, Y. Shin, A. Schirotzek and W. Ketterle (2008), arXiv:0802.0341v1.
29. T.-L. Dao, A. Georges, J. Dalibard, C. Salomon and I. Carusotto, *Phys. Rev. Lett.* **98**, 240402 (2007).
30. A. Damascelli, *Phys. Scr.* **T109**, 61 (2004).
31. J. P. Gaebler, J. T. Stewart, J. L. Bohn and D. S. Jin, *Phys. Rev. Lett.* **98**, 200403 (2007).

UNIVERSALITY IN STRONGLY INTERACTING FERMI GASES

PETER D. DRUMMOND[1], HUI HU[2,3], and XIA-JI LIU[2]

[1] *ARC Centre of Excellence for Quantum-Atom Optics, Center for Atom Optics and Ultrafast Spectroscopy*
Swinburne University of Technology, Melbourne, Victoria, Australia
[2] *ARC Centre of Excellence for Quantum-Atom Optics, Department of Physics, University of Queensland, Brisbane, Queensland 4072, Australia*
[3] *Department of Physics, Renmin University of China, Beijing 100872,China*

Experiments on ultra-cold atomic gases at nano-Kelvin temperatures are revo-
lutionizing many areas of physics. Their exceptional adaptability and simplicity
allows tests of many-body theory in areas long thought to be inaccessible, due
to strong interactions. Ultra-cold Fermi gases are now providing new insight
into the foundations of quantum theory. They are expected to exhibit a universal thermodynamic behaviour in the strongly interacting limit, independent
of any microscopic details of the underlying interactions. Here, we present a
systematic theoretical study of strong interacting fermions, using different field-
theoretic methods and comparisons with quantum Monte Carlo simulations.
Pioneering measurements have dramatically confirmed our theoretical predictions, giving the first known evidence for universal fermion thermodynamics.

Keywords: Strongly interacting Fermi gases; unitarity limit; many-body
T-matrix.

1. Introduction

The theory of strongly interacting fermions is of wide interest.[1,2] Interacting
fermions are involved in some of the most important unanswered questions
in condensed matter physics, nuclear physics, astrophysics and cosmology.
Though weakly-interacting fermions are well understood,[3,4] new approaches
are required to treat strong interactions. In these cases, one encounters a
"strongly correlated" picture which occurs in many fundamental systems
ranging from strongly interacting electrons to quarks.

The main theoretical difficulty lies in the absence of any small coupling
parameter in the strongly interacting regime. Although there are numerous efforts to develop strong-coupling perturbation theories of interacting

fermions, notably the many-body T-matrix fluctuation theories,[5-15] their accuracy is not well-understood. Quantum Monte Carlo (QMC) simulations are also less helpful than one would like, due to the sign problem for fermions[16] or, in the case of lattice calculations,[17,18] the need for extrapolation to the zero filling factor limit.

Recent developments in ultracold atomic Fermi gases near a Feshbach resonance with widely tunable interaction strength, densities, and temperatures have provided a unique opportunity to *quantitatively* test different strong-coupling theories.[19-23] In these systems, when tuned to have an infinite s-wave scattering length — the *unitarity* limit — a simple universal thermodynamic behaviour emerges.[24-26]

We give an overview of the current theoretical and experimental situation, including detailed quantitative comparisons of theory and several different experiments that establish the first evidence for universality. We also explore the open question of how to quantitatively distinguish between existing theories of strongly interacting Fermi gases.

2. Universality in strongly interacting Fermi gases

Dilute quantum Fermi gases were first considered by Lee, Huang and Yang (LHY) in 1950's.[3,4] They developed a perturbation theory in the weakly coupling regime, using a small gas parameter for interactions, $k_F a_s$, where k_F is the Fermi wavelength and a_s is the s-wave scattering length. The validity of their theory was, however, restricted to $|k_F a_s| \ll 1$. What will happen when the gas parameter increases to infinity? This fascinating and challenging theoretical problem has been studied intensively in recent years.

A brilliant idea, firstly proposed by Ho and co-workers,[24] is the universality hypothesis. This states that due to the infinitely large scattering length, the interatomic distance becomes the *only* relevant length scale in the problem. At this point, the gas is expected to show a universal thermodynamic behaviour, independent of any microscopic details of the underlying interactions. In particular, at zero temperature the homogeneous ground state energy U and the chemical potential μ scale with the Fermi energy ϵ_F as,

$$U/N\epsilon_{FG} = 1 + \beta, \qquad (1)$$

$$\mu/\epsilon_F = 1 + \beta, \qquad (2)$$

where β is a universal many-body parameter,[27] and $\epsilon_{FG} = (3/5)\epsilon_F$ is the mean energy of a non-interacting Fermi gas. At finite temperatures, dimensional analysis leads to an exact scaling identity between the pressure and the energy density at finite volume V,[24,25]

$$P = \frac{2}{3} \frac{E}{V}. \tag{3}$$

This simple equation relates the pressure P and energy E for a strongly interacting Fermi gas at unitarity in the same way as for its ideal, noninteracting counterpart. In the experimental situation with a harmonic trap, treating the gas as a collection of many uniform cells then leads to the simple result,[25]

$$E = 2N \langle E_P \rangle. \tag{4}$$

Hence, a strongly interacting Fermi gas at unitary obeys the same virial theorem as for an ideal quantum gas. As we shall see, this elegant theorem can be used to measure the total energy of a unitarity Fermi gas in harmonic traps, since the readily calibrated mean-square size of the cloud is proportional to the trapping potential energy $\langle E_P \rangle$.

3. Strong coupling theories

We first briefly review several commonly used strong coupling theories of a unitarity Fermi gas. These are mainly *approximate* many-body T-matrix theories, involving an infinite set of diagrams — the ladder sum in the particle-particle channel, since no exact results are known. It is generally accepted that this ladder sum is necessary for taking into account strong pair fluctuations in the strongly interacting regime, since it is the leading class of all sets of diagrams. Different many-body T-matrix theories are obtained from the approximations used for the diagrammatic structure of the T-matrix, which takes the form of $t(Q) = U/[1 + g\chi(Q)]$ in the normal state. Here and throughout, $Q = (\mathbf{q}, i\nu_n)$, $K = (\mathbf{k}, i\omega_m)$, while $g^{-1} = m/(4\pi\hbar^2 a_s) - \sum_{\mathbf{k}} 1/(2\epsilon_{\mathbf{k}})$ is the bare contact interaction renormalized in terms of the s-wave scattering length and atomic mass m. We use $\epsilon_{\mathbf{k}} = \hbar^2 k^2/(2m)$ and $\sum_K = k_B T \sum_m \sum_{\mathbf{k}}$, where $\mathbf{q}$ and $\mathbf{k}$ are wave vectors, while ν_n and ω_m are bosonic and fermionic Matsubara frequencies. Different T-matrix fluctuation theories differ in their choice of the particle-particle propagator $\chi(Q)$.

The simplest choice of $\chi(Q)$ was pioneered by Nozières-Schmidt-Rink above the critical temperature by using a thermodynamic potential.[5] This NSR approach was recently extended to the broken-symmetry superfluid phase by several authors,[8,9,12,13,29] using the mean-field 2×2 matrix BCS Green's function in the construction of $\chi(Q)$. In particular, we have considered the contributions of Gaussian fluctuations around the mean-field saddle point to the thermodynamic potential.[12] In the Nambu representation, it takes the form,

$$\delta\Omega = \frac{1}{2}\sum_Q \ln \det \begin{bmatrix} \chi_{11}(Q) & \chi_{12}(Q) \\ \chi_{12}(Q) & \chi_{11}(-Q) \end{bmatrix}, \tag{5}$$

where

$$\chi_{11} = \frac{m}{4\pi\hbar^2 a_s} + \sum_K \mathcal{G}_{11}(Q-K)\mathcal{G}_{11}(K) - \sum_{\mathbf{k}} \frac{1}{2\epsilon_{\mathbf{k}}}, \tag{6}$$

$$\chi_{12} = \sum_K \mathcal{G}_{12}(Q-K)\mathcal{G}_{12}(K), \tag{7}$$

are respectively the diagonal and off-diagonal parts of the pair propagator. Here, $\mathcal{G}_{11}$ and $\mathcal{G}_{12}$ are BCS Green's functions with a variational order parameter Δ. Together with the mean-field contribution

$$\Omega_0 = \sum_{\mathbf{k}} \left[\epsilon_{\mathbf{k}} - \mu + \frac{\Delta^2}{2\epsilon_{\mathbf{k}}} + 2k_B T f(-E_{\mathbf{k}}) \right] - \frac{m\Delta^2}{4\pi\hbar^2 a_s}, \tag{8}$$

where the excitation energy $E_{\mathbf{k}} = [(\epsilon_{\mathbf{k}} - \mu)^2 + \Delta^2]^{1/2}$ and the Fermi distribution function $f(x) = 1/(1 + e^{x/k_B T})$, we obtain the full thermodynamic potential $\Omega = \Omega_0 + \delta\Omega$. All the thermodynamic functions can then be calculated straightforwardly following the standard thermodynamic relations, once the chemical potential μ and the order parameter Δ are determined. It is important to note that in our formalism, the number conservation in the form of $n = -\partial\Omega/\partial\mu$ is strictly satisfied, reproducing exactly the scaling identity of $P = (2/3)(E/V)$. For consistency,[29] in our formalism we determine the order parameter at the level of mean field, using the gap equation $\partial\Omega_0/\partial\Delta = 0$. Part of our approach was also previously derived using a functional integral method.[8,13] In the case of the normal Fermi liquid with vanishing order parameter, the usual NSR formalism is recovered.[5]

This type of perturbation theory with *bare* BCS Green functions in the pair propagators, abbreviated as $\chi = G_0 G_0$, constitutes the simplest description of strongly interacting fermions, including the essential contribution from the low-lying collective Bogoliubov-Anderson (BA) modes. More sophisticated approximations with dressed Green functions in the pair propagators $\chi(Q)$, i.e., the fully self-consistent GG and an intermediate GG_0 schemes, have also been put forward. The self-consistent approximation was discussed in detail by Haussmann and co-workers,[7,15] both above the below the superfluid transition temperature T_c. Below T_c, an *ad hoc* renormalization of the interaction strength is required to obtain a gapless BA phonon spectrum, resulting in a slight violation of the exact scaling identity.

The intermediate GG_0 scheme was investigated in a series of papers by Levin and co-workers.[11,30] We note that, although the theory has been explored numerically to some extent,[30] a complete numerical implementation is yet to be reached. A simplified version of the GG_0 fluctuation theory was introduced based on a decomposition of the T-matrix $t(Q)$ in terms of a condensate part and a pseudogap part.[11] Thus, we refer to this simplified approach as the "pseudogap model".

We emphasize that there is no known *a priori* theoretical justification for which T-matrix theory is the most appropriate. Nevertheless, we will show that the NSR and self-consistent T-matrix calculations agree well with the lattice QMC simulations. In contrast, the prediction of the pesudogap model deviates substantially. The universal parameters predicted from the different theories are respectively, $\beta = -0.599$ (NSR), $\beta = -0.632$ (self-consistent T-matrix), and $\beta = -0.409$ (pseudogap model), compared to $\beta = -0.60 \pm 0.01$ as given by the latest zero temperature QMC simulations.[31]

The energy and entropy of a unitarity Fermi gas are the most useful benchmarks for comparisons, as they experimentally observable. For quantitative purposes, we calculate the thermodynamic potential from the chemical potential, using

$$\Omega\left(\mu, T = const\right) = -\int_{\mu_0}^{\mu} n\left(\mu'\right) d\mu' + \Omega\left(\mu_0, T\right) \tag{9}$$

at a given temperature. Here, the lower bound of the integral μ_0 is sufficiently small so that $\Omega\left(\mu_0, T\right)$ can be obtained accurately from a high temperature virial expansion. The energy and entropy can then be calculated from the rigorous scaling relations, $E = -3\Omega/2 = (3/2)PV$, and $S = (-5\Omega/2 - \mu N)/T$, valid at unitarity.

The energy and entropy obtained in this manner are given in Fig. 1, and compared to the predictions of QMC calculations. There is a reasonable agreement between T-matrix theories and the lattice QMC simulations. The pseudogap model appears to provide the least accurate predictions. At low temperatures the T-matrix entropies follow a T^3 scaling law, arising from the Bogoliubov-Anderson phonon modes.[15]

To include the effects of the trap, we employ the local density approximation by assuming that the system can be treated as locally uniform, with a position dependent local chemical potential $\mu\left(\mathbf{x}\right) = \mu - V\left(\mathbf{x}\right)$, where $V\left(\mathbf{x}\right)$ is the trapping potential. The local entropy and energy, calculated directly from the local thermodynamic potential using thermodynamic relations, are then summed to give the total entropy and energy.

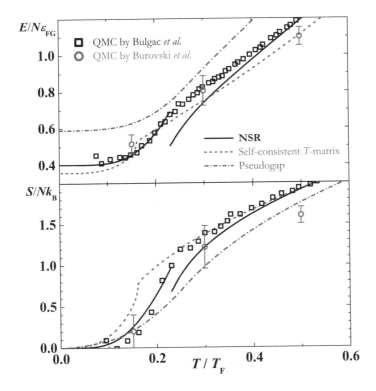

Fig. 1. Temperature dependence of the energy (upper panel) and of the entropy (lower panel) of a uniform Fermi gas at unitarity, obtained from different T-matrix approximations and QMC simulations as indicated.

4. Evidence for universal thermodynamics

The energy and entropy of a strongly interacting Fermi gas at unitarity can now be accurately measured, due to the pioneering efforts of experimentalists at Duke University (^{6}Li) and JILA (^{40}K) . We discuss first the ^{6}Li experiment in Duke.[23] Experimentally, the entropy of the gas is measured by an adiabatic passage to a weakly interacting region at field strength $B = 1200$ G, where $k_F a = -0.75$ and the entropy and temperature is known from the cloud size after the sweep. The energy E is determined model independently from the mean square radius of the strongly interacting fermion cloud $\langle z^2 \rangle_{840}$ measured at 840 G, according to the virial theorem,

$$\frac{E}{N E_F} = \frac{\langle z^2 \rangle_{840}}{z_F^2} \left(1 - \kappa \right), \tag{10}$$

where $E_F = (3N\omega_\perp^2\omega_z)^{1/3} = k_B T_F$ is the Fermi energy for an ideal harmonically trapped gas at the trap center, and z_F^2 is defined by $3m\omega_z^2 z_F^2 \equiv E_F$. The correction factor $1 - \kappa$ accounts for the anharmonicity in the shallow trapping potential $U_0 \simeq 10E_F$.

Calibration of the entropy from the measured mean square axial cloud size at 1200 G using the theoretically predicted dependence of the entropy on the size leads to the comparison for the entropy-energy relation, as shown in Fig. 2 in the blue squares. The agreement is very impressive indeed.

The procedure in the JILA experiment for ^{40}K atomic gases is essentially the same.[22] The strongly interacting gas prepared in a harmonic trap at the Feshbach resonance field is swept adiabatically to a zero scattering length field, and the potential energies at both fields are measured. As shown by the data in Fig (2) in the red circles, we find again the excellent agreement between the experimental data and NSR theoretical predictions.

In summary, Fig (2) illustrates the universal thermodynamic behavior of strongly interacting Fermi gases. This figure plots all the measured data in a single figure, in comparison with our NSR prediction for the entropy dependence of the energy of a harmonically trapped, strongly interacting Fermi gas. The agreement between theory and experiment is excellent for almost all the measured data. Exactly the same theory (NSR theory) is used in all cases, with results from three different laboratories with different densities and atomic species. This includes a single result from Rice University, which uses a different techniques to measure the temperature.

The universality of the thermodynamics of a strongly interacting Fermi gas is therefore clearly demonstrated, independent of which atomic species we compare with. Just above the critical entropy $S_c \simeq 2.2Nk_B$, for the superfluid-normal fluid phase transition, there is a slight discrepancy with these precise measurements. At this point the approximate NSR theory is least accurate.

A key feature of Duke and JILA experiments is that the lowest attainable entropy is around $S = 0.7Nk_B$, which corresponds to a temperature $0.10 - 0.15T_F$ at unitarity. This nonzero entropy or temperature affects significantly the precise determination of the many-body universal parameter. To remove the temperature dependence, we assume at the low entropy regime, a power law dependence of the energy on the entropy: $E - E_0 \propto S^\alpha$, in order to fit the experimental data. The fitting procedure leads to $E_0/N = 0.48E_F$ and $E_0/N = 0.47E_F$, for the Duke and JILA setups, respectively. Using the equation $E_0/(NE_F) = 3/4\sqrt{1+\beta}$ for a harmonic trap gives rise to $\beta \simeq -0.60$, which agrees fairly well with the most

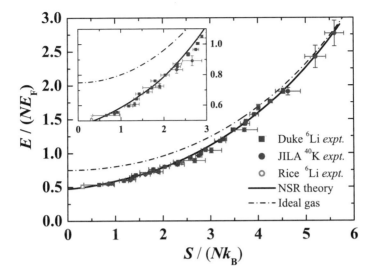

Fig. 2. Illustration of the universal thermodynamics of a strongly interacting Fermi gas.

accurate quantum Monte Carlo simulations, $E = -0.40 \pm 0.01$,[31] hence $\beta = -0.60 \pm 0.01$,[31] and our theoretical predictions, $\beta = -0.599$.[12]

5. Comparative study of strong-coupling theories

The accurate experimental measurement at Duke, at the level of a few percent, provides a very useful benchmark for *unbiased* test of different strong-coupling theories.[32] To better visualize the comparison, we subtract from the energy the ideal gas result E_{IG}. The resulting interaction energy vs entropy has been shown in Fig. 3.

The difference between NSR approach and the self-consistent T-matrix scheme, mostly of the order $0.05NE_F$, is relatively small. Despite this, the extraordinary precision of the measurements is able to discriminate between the different theories of the interaction energy (Fig. 3). The NSR approach is seen to give the best fit to the experimental data below T_c (corresponding to $S_c \approx 2.3Nk_B$) and above $T = 0.5T_F$ (corresponding to $S > 3.5Nk_B$). This indicates that the simplest T-matrix approximation captures the essential physics of strong pair fluctuations at both low (superfluid) and high (normal) temperatures. In the temperature region just above T_c, however, the NSR approach presumably does not fully capture the full effect of fluctuations, compared to the self-consistent T-matrix theory above T_c. Note

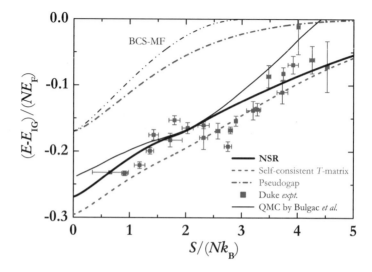

Fig. 3.   Interaction energy vs entropy in a shallow Gaussian trap as predicted by the strong-coupling theories in comparison with experimental data.

that, around T_c the experimental data shows evidence of what could be a first-order superfluid transition. Due to "critical slowing-down", systematic experimental errors cannot be ruled out in this regime, if the magnetic field sweep is not quite adiabatic.

There is a noticeable systematic difference between the QMC and experimental data at high entropy, which may be due to the improper use of an ideal gas approximation in the QMC estimates at high temperatures.[33] It is also clear that the pseudogap approximation[11] is in poor agreement with thermodynamic data, though it is better than BCS mean-field theory – which completely ignores the pairing fluctuations.

6. Conclusion

In conclusion, we have briefly reviewed the current theoretical and experimental status on the thermodynamic properties of a strongly interacting Fermi gas at unitarity, and show that, at this stage the anticipated universal thermodynamics has been clearly demonstrated experimentally, at an accuracy of a few percent. We have also attempted to quantitatively distinguish between existing strong-coupling theories, which may leads to useful insights on the development of more accurate theory of strongly interacting Fermi gases.

References

1. S. Giorgini, L. P. Pitaevskii and S. Stringari, arXiv:0706.3360; *Rev. Mod. Phys.* (in press 2008).
2. I. Bloch, J. Dalibard and W. Zwerger, *Rev. Mod. Phys.* **80**, 885 (2008).
3. T. D. Lee, K. Huang and C. N. Yang, *Phys. Rev.* **106**, 1135 (1957).
4. T. D. Lee and C. N. Yang, *Phys. Rev.* **105**, 1119 (1957).
5. P. Nozières and S. Schmitt-Rink, *J. Low Temp. Phys.* **59**, 195 (1985).
6. C. A. R. S. Melo, M. Randeria and J. R. Engelbrecht, *Phys. Rev. Lett.* **71**, 3202 (1993).
7. R. Haussmann, *Phys. Rev. B* **49**, 12975 (1994).
8. J. R. Engelbrecht, M. Randeria and C. A. R. S. Melo, *Phys. Rev. B* **55**, 15153 (1997).
9. Y. Ohashi and A. Griffin, *Phys. Rev. Lett.* **89**, 130402 (2002); *Phys. Rev. A* **67**, 063612 (2003).
10. A. Perali *et al.*, *Phys. Rev. Lett.* **92**, 220404 (2004).
11. For a review and references to earlier work, see Q. J. Chen *et al.*, *Phys. Rep.* **412**, 1 (2005).
12. H. Hu, X.-J. Liu and P. D. Drummond, *Europhys. Lett.* **74**, 574 (2006); *Phys. Rev. A* **73**, 023617 (2006).
13. R. B. Diener, R. Sensarma and M. Randeria, *Phys. Rev. B* **77**, 023626 (2008).
14. R. Combescot, X. Leyronas and M. Yu. Kagan, *Phys. Rev. A* **73**, 023618 (2006); Z. Nussinov and S. Nussinov, *Phys. Rev. A* **74**, 053622 (2006).
15. R. Haussmann *et al.*, *Phys. Rev. A* **75**, 023610 (2007).
16. V. K. Akkineni, D. M. Ceperley and N. Trivedi, *Phys. Rev. B* **76**, 165116 (2007).
17. A. Bulgac, J. E. Drut and P. Magierski, *Phys. Rev. Lett.* **96**, 090404 (2006).
18. E. Burovski *et al.*, *Phys. Rev. Lett.* **96**, 160402 (2006).
19. K. M. O'Hara *et al.*, *Science* **298**, 2179 (2002).
20. J. Kinast *et al.*, *Science* **307**, 1296 (2005).
21. G. B. Partridge *et al.*, *Science* **311**, 503 (2006).
22. J. T. Steward *et al.*, *Phys. Rev. Lett.* **97**, 220406 (2006).
23. L. Luo *et al.*, *Phys. Rev. Lett.* **98**, 080402 (2007).
24. T.-L. Ho, *Phys. Rev. Lett.* **92**, 090402 (2004).
25. J. E. Thomas, J. Kinast and A. Turlapov, *Phys. Rev. Lett.* **95**, 120402 (2005).
26. H. Hu, P. D. Drummond and X.-J. Liu, *Nat. Phys.* **3**, 469 (2007).
27. H. Heiselberg, *Phys. Rev. A* **63**, 043606 (2001).
28. D. T. Son and C. F. Wingate, *Ann. Phys. (NY)* **321**, 197 (2006).
29. Different realization of the NSR approaches in the superfluid phase below T_c are reviewed by Taylor and Griffin; see, E. Taylor, PhD thesis, University of Toronto (2007).
30. J. Maly, B. Janko and K. Levin, *Physica C* **321**, 113 (1999); *Phys. Rev. B* **59**, 1354 (1999).
31. A. Gezerlis and J. Carlson, *Phys. Rev. C* **77**, 032801(R) (2008).
32. H. Hu, X.-J. Liu and P. D. Drummond, *Phys. Rev. A* **77**, 061605(R) (2008).
33. A. Bulgac, J. E. Drut and P. Magierski, *Phys. Rev. Lett.* **99**, 120401 (2007).

MAPPING THE PHASE DIAGRAM OF A TWO-COMPONENT FERMI GAS WITH STRONG INTERACTIONS

Y. SHIN*, A. SCHIROTZEK, C. H. SCHUNK and W. KETTERLE

*MIT-Harvard Center for Ultracold Atoms, Research Laboratory of Electronics,
Department of Physics, Massachusetts Institute of Technology,
Cambridge, MA 02139, USA*
* E-mail: yishin@mit.edu

We describe recent experimental studies of a spin-polarized Fermi gas with strong interactions. Tomographically resolving the spatial structure of an inhomogeneous trapped sample, we have mapped out the superfluid phases in the parameter space of temperature, spin polarization, and interaction strength. Phase separation between the superfluid and the normal component occurs at low temperatures, showing spatial discontinuities in the spin polarization. The critical polarization of the normal gas increases with stronger coupling. Beyond a critical interaction strength all minority atoms pair with majority atoms, and the system can be effectively described as a boson-fermion mixture. Pairing correlations have been studied by rf spectroscopy, determining the fermion pair size and the pairing gap energy in a resonantly interacting superfluid.

Keywords: Superfluidity; Phase separation; Bose-Fermi mixture; RF spectroscopy.

1. Introduction

Below a critical temperature, an equal mixture of two fermionic gases with attractive interactions undergoes a phase transition to the Bardeen-Cooper-Schrieffer (BCS) superfluid state via Cooper pairing. Since pairing occurs preferably at the Fermi surface, pairing becomes energetically less favorable if the two Fermi surfaces don't overlap. Eventually superfluidity will break down when the difference in Fermi energies exceeds the energy gain from pairing. This is the so-called Chandrasekhar-Clogston (CC) limit of superfluidity.[1,2] Pairing and superfluidity in an imbalanced Fermi mixture has been an intriguing issue for many decades, especially because of the possibility of new exotic ground states such as the Fulde-Ferrell-Larkin-

Ovchinnikov (FFLO) state[3,4] in which either the phase or the density of the superfluid has a spatial periodic modulation.

Mismatched Fermi surfaces can be created in electron gases by applying magnetic field. However, the situation in conventional superconductors is more complicated due to spin-orbit coupling, i.e., the field is shielded by the Meissner effect. On the other hand, in atomic Fermi gases one can prepare a mixture with an arbitrary population ratio, since collisional relaxation processes are very slow. This unique feature, together with tunable interactions using Feshbach resonances, allows the ultracold atomic Fermi gas system to be a highly controllable and clean model system for studying interacting Fermi mixtures. With balanced mixtures near a Feshbach resonance the crossover from a Bose-Einstein condensate (BEC) to a BCS superfluid has been investigated.[5] Recently with population-imbalanced mixtures, the behavior consistent with the CC limit has been observed,[6,7] i.e., a superfluid becomes more robust against imbalance with stronger coupling. The apparent absence of the CC limit in mesoscopic, highly elongated samples[8,9] is not understood and seems to depend on the aspect ratio of the cloud shape.

In this paper, we present the phase diagram of a two-component Fermi gas of ^{6}Li atoms with strong interactions. We have identified and/or determined several important critical points including a tricritical point where the superfluid-to-normal phase transition changes from first-order to second-order, critical spin polarizations of a normal phase, and a critical interaction strength for a stable fermion pair in a Fermi sea of one of its constituents.[10–12] We also present recently measured rf spectra, where we have determined the fermion pair size and the superfluid gap energy in a resonantly interacting Fermi gas.[13,14]

2. Two-Component Fermi Mixture in a Harmonic Potential

In our experiments, we prepared a two-component spin mixture of ^{6}Li atoms, using two states of the three lowest hyperfine states, around a Feshbach resonance. The population imbalance between the two components was controlled by a radio frequency (rf) sweep with an adjustable sweep rate. The atom cloud was confined in a three-dimensional harmonic trap with cylindrical symmetry, thus having an inhomogeneous density distribution. Due to the population imbalance, the chemical potential ratio of the majority (labeled as spin ↑) and the minority (spin ↓) components varies spatially over the trapped sample. Under the local density approximation (LDA), each sample represents a line in the phase diagram. Using spatially resolved measurements, we have mapped out the phase diagram of the sys-

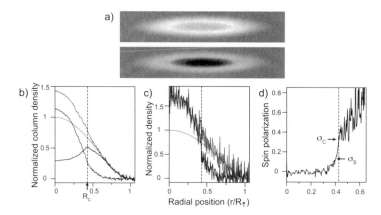

Fig. 1.   Spatial structure of a trapped Fermi mixture with population imbalance. (a) The *in situ* column density distributions are obtained using a phase contrast imaging technique.[10] The probe frequencies of the imaging beam are different for two images so that the first image measures the density difference $n_\uparrow - n_\downarrow$ and the second image measures the weighted density difference $0.76n_\uparrow - 1.43n_\downarrow$. (b) The smooth column density profiles are obtained from the elliptical averaging of the images under the local density approximation (red: majority, blue: minority, black: difference). (c) The reconstructed three-dimensional density profiles. (d) The spin polarization profile shows a sharp increase, indicating the phase separation between a core superfluid and a outer normal region. The vertical dashed line marks the location of the phase boundary.

tem. The temperature was controlled by adjusting the trap depth, which determined the final temperature of evaporative cooling.

For typical conditions, the spatial size of our sample was ~ 150 μm $\times 150$ μm $\times 800$ μm with a total atom number of $\sim 10^7$ and a radial (axial) trap frequency of $f_r = 130$ Hz ($f_z = 23$ Hz). Our experiments benefit from the big size of the sample. Using a phase-contrast imaging technique, we obtained the *in situ* column density distributions of the two components $\tilde{n}_{\uparrow,\downarrow}(r)$, and the three-dimensional density profiles $n_{\uparrow,\downarrow}(r)$ were tomographically reconstructed from the averaged column density profiles (Fig. 1). The imaging resolution of our setup was ~ 2 μm.

At low temperature, the outer part of the sample is occupied by only the majority component, forming a non-interacting Fermi gas. This part fulfills the definition of an ideal thermometer, namely a substance with exactly understood properties in contact with a target sample. We determined temperature from the in situ majority wing profiles. This *in situ* method provides a clean solution for the long-standing problem of measuring the temperature of a strongly interacting sample.

The parameter space of the system can be characterized by three dimensionless quantities: reduced temperature $T/T_{F\uparrow}$, interaction strength $1/k_{F\uparrow}a$ and spin polarization $\sigma = (n_\uparrow - n_\downarrow)/(n_\uparrow + n_\downarrow)$, where $T_{F\uparrow}$ and $k_{F\uparrow}$ are the Fermi temperature and wave number of the majority component, respectively, and a is the scattering length of the two components. The BCS-BEC crossover physics has been studied in the $\sigma = 0$, equal-mixture plane.

3. Phase Diagram at Unitarity

In the case of fixed particle numbers, it has been suggested that unpaired fermions are spatially separated from a BCS superfluid of equal densities due to the pairing gap energy in the superfluid region.[15–17] At low temperature, we have observed such a phase separation between a superfluid and a normal component in a trapped sample. A spatial discontinuity in the spin polarization clearly distinguishes two regions (Fig. 1). By correlating a non-zero condensate fraction[6] with the existence of the core region, we verified that the inner core is superfluid.[10] At the phase boundary two critical polarizations σ_s and σ_c are determined for a superfluid and normal phase, respectively. $\sigma_s \neq \sigma_c$ means that there is a thermodynamically unstable window, $\sigma_s < \sigma < \sigma_c$, leading to a first-order superfluid-to-normal phase transition. As the temperature increases, the discontinuity reduces with decreasing σ_c and increasing σ_s, and eventually disappears above a certain temperature. This is a tricritical point where the nature of the phase transition changes from first-order to second-order.[18] Above the tricritical point, the system shows smooth behavior across the superfluid-to-normal phase transition in density profiles and condensate fraction, which is characteristic of a second-order phase transition.

The phase diagram with resonant interactoins ($1/k_{F\uparrow}a = 0$) is presented in Fig. 2(a),[11] characterized by three distinct points: the critical temperature T_{c0} for a balanced mixture, the critical polarization σ_{c0} of a normal phase at zero temperature and the tricritical point (σ_{tc}, T_{tc}). From linear interpolation of the measured critical points, we have estimated $T_{c0}/T_{F\uparrow} \approx 0.15$, $\sigma_{c0} \approx 0.36$ and $(\sigma_{tc}, T_{tc}/T_{F\uparrow}) \approx (0.20, 0.07)$. The quantitative analysis of the *in situ* density profiles at the lowest temperature reveals the equation of state of a polarized Fermi gas,[19] showing that the critical chemical potential difference is $2h_c = 2 \times 0.95\mu$, where $\mu = (\mu_\uparrow + \mu_\downarrow)/2$. The pairing gap energy Δ of a superfluid has been measured to be $\Delta \gtrsim \mu$,[14] and the observation of $h_c < \Delta$ excludes the existence of a polarized superfluid at zero temperature. A polarized superfluid at finite temperature results from thermal population of spin-polarized quasiparticles.[18]

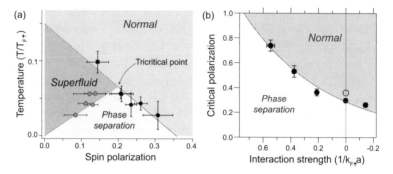

Fig. 2.   Phase diagram of a two-component Fermi gas with strong interactions. (a) With resonant interactions ($1/k_{F\uparrow}a = 0$). At low temperature, the system shows a first-order superfluid-to-normal phase transition via phase separation, which disappears at a tricritical point where the nature of the phase transition changes from first-order to second-order. (b) The critical polarization σ_c of a partially-polarized normal phase increases with stronger interactions. Above a critical interaction strength ($1/k_{F\uparrow}a \approx 0.7$, $\sigma_c = 1$), all minority atoms can pair up to form a superfluid.

4. Strongly Interacting Bose-Fermi Mixture

On the BEC side, two different fermions in free space have a stable bound state, forming a bosonic dimer which undergoes Bose-Einstein condensation at low temperature. Therefore, in the BEC limit a two-component Fermi gas with population imbalance will evolve into a binary mixture of bosonic dimers and unpaired excess fermions. Strong interactions and high degeneracy pressure can affect the structure of the composite boson and eventually destabilize it. This is the reason why we have a partially-polarized normal phase near resonance even at zero temperature. With stronger coupling, the critical polarization σ_c of a partially-polarized normal phase increases, and becomes unity at a critical interaction strength of $1/k_{F\uparrow}a \approx 0.7$.[12] This means that beyond the critical coupling all minority atoms pair up with majority atoms and form a Bose condensate. This is the regime where a polarized Fermi gas can be effectively described as a Bose-Fermi mixture.

In the limit of a BF mixture,[20] we have observed repulsive interactions between the fermion dimers and unpaired fermions. They are parameterized by an effective dimer-fermion scattering length of $a_{\mathrm{bf}} = 1.23(3)a$. This value is in reasonable agreement with the exact value $a_{\mathrm{bf}} = 1.18a$ which has been predicted over 50 years ago for the three fermion problem,[21] but has never been experimentally confirmed. Our finding excludes the mean-field prediction $a_{bf} = (8/3)a$. The boson-boson interactions were found to be stronger than the mean-field prediction in agreement with the Lee-Huang-Yang prediction.[22] Including the LHY correction, the effective dimer-dimer scattering length was determined to be $a_{bb} = 0.55(1)a$, which is close to the exact value for weakly bound dimers $a_{bb} = 0.6a$.

5. Tomographic RF Spectroscopy with a New Superfluid

RF spectroscopy of a two-component Fermi gas measures a single-particle excitation spectrum by flipping the spin state of an atom to a third spin state. Since a fermion pair can be dissociated via spin flip, RF spectroscopy provides valuable information about the pair such as binding energy and size. In early experiments,[23,24] a spectral shift has been observed in a Fermi gas at low temperature and interpreted as a manifestation of pairing. However, it turned out that the spectral line shape is severely affected by the strong interactions of the third, final spin state and broadened due to the inhomogeneous density distribution of a trapped sample, preventing clear comparison of the experimental results to theory. Recently, we have developed several experimental techniques to overcome these problems. In order to minimize final state effects we have exploited a new spin mixture of states $|1\rangle$ and $|3\rangle$ of ^{6}Li atoms[13] (corresponding to $|F = 1/2, m_F = 1/2\rangle$ and $|F = 3/2, m_F = -3/2\rangle$ at low field), and using a tomographic technique, we have obtained local RF spectra from an inhomogeneous sample.[25]

Figure 3 shows the RF spectra of the various phases in a trapped sample with population imbalance. For a balanced superfluid, the majority and the minority spectra completely overlap, showing the characteristic behavior of pair dissociation, i.e. a sharp threshold and a slow high-energy tail. From the spectral width, we have determined the pair size to be $2.6(2)/k_F$ at unitarity, about 20% smaller than the interparticle spacing.[13] These are the smallest pairs so far observed in fermionic superfluids, highlighting the importance of small fermion pairs for superfluidity at high critical temperature.[26]

Excess fermions in a low-temperature superfluid constitute quasiparticles populating the minimum of the dispersion curve. The RF spectrum of a superfluid with such quasiparticles shows two peaks, which, in the BCS limit, would be split by a superfluid gap Δ. Therefore, RF spectroscopy of quasiparticles is a direct way to observe the superfluid gap in close analogy with tunneling experiments in superconductors. In a polarized superfluid near the phase boundary, we have obtained a local majority spectrum of a double-peak structure, from which the superfluid gap has been deter-

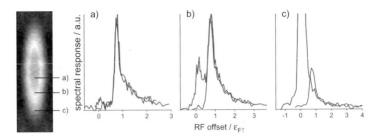

Fig. 3. Tomographic RF spectroscopy of strongly interacting Fermi mixtures. A trapped, inhomogeneous sample has various phases in spatially different regions. The spectra of each region (red: majority, blue: minority) reveals the nature of pairing correlation of the corresponding phase. (a) Balanced superfluid. (b) Polarized superfluid. The additional peak in the majority spectrum is the contribution of the excess fermions, which can be identified as fermionic quasiparticles in a superfluid. From the separation of the two peaks, the pairing gap energy of a resonantly interacting superfluid has been determined.[14] (c) Highly polarized normal gas. The minority peak no longer overlaps with the majority spectrum, indicating the transition to polaronic correlations.

mined to be $\Delta = 0.44(3)E_{F\uparrow}$ at unitarity.[14] In addition, a Hartree term of $-0.43(3)E_{F\uparrow}$ is necessary to explain the observed spectral behavior.

The peak positions of the majority and the minority spectra become different in the partially-polarized normal phase, but still overlap in the high-energy tail. At large spin polarization, the limit of a single minority immersed in a majority Fermi sea is approached, where several theoretical studies suggest a polaron picture, associating the minority with weakly interacting quasiparticles in a normal Fermi liquid.[27–29] We found that these different kinds of pairing correlations are smoothly connected across the superfluid-to-normal phase transition at finite temperature.

6. Summary and Discussion

In a series of experiments with population-imbalanced Fermi mixtures near Feshbach resonances, we have established the phase diagram of a two-component Fermi gas with strong interactions. This includes the identification of a tricritical point at which the critical lines for first-order and second-order phase transitions meet, and the verification of a zero-temperature quantum phase transition from a balanced superfluid to a partially-polarized normal gas at unitarity. The observed critical points such as the critical polarization of a normal phase and the critical interaction strength of a composite boson in a Fermi sea provide quantitative tests of theoretical calculations on the stability of fermionic superfluidity.

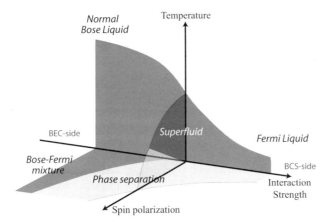

Fig. 4. Various phases of a two-component Fermi gas. The structure of the phase diagram is illustrated in the parameter space of temperature, interaction strength and spin polarization.

Figure 4 sketches the structure of the phase diagram of the system in a 3D parameter space (vs. temperature, spin polarization and interaction strength). For a complete understanding, this macroscopic characterization of the different phases should be complemented by an investigation of their microscopic properties. Currently, we understand the observed polarized superfluid as a result of thermal population of spin-polarized quasiparticles at finite temperature. However, the behavior at higher temperature or/and in a stronger coupling regime is not yet completely understood. Resolving the momentum distribution of the excess fermions might reveal a gapless region ($h > \Delta$) in the parameter space. The nature of a partially polarized normal phase near the resonance is also an interesting subject. Measurement of the binding energy and the effective mass of a minority atom might be helpful to test the polaron picture and to observe the polaron-to-molecule transition near a critical interaction strength. However, it is an open question whether the Fermi liquid description is still valid for high minority concentrations, where the Pauli blocking effect of the minority Fermi sea might play an important role. Furthermore, it has been speculated that exotic pairing states might exist in the partially-polarized phase.[30] So far, predicted exotic superfluid states such as the breached-pair state in a stronger coupling regime and the FFLO state in a weaker coupling regime have not been observed. The novel methods developed in our experiments such as tomography and thermometry will be important tools in the search for these states.

Acknowledgements

This work was supported by NSF, ONR, MURI and ARO Award W911NF-07-1-0493 (DARPA OLE Program).

References

1. B. S. Chandrasekhar, *Appl. Phys. Lett.* **1**, 7 (1962).
2. A. M. Clogston, *Phys. Rev. Lett.* **9**, 266 (1962).
3. P. Fulde and R. A. Ferrell, *Phys. Rev.* **135**, A550 (1964).
4. A. I. Larkin and Y. N. Ovchinnikov, *Sov. Phys. JETP* **20**, 762 (1965).
5. M. Incuscio, W. Ketterle and C. Salomon (eds.), *Ultra-cold Fermi Gases*, Proceedings of the International School of Physics "Enrico Fermi" Course CLXIV, (IOS Press, 2007).
6. M. W. Zwierlein, A. Schirotzek, C. H. Schunck and W. Ketterle, *Science* **311**, 492(2006).
7. M. W. Zwierlein, C. H. Schunck, A. Schirotzek and W. Ketterle, *Nature* **442**, 54 (2006).
8. G. B. Partridge, W. Li, R. I. Karmar, Y. Liao and R. G. Hulet, *Science* **311**, 503 (2006).
9. G. B. Partridge, W. Li, R. I. Karmar, Y. Liao and R. G. Hulet, *Phys. Rev. Lett.* **97**, 190407 (2006).
10. Y. Shin, M. W. Zwierlein, C. H. Schunck, A. Schirotzek and W. Ketterle, *Phys. Rev. Lett.* **97**, 030401 (2006).
11. Y. Shin, C. H. Schunck, A. Schirotzek and W. Ketterle, *Nature* **451**, 689 (2008).
12. Y. Shin, C. H. Schunck, A. Schirotzek and W. Ketterle, *Phys. Rev. Lett* **101**, 070404 (2008).
13. C. H. Schunck, Y. Shin, A. Schirotzek and W. Ketterle, *Nature* **454**, 739 (2008).
14. A. Schirotzek, Y. Shin, C. H. Schunck and W. Ketterle, arXiv:0808.0026.
15. P. F. Bedaque, H. Caldas and G. Rupak, *Phys. Rev. Lett.* **91**, 247002 (2003).
16. J. Carlson and S. Reddy, *Phys. Rev. Lett.* **95**, 060401 (2005).
17. D. E. Sheehy and L. Radzihovsky, *Phys. Rev. Lett.* **96**, 060401 (2006).
18. M. M. Parish, F. M. Marchetti, A. Lamacraft and B. D. Simons, *Nature Phys.* **3**, 124 (2007).
19. Y. Shin, *Phys. Rev. A* **77**, 041603(R) (2008).
20. P. Pieri and G. C. Strinati, *Phys. Rev. Lett.* **96**, 150404 (2006).
21. G. V. Skorniakov and K. A. Ter-Martirosian, *Sov. Phys. JETP* **4**, 648 (1957).
22. T. D. Lee, K. Huang and C. N. Yang, *Phys. Rev.* **106**, 1135 (1957).
23. C. Chin, M. Bartenstein, A. Altmeyer, S. Riedl, S. Jochim, J. H. Denschlag and R. Grimm, *Science* **305**, 1128 (2004).
24. C. H. Schunck, Y. Shin, A. Schirotzek, M. W. Zwierlein and W. Ketterle, *Science* **316**, 867 (2007).
25. Y. Shin, C. H. Schunck, A. Schirotzek and W. Ketterle, *Phys. Rev. Lett.* **99**, 090403 (2007).
26. F. Pistolesi and G. C. Strinati, *Phys. Rev. B* **49**, 6356 (1994).

27. C. Lobo, A. Recati, S. Giorgini and S. Stringari, *Phys. Rev. Lett.* **97**, 200403 (2006).
28. R. Combescot, A. Recati, C. Lobo and F. Chevy, *Phys. Rev. Lett.* **98**, 180402 (2007).
29. N. Prokof'ev and B. Svistunov, *Phys. Rev. B* **77**, 020408(R) (2008).
30. A. Bulgac, M. M. Forbes and A. Schwenk, *Phys. Rev. Lett.* **97**, 020402 (2006).

EXPLORING UNIVERSALITY OF FEW-BODY PHYSICS BASED ON ULTRACOLD ATOMS NEAR FESHBACH RESONANCES

NATHAN GEMELKE, CHEN-LUNG HUNG, XIBO ZHANG and CHENG CHIN*

*James Franck Institute and Physics Department, University of Chicago,
Chicago, IL 60637, USA*
* E-mail: cchin@uchicago.edu
Website: http://ultracold.uchicago.edu

A universal characterization of interactions in few- and many-body quantum systems is often possible without detailed description of the interaction potential, and has become a defacto assumption for cold atom research. Universality in this context is defined as the validity to fully characterize the system in terms of two-body scattering length. We discuss universality in the following three contexts: closed-channel dominated Feshbach resonance, Efimov physics near Feshbach resonances, and corrections to the mean field energy of Bose-Einstein condensates with large scattering lengths. Novel experimental tools and strategies are discussed to study universality in ultracold atomic gases: dynamic control of interactions, run-away evaporative cooling in optical traps, and preparation of few-body systems in optical lattices.

Keywords: Universality, Bose-Einstein condensation, Feshbach, Efimov, mean-field interaction.

1. Introduction

Quantum gases of ultracold atoms distinguish themselves from other quantum systems in two unique and useful ways. First of all, the diluteness of the gases permits a very simple and accurate description of the effect of interactions. Degenerate gases of atoms can be described well by textbook models of fundamental and general interest. Extending beyond these, complexity can be built in slowly to study novel quantum phases, and even intractable mathematical models. This aspect has inspired new research in the vein of *quantum simulation*, promising far-reaching impact on the understanding of other quantum systems in nature, including condensed and nuclear matter.

A second gainful aspect of ultracold atomic gases lies in the ability to tune interactions via Feshbach resonant scattering. Exploitation of this feature, first observed in 1998,[1] has only fully matured in recent years, and promises numerous future applications. Full control of interaction in a quantum gas not only allows for an easy exploration of the quantum system in different interaction regimes, but also leads to new methods to observe dynamic evolution and to scrutinize quantum states in previously unimaginable ways. For example, projecting a complex many-body state onto a non-interacting single particle basis can be realized by diabatically switching off atomic interactions.

The majority of quantum gas systems studied to date admit a universal description of the effect of interaction. In this paper, we describe our approach to explore situations in which universality requires nontrivial extensions. Our experimental platform, based on optically trapped cesium atoms, exploits both of the aforementioned features, allowing us to address long-standing questions concerning the universality of an interacting gas. In particular, we will focus on three topics: universality and its minimal extensions in the study of dimer molecules, three-body Efimov states near a Feshbach resonance, and beyond mean-field interactions in Bose-Einstein condensates. Finally, we will outline our approach to study few-body interactions and our experimental progress.

2. Universality in N-body physics

The connection of quantum degenerate atomic gases to other physical systems is made possible by the expected universality of physics at low temperatures. Here, universality arises when the quantum system is fully described by a single parameter, the two-body scattering length a.[2] Universality is well established in two-body, low energy scattering theory, where the s-wave scattering phase shift is $\eta = -\tan^{-1} ka$, with k the scattering wave number. In the many-body regime, universal behavior of dilute Bose-Einstein condensates (BECs) of different bosonic atomic species is expected for small and positive scattering lengths. Universality is further expected and verified in two-component degenerate Fermi gases with large scattering lengths, as in the BEC-BCS (Bardeen-Cooper-Schrieffer superfluid) crossover regime.[3–5]

Non-universal parameters, however, can play an important role in certain low energy few- and many-body systems, and represent the entrance of a richer underlying scattering physics. For example, binding energies of Efimov states in three-body systems[6] and three-body interactions in BECs with large scattering lengths[7] are expected to be non-universal. Both cases strongly depend on the three-body scattering phase shifts, which likely cannot be universally derived from a.[2]

3. Feshbach resonances

3.1. *Origin of Feshbach resonance*

In cold atom experiments, Feshbach resonances occur when two free atoms interact in the scattering channel and resonantly couple to a bound molecular state in a closed channel.[8] In many cases, the bound state can have a different magnetic moment from that of the scattering atoms, and resonant coupling between the channels can be induced by tuning the bound state energy with an external magnetic field.

Near a Feshbach resonance, the scattering phase shift η follows the Breit-Wigner formula[9]:

$$\eta = \eta_{\mathrm{bg}} - \tan^{-1}\frac{\Gamma/2}{E - E_c - \delta E}, \tag{1}$$

where η_{bg} is the background, or off-resonant phase shift, $E = \hbar^2 k^2/m$ is the scattering energy, k is the scattering wave number, m is twice the reduced mass, $\Gamma \propto k$ is the coupling strength between the scattering and bound states, E_c is the energy of the bare bound state, and δE is the self-energy shift.

At low scattering energies $E \to 0$, the (background) scattering length is given by $a_{(\mathrm{bg})} = -\tan\eta_{(\mathrm{bg})}/k$.[9] We further assume a linear Zeeman shift to the bound state $E_c = \delta\mu(B - B_c)$, where $\delta\mu$ is the relative magnetic moment between open and closed channels, and the bound state is shifted to the continuum when $B = B_c$. These allow us to derive scattering length in the standard resonance form $a = a_{\mathrm{bg}}[1 - \Delta/(B - B_0)]$. Here $\Delta = \lim_{k \to 0}\Gamma/(2ka_{\mathrm{bg}}\delta\mu)$ is the resonance width, $B_0 = B_c - \delta E/\delta\mu$ is the resonance position. Note that a diverges when $B = B_0$, or equivalently, $\eta = (N + \frac{1}{2})\pi$, where N is the number of molecular states below the continuum.

It is important to note that Feshbach resonance occurs not exactly when the bare state is tuned to the continuum $B = B_c$. From the van der Waals potential model, the resonance position offset is[10]

$$B_0 - B_c = -\frac{\delta E}{\delta\mu} = -\frac{r^2 - r}{r^2 - 2r + 2}\Delta, \tag{2}$$

where $r = a_{\mathrm{bg}}/\bar{a}$ and $\bar{a}$ is the mean scattering length of the van der Waals potential.[11]

Equation 2 shows that the difference in magnetic field between the bare state crossing and the resonance position $B_0 - B_c$ is on the order of the resonance width Δ when $|a_{\mathrm{bg}}| > \bar{a}$.

3.2. *Non-universality of Feshbach molecules*

The two-channel nature of the interaction potential described in the previous section implies that, in the absence of Feshbach coupling, the scattering length of atoms in the entrance channel does not reveal the properties of the closed channel bound state. Thus, the properties of the molecular state are clearly non-universal. This point can also been seen in Fig. 1. When the molecular state is well below the continuum, the molecular energy approaches the bare state value $E_c = \delta\mu(B - B_c)$, which cannot be universally derived from mere knowledge of a.

When the magnetic field is tuned sufficiently near the Feshbach resonance, Feshbach coupling strongly modifies the nature of the bound state, whose character is now dominated by the open channel. In this regime, the molecular state does develop a universal behavior with a binding energy of $E_b = \hbar^2/ma^2$. The Universal regime can be seen in the Fig. 1(b) inset.

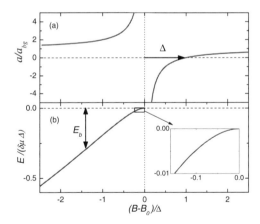

Fig. 1. Scattering length a in (a) and molecular state energy E in (b) near a magnetically tuned Feshbach resonance. $E_b > 0$ is the binding energy. The inset shows the universal regime where $E_b = \hbar^2/ma^2$.

The transition between the non-universal and universal regimes differs for different resonances. One instructive way to see the transition behavior is to expand the molecular binding energy near the resonance, which, from a simplified two-channel model potential,[12] gives

$$E_b \approx \frac{\hbar^2}{m(a - \bar{a} - R^*)^2}. \tag{3}$$

Here $E_{bg} = \hbar^2/ma_{bg}^2$. $\bar{a}$[11]and $R^* = a_{bg}E_{bg}/(\delta\mu\Delta)$[13] are two leading order non-universal length scales associated with the finite interaction range and the coupling strength to the closed channel bound state, respectively. Universality is valid only when $a \gg \bar{a}$ and R^*. In particular, R^*, can be uncharacteristically large, $R^* \gg \bar{a}$, for narrow resonances. Resonances of this type are deemed closed-channel dominated and can have strong non-universal behavior.

4. Efimov physics and Efimov states

Efimov states are a set of three-body, long-range bound states which emerge when the pairwise interactions in a three-particle system are resonantly enhanced. These states are supported by the Efimov potential which scales like $-1/R^2$ for $R < |a|$, where the hyperspherical radius R characterizes the geometric size of the system.[6]

The connection between the three-body Efimov potential and scattering length a can be understood using a hand-waving picture. Assume two bosonic atoms are separated by R, the wave function of the third atom is scattered by each atom, with $|a|$ characterizing the length scale of the scattered waves. The total wave function ψ, after Bose-symmetrization, can be significantly enhanced when the two scattered waves overlap. Using Schrödinger's equation, we can model the wave function enhancement $\phi(R) = \delta\psi(R)$ as a result of an effective Efimov potential $V_{efm}(R)$, which satisfies Schrodinger's equation:

$$-\frac{\hbar^2}{m}\phi''(R) + V_{efm}(R)\phi(R) = E\phi(R) \quad \text{for } R < a. \tag{4}$$

To evaluate the curvature term, we note that ϕ is localized with a length scale of system size R. The curvature is thus negative and we can rewrite $\phi''(R) = -\alpha\phi(R)/R^2$, where $\alpha > 0$ is a proportionality constant. In the low energy collision limit $E \to 0$, we get

$$V_{efm}(R) = -\frac{\alpha\hbar^2}{mR^2} \quad \text{for } R < a. \tag{5}$$

A rigorous calculation performed by V. N. Efimov shows that with R identified as the three-body hyperspherical radius, we have $\alpha = s_0^2 + \frac{1}{4}$ and $s_0 = 1.00624...$ is a constant.[6] For $R > |a|$, the effective potential is no longer attractive.[14]

Right on two-body resonance $a \to \pm\infty$, the $-1/R^2$ Efimov potential extends to infinity and can support an infinite number of three-body bound states (Efimov states); simple scaling laws for the spatial extent A_N and binding energy E_N of the N-th lowest Efimov state have been derived as

$$A_N = \beta^N \times A^* \qquad (6)$$

$$E_N = \beta^{-2N} \times E^*, \qquad (7)$$

where $\beta = e^{\pi/s_0} \approx 22.7$ is a universal constant.[6] These size and energy scaling laws are among the most prominent universal features of Efimov's predictions. Constants A^* and E^* depend on the three-body potential at short range and are thus expected to be non-universal.[2]

4.1. *Universality of Efimov physics near different Feshbach resonances*

Recent observation of an Efimov resonance in the three-body recombination process[15] of ultracold cesium atoms represents a major breakthrough in few-body physics.[16]

Here we suggest a new scheme to check the "defacto" universality of Efimov physics implied by the expected slow variation of the short-range three-body potential with magnetic field tuning. By monitoring recombination loss near different, isolated *open-channel dominated* Feshbach resonances, we expect that Efimov resonances of the same order can occur at the same scattering lengths. Here we point out that the application of magnetic field barely changes the interatomic potential in the entrance scattering channel. We thus expect that, in the three-body sector, systems have nearly identical off-resonant phase shift near different Feshbach resonances.

To estimate the insensitivity of the open channel potential to magnetic field, we note that the two-body background scattering length varies less than 1 a_0 over 100 G at $a = 2400\, a_0$. (This estimation is based on numerical calculation of cesium atom scattering length in the highest triplet scattering channel, in which Feshbach resonances do not exist.) This small variation can be translated into a small fractional change of the scattering phase shift by $|\delta\eta/\eta| < 3 \times 10^{-10}$ per Gauss.[17] This result suggests that the non-universal effects of the three-body potential can potentially remain nearly unchanged when the magnetic field is tuned to different Feshbach resonances.

5. Universality in a dilute BEC with large scattering length: Lee-Huang-Yang Corrections and beyond

In a dilute Bose-Einstein condensate, the energy per particle is given as $2\pi na\hbar^2/m$, which describes the fluid on length scales longer than the coherence length $l = (16na)^{-1/2}$. Due to the weak coupling, corrections to the mean field term can be calculated as expansions of a dimensionless parameter a/l, which is in turn proportional to the diluteness parameter $\sqrt{na^3}$. The energy per particle in a dilute homogeneous BEC is given by[7]

$$\frac{E}{N} = \frac{2\pi\hbar^2 na}{m}[1 + \frac{128}{15\sqrt{\pi}}\sqrt{na^3} + \frac{8(4\pi - 3\sqrt{3})}{3}na^3 \ln na^3 + Cna^3 + ...], \quad (8)$$

where the lowest order contributions $\sqrt{na^3}$, called the Lee-Huang-Yang (LHY) correction,[18] and $na^3 \ln na^3$ term[19] result from universal two- and three-body correlations, and C is a three-body parameter which depends on three-body interactions and Efimov physics.[7] Although Eq. 8 was originally derived based on a hard-sphere potential, the validity of the LHY term for soft-sphere and short-ranged attractive potentials has been numerically verified.[20]

Beyond mean-field effects can be amplified by tuning the scattering length to large values. Previous approaches along this line with ^{85}Rb reached $na^3 = 0.1$, but were complicated by limited lifetimes due to three-body inelastic collisions. Here we point out that a careful choice of scattering length and a fast measurement can allow for a detectable beyond mean-field signal.

To see this, we first note that the LHY term, on the order of $(na^3)^{1/2}$, is a lower order process than is the three body process of na^3. Measurement of the former effect can be immune from three-body loss when na^3 is low. For example, the typical mean-field energy of a BEC is $U = h \times 1$ kHz and the scattering length can be tuned such that $na^3 = 0.01$. In this case, the LHY term is about $(na^3)^{1/2} = 10\%$ of the mean-field energy and is 10 times larger than the three-body energy scale. The associated three-body time scale is $na^3 U/\hbar \approx (10 \text{ ms})^{-1}$. Determination of interaction energy of a condensate within 10 ms can be realized by promptly releasing the condensate into free space. The expansion of the condensate thus converts the interaction energy into detectable atomic kinetic energy.

6. Experimental approach

Two powerful experimental tools will be employed to explore few-body physics: optical lattices to confine and isolate few atoms at each lattice site in the Mott insulator phase, and magnetic Feshbach resonances to control atomic interactions. Both can lead to precise control of the few-body samples in different interaction regimes.

6.1. *Scattering properties of Cesium atoms*

Cesium-133 is chosen in the experiment for their convenient tuning of interaction. In the range of 0 to 50 G, the s-wave scattering length in the lowest hyperfine ground state $|F = 3, m_F = 3\rangle$ can be smoothly tuned from $-2500\ a_0$ to $1000\ a_0$. Here, F is the total angular momentum quantum number and m_F its projection along the magnetic field. At higher fields, two more broad resonances exist at 547 G and 800 G. See Fig. 2.

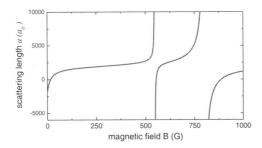

Fig. 2. Calculated s-wave Feshbach resonances in collisions of ground state cesium atoms. Three broad resonances at -11.7 G, 547 G and 800 G allow for tuning of the scattering length. Other higher-order resonances are omitted here for simplicity. The numerical calculation code is provided by Eite Tiesinga, NIST.

The existence of multiple broad s-wave Feshbach resonances permits tests of universality by probing the cold atoms sample at different scattering lengths. As discussed in Secs. 3.2 and 4.1, unique tests of universality in two- and three-body systems can be performed by tuning the scattering length to the same value, but near different Feshbach resonances.

6.2. *Fast evaporation to Bose-Einstein condensation in optical traps*

We employ a novel scheme to achieve fast, runaway evaporative cooling of cesium atoms in optical traps. This is realized by tilting the optical potential with a magnetic field gradient. Runaway evaporation is possible in this trap geometry due to the very weak dependence of vibration frequencies on trap depth, which preserves atomic density during the evaporation process. When the trap depth is reduced by a large factor of 100, the geometric mean of the trap frequencies is only reduced by a factor of 2 and thus preserves the high collision rate.[21]

Using this scheme, we show that Bose-Einstein condensation with $\sim 10^5$ cesium atoms can be realized in $2 \sim 4$ s of forced evaporation.[21] The evaporation speed and energetics are consistent with the three-dimensional evaporation picture, despite the fact that atoms can only leave the trap in the direction of tilt.

6.3. Preparation of few-atom systems in optical lattices

Few-body experiments will begin with segmentation of a bulk condensed superfluid into the ground states of isolated optical lattice sites. Each site will be populated with a small and in general indefinite number, $1 < N < 10$, of atoms.

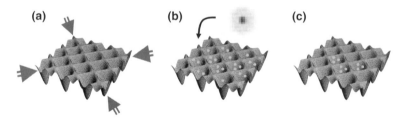

Fig. 3.   Loading an optical lattice and preparation of lattice sites with three atoms. (a) Optical lattices are formed by the interference pattern of intersecting laser beams. For a red-detuned lattice, the potential minima are defined by the anti-nodes of the standing waves. (b) Condensed atoms are loaded into the optical lattices. (c) To prepare lattice sites with three and only three atoms, atoms in lattice sites with other occupancies may be removed by precision radio-frequency excitation.

We have constructed a novel optical lattice configuration - a thin layer, 2D optical lattice, which permits direct imaging of atomic density by sending an imaging beam perpendicular to the lattice plane. The optical lattice is defined by interfering four laser beams derived from a single frequency fiber laser operated at a wavelength $\lambda = 1.06 \mu$m. Two counterpropagating beam pairs on the horizontal plane form a square optical lattice. In the vertical direction, confinement is provided by a single CO_2 laser beam focused to 50 μm vertically and 2 mm in the radial direction. This tight vertical confinement holds atoms against gravity without need for a magnetic field gradient, and provides an ideal mode-matching potential for transferring condensates into the 2D lattice, see Fig. 3.

Probing of few-body energies will be performed through the combined methods of precision radio-frequency spectroscopy, collective mode excitation and dynamic evolution of matter wave coherence. In particular, methods which allow precise determination of the variation of few-body energies with atom number will permit direct investigation of various interactions, including two- and three-body scattering of free atoms, strong correlations, unitarity at strong confinement, atom-dimer interactions, and three-body recombination. Working in a tightly bound optical lattice allows quantum pressure to determine atomic density profiles and permits accurate extrapolation of single-particle measurements to interacting few-body systems. In addition, methods of adiabatically preparing specific few-body systems as combinations of free and bound states (e.g. atom+dimer) will be explored, providing a basis for directly studying the universality of higher complexity interactions.

7. Conclusion

We describe key issues in three- and many-body physics including Efimov physics and beyond mean-field effects in the context of Feshbach tuning in quantum gases. In both cases, Bose-condensed cesium atoms provide unique opportunities to investigate universal behavior of energy shifts and energy structure. In particular, we point out possible non-universal parameters, including the finite interaction range $\bar{a}$, Feshbach coupling length scale R^* and three-body phase shift.

We propose a brand new approach to prepare and study few-body systems in optical lattices by inducing superfluid-Mott insulator transitions in a single-layer 2D optical lattice. This system provides complete and independent control over the filling factor, on site interaction and tunneling. Together with the rich interaction properties of cesium atoms and fast evaporation, one expects a new level of few-body physics can be explored in this lattice setting. We anticipate that a firm understanding of universality in finite systems will provide practical applications in quantum simulation of few-body systems in nuclear physics, helium physics, physical chemistry and the physics of atom clusters.

Acknowledgement

The authors acknowledge support from the NSF-MRSEC program under No. DMR-0820054, ARO Grant No. W911NF0710576 from the DARPA OLE program and Packard foundation. N. Gemelke acknowledges support from the Grainger Foundation.

References

1. S. Inouye, M. R. Andrews, J. Stenger, H.-J. Miesner and D. M. Stamper-Kurn and W. Ketterle, *Nature* **392**, 151 (1998).
2. E. Braaten and H.-W. Hammer, *Phys. Rep.* **428**, 259 (2006).
3. C. A. Regal, M. Greiner and D. S. Jin, *Phys. Rev. Lett.* **92**, 040403 (2004).
4. M. Bartenstein, A. Altmeyer, S. Riedl, S. Jochim, C. Chin, J. Hecker Denschlag and R. Grimm, *Phys. Rev. Lett.* **92**, 120401 (2004).
5. M. W. Zwierlein, C. A. Stan, C. H. Schunck, S. M. F. Raupach, A. J. Kerman and W. Ketterle, *Phys. Rev. Lett.* **92**, 120403 (2004).
6. V. Efimov, *Phys. Lett. B* **33**, 563 (1970).
7. E. Braaten, H.-W. Hammer and T. S. Mehen, *Phys. Rev. Lett.* **88**, 040401 (2002).
8. E. Tiesinga, B.J. Verhaar, B. J. and H.T.C. Stoof, *Phys. Rev. A*, **47**, 4114 (1993).
9. N. F. Mott and H. D. W. Massey, *Theory of Atomic Collisions* (Oxford University Press, London, 1965).
10. P. S. Julienne and B. Gao, *Atomic Physics 20*, eds. C. Roos, H. Häffner and R. Blatt (AIP, Melville, New York, 2006), pp. 261–268.
11. G. F. Gribakin and V. V. Flambaum, *Phys. Rev. A* **48**, 546 (1993).
12. C. Chin, R. Grimm, E. Tiesinga and P. S. Julienne, cond-mat/0812.1496 submitted to *Rev. Mod. Phys.*
13. D. S. Petrov, *Phys. Rev. Lett.* **93**, 143201 (2004).
14. J. P. D'Incao and B. D. Esry, *Phys. Rev. Lett.* **94**, 213201 (2005).
15. B. D. Esry, C. H. Greene and J. P. Burke, *Phys. Rev. Lett.* **83**, 1751 (1999).
16. T. Kraemer, M. Mark, P. Waldburger, J. G. Danzl, C. Chin, B. Engeser, A. D. Lange, K. Pilch, A. Jaakkola, H.-C. Nägerl and R. Grimm, *Nature* **440**, 315 (2006).
17. C. Chin and V. V. Flambaum, *Phys. Rev. Lett.* **96**, 230801 (2006).
18. T. D. Lee, K. Huang, and C. N. Yang, *Phys. Rev.* **106**, 1135 (1957).
19. T. T. Wu, *Phys. Rev.* **115**, 1390 (1959).
20. S. Giorgini, J. Boronat and J. Casulleras, *Phys. Rev. A* **60**, 5129 (1999).
21. C.-L. Hung, X. Zhang, N. Gemelke and C. Chin, *Phys. Rev. A* **78**, 011604 (2008).

ATOM INTERFEROMETRY WITH A WEAKLY INTERACTING BOSE-EINSTEIN CONDENSATE

M. FATTORI[1], B. DEISSLER[1], C. D'ERRICO[1], M. JONA-LASINIO[1],

M. MODUGNO[1], G. ROATI[1], L. SANTOS[2], A. SIMONI[3], M. ZACCANTI[1],

M. INGUSCIO[1], and G. MODUGNO[1]

[1] *LENS and Dipartimento di Fisica, Università di Firenze, INFN and CNR-INFM*
Via Nello Carrara 1, 50019 Sesto Fiorentino, Italy

[2] *Institut für Theoretische Physik, Leibniz Universität, D-30167 Hannover, Germany*

[3] *Laboratoire de Physique des Atomes, Lasers, Molécules et Surfaces*
UMR 6627 du CNRS and Université de Rennes, 35042 Rennes Cedex, France

Bose-Einstein condensates have long been considered the most appropriate source for interferometry with matter waves, due to their maximal coherence properties. However, the realization of practical interferometers with condensates has been so far hindered by the presence of the natural atom-atom interaction, which dramatically affects their performance. We describe here the realization of a lattice-based interferometer based on a Bose-Einstein condensate where the contact interaction can be tuned by means of a Feshbach resonance, and eventually reduced towards zero. We observe a strong increase of the coherence time of the interferometer with vanishing scattering length, and see evidence of the effect of the weak magnetic dipole-dipole interaction. Our observations indicate that high-sensitivity atom interferometry with Bose-Einstein condensates is feasible, via a precise control of the interactions.

Keywords: Atom interferometry, Bose-Einstein condensates.

1. Introduction: atom interferometry with quantum gases

Interferometry with atoms allows one to perform measurements that are complementary to those achievable with photons: besides accelerations and rotations, atoms allow detection of gravitational, magnetic and electric forces. Many atom interferometers capable of performing high-accuracy measurements and tests of physical laws have been demonstrated in recent years.[1] The field has however not yet reached the stage of light interferometers, where coherent sources are routinely employed. Most interferometers are so far operated with ultracold but nondegenerate samples of atoms, the main requirement being a momentum spread smaller than the momentum associated to the photons that are used to manipulate the atoms. Bose-Einstein condensates have actually long been considered the most appropriate source for interferometry, due to their maximal coherence properties, and various interferometric schemes have been demonstrated in recent years,[2-7] but not yet employed in precision measurements because of the presence of atom-atom interactions. This is a serious roadblock that is not present in light interferometers: the strong nonlinearity arising from interaction degrades the performances of atom interferometers by adding a phase shift and, even worse, a decoherence term. The effects of interactions become dramatic when one wants to work with a trapped, high-density sample.

The leading contact interaction of atoms in a condensate is described by the s-wave scattering length a, which is typically of the order of 100 a_0. This translates into an energy per atom $U^c = (4\pi\hbar^2\, a\, n)/m$ that for a typical density $n = 10^{13} \text{cm}^{-1}$ is of the order of $h \times 1$ kHz. Let us now consider a two-arm, e.g. a Mach-Zehnder, interferometer. When a condensate is split into the two arms of the interferometer, the two separate condensates can be described by two superpositions of numbers states with a finite variance and with possibly different mean values. In the presence of a total interaction energy that depends quadratically on the atom number, such superpositions give rise to both a phase diffusion and a phase shift that can dramatically degrade the performance of the interferometer. In particular, the unavoidable phase diffusion[8] has a rate of the order of U^c/h, which limits the phase coherence time to few ms. This fundamental problem can be partially solved by employing the same interaction to squeeze the atom number fluctuations during the splitting procedure, as it has been recently demonstrated in experiments.[9,10] Number squeezing does however degrade the phase coherence of the condensate, and hence the ultimate sensitivity of the interferometer. In addition, other issues related to the possible phase shifts caused by fluctuations of the splitting procedure and to the recombination process are not easily solved.

One possibility is of course to use a non-interacting sample. A single-species Fermi gas provides an example of a very weakly interacting quantum gas. We recently studied the operation of a specific lattice-based interferometer based on a Fermi gas, which appeared to perform much better than an analogous interferometer based on a standard interacting condensate.[11] The Fermi gas however suffers from a rather poor phase coherence that arises from its broad momentum distribution. The best solution would therefore be to use a non-interacting Bose gas. We explore here the performance of an interferometer based on a Bose gas where the natural contact interaction has been reduced via magnetic tuning in the neighborhood of a Feshbach resonance. We employ a ^{39}K gas, which turns out to be very appropriate for this kind of application, because of the high degree of control of the s-wave scattering length around zero.[12] We find that the performance of an interferometer based on a Bose gas greatly improve when the system is brought into the weakly interacting regime. One can have at the same time a good phase coherence and a long coherence time, which result in an over-all large phase sensitivity. The control of the scattering length around zero is so good that we are able to detect the next order interaction term, i.e. the magnetic dipole-dipole interaction, which in a standard alkali condensate is more than two orders of magnitude smaller that the contact interaction. We study the interplay of the two different interactions, and find that they partially compensate each other. We finally discuss prospects for employing such a system for high-sensitivity atom-interferometry schemes.

2. Weakly interacting potassium condensate

A promising atomic system for interferometry with weakly interacting gases is potassium, which has a bosonic isotope (^{39}K) where the s-wave scattering length can be conveniently controlled through Feshbach resonances.[12,13] We have in particular focused our attention on the absolute ground state $F = 1$, $m_F = 1$. The molecular potential of ^{39}K$_2$ has a relatively deeply bound state which is responsible for a small background scattering length $a_{bg} = -29\,a_0$ and a low background three-body recombination rate $K_3 = (1.3 \pm 0.5) \times 10^{-29}$ cm^{6}s^{-1}. A molecular state with a relative magnetic moment of 1.5 μ_B crosses the atomic threshold around 402 G, giving rise to a 52-G broad Feshbach resonance,[13] as shown in Fig. 1a. As a result of this, around 350 G the s-wave scattering length crosses zero, i.e. the low-temperature phase shift associated with a two body collision is zero, and the particles effectively do not interact. The slope of $a(B)$ in this region is about 0.6 a_0/G.

In a first experiment, described in more detail in Ref. 12, we have tested our capability of changing the interaction in a ^{39}K condensate by measuring the release energy and the stability properties in the magnetic field region 350–402 G where the scattering length is positive. The condensate is produced by evaporative cooling in an optical trap at 395 G, where $a = 180\,a_0$, and the magnetic field is then adiabatically tuned in about 100 ms to a different value. The trap, which has a mean frequency of 90 Hz, is then switched off and the condensate is imaged after a long ballistic expansion of 31 ms. From the rms size of the cloud we extract the kinetic energy the atoms, which in general is the sum of kinetic energy and interaction energy in the trap. The data reported in Fig. 1b clearly show how the interaction energy decreases with vanishing a, until in the zero-crossing region the energy is purely kinetic, due to the zero point motion in the trap. For sufficiently large negative a the condensate collapses, and the total energy of the system rises abruptly. Note that the finite precision of our collisional model[13] allows us to predict the scattering length with an accuracy of about 0.4 a_0 around 350 G (error bars in the inset).

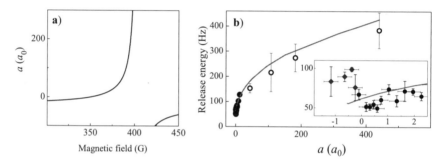

Fig. 1.   Control of the interaction in a ^{39}K condensate. a) The broad Feshbach resonance in the absolute ground state employed in this experiment. b) Release energy per atom in the condensate, as extracted from the rms size of the cloud after a long free expansion. The increase in energy for negative scattering lengths signals a collapse of the condensate. The continuous line is the prediction of the Gross-Pitaevskii theory.

3. Control of the interaction in an atom interferometer

The interferometer we adopted is based on a multiple well scheme[3,11] realized with an optical lattice. This kind of interferometer is particularly sensitive to interactions, because it is hard to avoid a large number unbalance between the several arms. The condensate is adiabatically loaded into a sinusoidal potential with period $\lambda/2$, realized with an optical standing wave of wavelength λ. In the presence of an external force F (which in our case is gravity), the macroscopic wavefunction ψ of the condensate can be described as a coherent superposition of Wannier-Stark states ϕ_i,[14] parametrized with the lattice site index i, characterized by complex amplitudes of magnitude $\sqrt{\rho_i}$ and phase θ_i, $\psi = \sum_i \sqrt{\rho_i} exp(j\theta_i)\phi_i$.[17] In the absence of interaction the phase of each state evolves according to the energy shift induced by the external potential, i.e. $\theta_i = F\lambda it/2$. By releasing the cloud from the lattice, the macroscopic interference between different Wannier-Stark states gives rise to the well known Bloch oscillations of the density pattern, with period $t_{bloch} = 2h/F\lambda$. A measurement of the frequency of such oscillations allows a direct measurement of the external force.

The presence of interactions and of the spatial inhomogeneity of the trapped condensate shifts the energy of individual Wannier-Stark states, as shown in the cartoon in Fig. 2(a). Generally, interactions give rise to a complex system of non linear equations for ρ_i and θ_i. In the weakly interacting limit the ρ_i don't change and, in addition to θ_i, extra phase terms θ'_i, proportional to the local interaction energy, are accumulated, i.e. $\theta'_i \propto g\rho_i t/h$, where g is the interaction strength. This causes a phase broadening, hence a progressive destruction of the interference pattern, as shown by the absorption images of a condensate with $a = 100\,a_0$ in Fig. 2(b). After two Bloch periods the interference pattern is drastically broadened. Fig. 2(b) shows also an analogous experiment performed with a very weakly interacting condensate with $a = 1\,a_0$. No broadening is discernible on the same time scale.

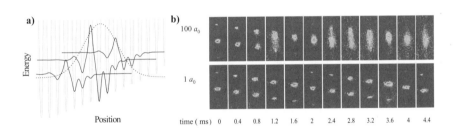

Fig. 2. (a) modification of the regular Wannier-Stark ladder, due to the interaction energy in an inhomogeneous condensate (only every third state is shown, for clarity). (b) Bloch oscillations of a Bose gas with standard interaction strength, $a = 100\,a_0$ and of a weakly interacting Bose gas, $a = 1\,a_0$.

To analyze quantitatively the effect of the interactions on the dephasing of the interferometer we repeated the same experimental sequence for different a, measuring the width of the central peak at integer times of the Bloch period. Initially the widths increase linearly with time, as a direct consequence of the phase terms $\theta'_i \propto g\rho_i t/h$, with $\rho_i =$const.[17] Later on, when the momentum distribution of the condensate occupies the whole first Brillouin zone, the widths saturate. In Fig. 3 we compare the measured decoherence rate, defined as the the slope at short times of the width of the central interference peak in units of $2\hbar k$, with theory. The theoretical curve is derived from a numerical calculation of ρ_i and the analysis described in.[17]

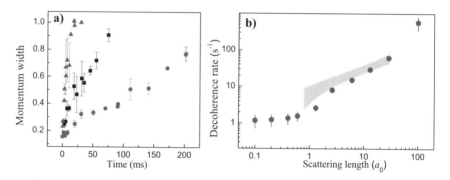

Fig. 3.   Decoherence in the interferometer with tunable interactions. (a) Time evolution of the momentum width (in units of $2\hbar k$) for three different values of a: 29 a_0 (triangles), 6 a_0 (squares), 2.5 a_0 (circles). (b) Decoherence rate, defined as the slope of the curves in a), vs the scattering length. The grey region is the theoretical prediction of the model in Ref. 17. Below 1 a_0 the decoherence rate is dominated by laser noise.

The experimental data feature an almost linear decrease of the decoherence rate with decreasing a. The rate passes from about 500 s^{-1} for $a = 100\,a_0$ to about 2 s^{-1} for $a = 1\,a_0$. Below 1 a_0 we find that noise in the lattice laser starts to significantly contribute to the decoherence, preventing a quantitative comparison of the observation with theory. We have however further investigated the effect of interactions on the decoherence of the interferometer around the zero crossing. In this region we have used a cloud dense enough to make the effect of interactions visible but sufficiently diluted in order to exclude the effect of three body losses and to prevent the condensate from collapse for small negative a.[12] A condensate with $a = a_{in} \neq 0$ is initially prepared. Right after the beginning of Bloch oscillations the external magnetic field is tuned to a final value in 2 ms. This value is kept constant for 180 ms of Bloch oscillations and for 12 ms of expansion from the lattice. The width of the interference peak reveals a minimum at 350.0(1) G, close to the expected position of the zero crossing, 350.4(4) G. The symmetric trend of the data confirms that the decoherence depends on the magnitude and not on the sign of the scattering length. More details on the experiment can be found in Ref. 15. Note that an analogous experiment has been performed with Cs atoms[16] in Innsbruck.

4. Interplay of contact and magnetic dipole-dipole interactions

In this magnetic field region we have $a \approx 0.1\,a_0$, which corresponds to a contact energy so small that the next order interaction term, i.e. the magnetic dipole-dipole interaction, can start to contribute significantly.[18] We have actually investigated the role of such a dipolar interaction in further experiments.[19] The two-body dipolar interaction has a long-range and anisotropic nature:

$$U^d(\vec{r}) = \frac{\mu_0\mu^2}{4\pi}\left(\frac{1 - 3\cos^2\theta}{r^3}\right), \tag{1}$$

where μ is the magnetic moment of the atoms ($\mu \approx 0.95\,\mu_B$ for ground-state ^{39}K atoms at 350 G, with all dipoles aligned along the Feshbach field), and θ is the angle between $\hat{\mu}$ and $\hat{r}$. For two atoms separated by $r = 0.1\,\mu m$ the interaction strength is about 10 Hz, which is comparable to the strength of the contact interaction for $a = 0.1\,a_0$. If one notes that from Eq. (1) two dipoles aligned with $\theta = 0$ attract each other with a strength that is double the repulsion they experience for $\theta = \pi/2$, it is easy to evaluate the character of the dipolar interaction for our experimental configuration. As summarized in the cartoon in Fig.4a, if we model the condensate in each lattice site as a quasi-2D pancake, the interaction within each site is purely repulsive. A weaker attractive contribution however comes from distant sites, since for them $\theta \approx \pi/2$. Note that the total dipolar interaction will change along the extension of the condensate, hence giving rise to a decoherence term in analogy with the contact interaction. One could therefore expect that the minimum of decoherence we find in the experiment corresponds to a slightly negative value of a, for which an attractive contact interaction compensates the repulsive dipole interaction. To test this conjecture, we have repeated the experiment in a different geometry, where now the lattice axis is oriented perpendicular to the magnetic field, as depicted in Fig.4b. The on-site dipolar interaction is now attractive, while the small correction from distant sites is repulsive. One could therefore expect that a minimum of decoherence should now be found for a slightly positive value of a. This is actually what we observe in the experiment, when we drive Bloch oscillations with an appropriate magnetic field gradient.

We have developed a simple model for dipolar interaction in a deep lattice to quantitatively check the observations. The model is based on a

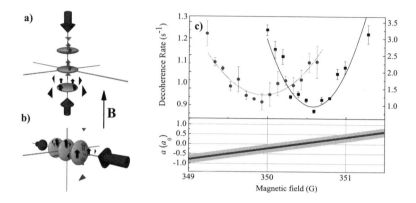

Fig. 4. Decoherence of the interferometer around the zero crossing due to the interplay of contact and dipolar interactions. The character of the dipolar interaction depends on the relative orientation of the lattice and magnetic field: (a) prevalently repulsive interaction; (b) prevalently attractive interaction. (c) The width of the momentum distribution after a few 100 ms of Bloch oscillations shows a minimum when the contact interaction compensates the dipolar one: circles are for the case (a), while squares are for the case (b). The lines are parabolic fits to the data, which constrain the position of the zero-crossing in a comparison with theory (black region in the lower panel) better than Feshbach spectroscopy (gray region).

discrete, non-linear Schrödinger equation that assumes a Gaussian density distribution of each condensate trapped in individual sites of the lattice:

$$i\hbar\frac{\partial}{\partial t}\psi_j = -J(\psi_{j+1} + \psi_{j-1}) + \Delta j\psi_j + NU^c(a_s)|\psi_j|^2\psi_j +$$

$$+NU^d_{j,j}|\psi_j|^2\psi_j + N\sum_{\delta\neq0}U^d_{j,j+\delta}|\psi_{j+\delta}|^2\psi_j \qquad (2)$$

Here, J is the tunnelling energy between neighbouring lattice sites and $U^d_{j,j'}$ is the dipolar interaction energy between the lattice sites j and j'. Note that the inter-site contribution of the dipolar interaction $U^d_{j,j'\neq j}$ does not have the same spatial dependence as the on-site contribution $U^d_{j,j}$, and therefore it cannot be completely compensated by the contact interaction. The model indicates that the minimal decoherence is reached when the spatial variance of the total interaction energy (contact interaction plus on-site and inter-site dipolar interaction) is minimal. This happens close to the minimum for the total energy. The predicted positions of the two minima of decoherence for our experimental parameters are $a = -0.32\,a_0$ and $a = +0.15\,a_0$, with a residual decoherence rate of the order of 0.05 Hz at the minima. Since our uncertainty in the predicted position of the zero-crossing is large, we can only compare the separation of the two minima in experiment and theory. The experimental observation is $\Delta a = 0.34(10) \, a_0$, which is close to the theoretical value $\Delta a = 0.47 \, a_0$.

In the present experiment we cannot test the theory prediction for the residual decoherence rate due to the uncompensated dipolar interaction, because decoherence due to technical noise is one order of magnitude larger. We plan to study this fundamental limit to the interferometer's coherence with an optimized experimental apparatus in the near future. A higher sensitivity to interaction-induced decoherence would also allow to verify the presence of second-order effects that cannot be taken into account by our simple model, such as the possible formation of density waves due to the dipolar interaction. Note that in principle the dipolar interaction could be canceled by choosing a "magic angle" between the dipoles and the lattice axis $\bar{\theta} = 54.7°$. At such angle, the $(1 - 3\cos^2 \theta)$ factor for two dipoles on the lattice axis in different sites is zero, while it is ranges from $+1$ to -1 for two dipoles in the same site. Given the cylindrical symmetry of our system, both on-site and inter-site terms will therefore on average be zero.

5. Microscopic atom interferometer

Tuning the scattering length has another important consequence. During Bloch oscillations, the in-trap extension of the sample results from the spatial interference of different Wannier-Stark states ϕ_i. Therefore the size of the cloud can have at most a variation on the order of the extension l of the single ϕ_i. In our case, for $s = 6$, $l \sim 2 \, \mu$m.[14] As a consequence the spatial resolution of our interferometer depends on the initial size of the condensate if this is larger than l. Tuning a to zero allows the condensate to occupy the ground state of the trapping potential and by an appropriate choice of the external confinement we can prepare very small samples. We have verified that our condensate occupies only about 10 lattice sites, in agreement with the 4.5 μm $1/e^2$ spatial width of the ground state of our combined potential with 100 Hz trapping frequency. The possibility of reaching the ground state of the trapping potential has clearly a great importance in view of using an atom interferometer for local phase or force sensing. One example is the application of atom interferometers to the measurement of Casimir-Polder and possible non-Newtonian gravitational forces in the proximity of surfaces, where an atomic sample with size of the order of 1 μm would allow to access unexplored regions.[20]

One should keep in mind that for very small values of a the peak density of the condensate in the presence of the lattice can approach 10^{15} atoms/cm^3. Having a system with a low K_3 coefficient, like potassium, is therefore of crucial importance for interferometric applications where the observation time needs to be long. Note however that three-body recombination in a non interacting condensate only causes losses, but is not accompanied by heating.

6. Outlook

In conclusion, we have shown how the coherence time of an atom interferometer based on Bose-Einstein condensates can be greatly increased by reducing the contact interaction with an appropriate Feshbach resonance. The interaction energy U/h can be reduced from a few kHz down to the Hz level, where the weaker magnetic dipolar interaction comes into play. Our study also demonstrates that also the latter can be partially canceled by a weak contact interaction of opposite sign. We speculate that the residual sub-Hz inhomogeneous interaction energy will not be a major obstacle on the way to the realization of high-sensitivity condensate-based interferometers, since it can also be canceled by a proper choice of the system geometry. A potential issue is of course the high stability of the magnetic field gradients that are unavoidably associated with the Feshbach field. In this respect, we note that other Feshbach resonances at lower magnetic field in excited states of potassium[13] might provide an analogous fine-tuning of the scattering length with weaker requirements for the field stability.

The possibility of a dynamical tuning of the interaction we demonstrated here is of course also very appealing for the implementation of schemes aiming at sensitivities below the shot-noise. It is well known that the ultimate phase sensitivity of an atom interferometer fed with a Bose-Einstein condensate, i.e. a coherent state, scales as $1/\sqrt{N}$, where N is the number of particles. In principle it is possible to feed the interferometer with squeezed or entangled states, which should give phase sensitivities that scale as $1/N$ (for a general review, see[21]). All the various proposals that have been made in this direction rely on relatively strong interactions to create squeezed or entangled states in condensates during a preparation phase,[22] but of course the interaction must be canceled during the measurement phase. The potassium condensate is a good candidate system for future experiments in this direction.

Acknowledgements

We acknowledge useful discussions with A. Trombettoni, A. Smerzi and S. Stringari. M. Fattori is also at Museo Storico della Fisica e Centro Studi e Ricerche 'Enrico Fermi', Compendio del Viminale, 00184 Roma, Italy, and M. Modugno is also at BEC-INFM Center, Università di Trento, I-38050 Povo, Italy. This work was supported by INFN, by MIUR (PRIN 2006) and by Ente CRF, Firenze.

References

1. A. Peters, K. Y. Chung and S. Chu, *Nature* **400** 849 (1999); T. L. Gustavson, P. Bouyer and M. A. Kasevich, *Phys. Rev. Lett.* **78**, 2046 (1997); R. Battesti, P. Cladè, S. Guellati-Khèlifa, C. Schwob, B. Grèmaud, F. Nez, L. Julien and F. Biraben, *Phys. Rev. Lett.* **92**, 253001 (2004); J. B. Fixler, G. T. Foster, J. M. McGuirk and M. A. Kasevich, *Science* **315**, 74 (2007).
2. M. R. Andrews, C. G. Townsend, H.-J. Miesner, D. S. Durfee, D. M. Kurn and W. Ketterle, *Science* 275 637-641 (1997).
3. B. P. Anderson and M. A. Kasevich, *Science* **282**, 1686 (1998).
4. Y. Shin, M. Saba, T. A. Pasquini, W. Ketterle, D. E. Pritchard and A. E. Leanhardt, *Phys. Rev. Lett.* **92**, 050405 (2004).
5. T. Schumm, S. Hofferberth, L. M. Andersson, S. Wildermuth, S. Groth, I. Bar-Joseph, J. Schmiedmayer and P. Krueger, *Nat. Phys.* **1**, **57** (2005).
6. O. Garcia, B. Deissler, K. J. Hughes, J. M. Reeves and C. A. Sackett, *Phys. Rev. A* **74**, 031601(R) (2006).
7. A. Guenther, S. Kraft, C. Zimmermann and J. Fortagh, *Phys. Rev. Lett.* **98** 140403 (2007).
8. J. Javanainen and M. Wilkens, *Phys. Rev. Lett.* **78**, 4675 (1997).
9. G. B. Jo, Y. Shin, S. Will, T. A. Pasquini, M. Saba, W. Ketterle, D. E. Pritchard, M. Vengalattore and M. Prentiss, *Phys. Rev. Lett.* **98**, 030407 (2007)
10. W. Li, A. K. Tuchman, H. Chien and M. Kasevich, *Phys. Rev. Lett.* **98** 040402 (2007).
11. G. Roati, E. de Mirandes, F. Ferlaino, H. Ott, G. Modugno and M. Inguscio, *Phys. Rev. Lett.* **92** 230402 (2004).
12. G. Roati, M. Zaccanti, C. D'Errico, J. Catani, M. Modugno, A. Simoni, M. Inguscio and G. Modugno, *Phys. Rev. Lett.* **99** 010403 (2007).
13. C. D'Errico, M. Zaccanti, M. Fattori, G. Roati, M. Inguscio, G. Modugno and A. Simoni, *New J. Phys.* **9**, 223 (2007).
14. J. Zapata, A. M. Guzmán, M. G. Moore and P. Meystre, *Phys. Rev. A* **63**, 023607 (2001).
15. M. Fattori, C. D'Errico, G. Roati, M. Zaccanti, M. Jona-Lasinio, M. Modugno, M. Inguscio and G. Modugno, *Phys. Rev. Lett.* **100**, 080405 (2008).
16. M. Gustavsson, E. Haller, M. J. Mark, J. G. Danzl, G. Rojas-Kopeinig and H.-C. Nägerl, *Phys. Rev. Lett.* **100**, 080404 (2008).

17. D. Witthaut, M. Werder, S. Mossmann and H. J. Korsch, *Phys. Rev. E* **71** 036625 (2005).
18. A. Griesmaier, J. Stuhler, T. Koch, M. Fattori, T. Pfau and S. Giovanazzi, *Phys. Rev. Lett.* **97**, 250402 (2006).
19. M. Fattori, G. Roati, B. Deissler, C. D'Errico, M. Zaccanti, M. Jona-Lasinio, L. Santos, M. Inguscio and G. Modugno, *Phys. Rev. Lett.* **101**, 190405 (2008).
20. S. Dimopoulos and A. A. Geraci, *Phys. Rev. D* **68** 124021 (2003); I. Carusotto, L. Pitaevski, S. Stringari, G. Modugno and M. Inguscio, *Phys. Rev. Lett.* **95**, 093202 (2005); J. Obrecht, R. Wild, M. Antezza, L. P. Pitaevskii, S. Stringari and E. Cornell, *Phys. Rev. Lett.* **98**, 063201 (2007).
21. V. Giovannetti, S. Loyd and L. Maccone, *Science* **306**, 1330 (2004).
22. P. Bouyer and M. A. Kasevich, *Phys. Rev. A* **56**, R1083 (1997); L. Pezzè, L. A. Collins, A. Smerzi, G. P. Berman and A. R. Bishop, *Phys. Rev. A* **72**, 043212 (2005).

AN OPTICAL PLAQUETTE: MINIMUM EXPRESSIONS OF TOPOLOGICAL MATTER

B. PAREDES*

Institut fur Physik, University of Mainz,
Mainz, 55128, Germany
* E-mail: paredes@uni-mainz.de

Topological matter is an unconventional form of matter: it exhibits a global hidden order which is not associated with the spontaneous breaking of any symmetry. The defects of this exotic type of order are anyons, quasiparticles with fractional statistics. Moreover, when living on a surface with non-trivial topology, like a plane with a hole or a torus, this type of matter develops a number of degenerate states which are locally indistinguishable and could be used to build a quantum memory naturally resistant to errors. Except for the fractional quantum Hall effect there is no experimental evidence as to the existence of topologically ordered phases, and it remains a huge challenge to develop theoretical techniques to look for them in realistic models and find them in the laboratory. Here we show how to use ultracold atoms in optical lattices to create and detect different instances of topological order in the minimum non-trivial system: four spins in a plaquette. By combining different techniques we show how to prepare these spins in mimimum versions of topical topological liquids like resonant valence bond or Laughlin states, probe their fractional quasiparticle excitations, and exploit them to build a mini-topological quantum memory.

Keywords: Topological matter; plaquette; anyons; optical lattices.

1. Introduction

Strong correlations between particles can lead to unconventional states of matter that break the traditional paradigms of condensed matter physics.[1] Among these exotic phases, topological liquids[2] are at the frontier of current theoretical and experimental research. They are disordered states that do not break any symmetries when cooled to zero temperature. Surprisingly, they exhibit some kind of exotic order, dubbed topological order,[2,3] which cannot be understood in terms of a local order parameter. This global hidden pattern is revealed in the peculiar behavior both of the ground state, with a degeneracy that depends on the topology of the system, and of the elementary excitations, which are anyons with fractional statistics.[4]

The interest in topological liquids started in connection with two landmark phenomena in condensed matter physics: the fractional quantum Hall effect[5] and high temperature superconductivity.[6] In fractional quantum Hall systems electrons organize themselves in topological liquids, like the Laughlin state,[7] following a global pattern that cannot be locally destroyed. High temperature superconductivity was proposed by Anderson[8] to occur when doping a topological spin liquid: a resonating valence bond (RVB) state in which the system fluctuates among many singlet bond configurations.[8,9] Recently, the study of topological states of matter has received special attention in the context of topological quantum computation,[3,10] which seeks to exploit them to encode and manipulate information in a manner which is resistant to errors. Moreover, understanding topological order may help us to understand the origin of elementary particles. According to Wen's theory,[11] fundamental particles, like photons and electrons, may be indeed collective excitations that emerge from a topologically ordered vacuum, a string-net condensate.[11]

Except for the fractional quantum Hall effect, there is no experimental evidence as to the existence of topologically ordered phases. It remains a huge challenge to develop theoretical techniques to look for topological liquids in realistic models and find them in the laboratory. In this direction, artificial design of topological states in the versatile and highly controllable atomic systems in optical lattices[12] appears to be a very promising possibility.[13–19]

Here we show how to use ultracold atoms in optical lattices to create and detect different instances of topological order in the minimum non-trivial lattice system: four spins in a plaquette. Using a superlattice structure[20–24] it is possible to devise an array of disconnected plaquettes, which can be controlled and detected in parallel. When the hopping amplitude between plaquette sites is very small, atoms are site localized and the physics is governed by the remaining spins. By combining different techniques we show how to prepare these spins in minimum versions of topical topological liquids like resonant valence bond states or Laughlin states, probe their fractional quasiparticle excitations and exploit them to build a mini-topological quantum memory.

We will start by briefly discussing the concept of topological order and its manifestations within the example of fractional quantum Hall systems. Next we will show that a plaquette can exhibit the two marks of topological order, namely, existence of locally indistinguishable ground states and any-onic excitations. In particular we will show the existence of two degenerate ground states and discuss how they could be used to encode a qubit in a manner resistant to local errors. We will then present experimental schemes with atoms in a superlattice optical structure to realize such a mini topolog-ical quantum memory. Finally, we will propose a way to design a four-body plaquette interaction, a Hamiltonian exhibiting anyonic excitations.

2. Topological order with an example

Fractional quantum Hall systems[5] are the only ones in nature in which we can find topologically ordered phases. They are systems of electrons living at the interface of two semiconductors and subject to a very strong perpendicular magnetic field. For certain magic ratios $\nu = 1/m$ between the density of electrons and the magnetic flux piercing the two-dimensional sample (with m being an odd integer) the ground state of the system is a very special quantum liquid. It can be described by a wave function proposed by Laughlin,[7] which, in the case, for example, of $m = 3$ has the form:

$$\psi \propto \prod_{i<j}(z_i - z_j)^3, \qquad (1)$$

with the z's being complex coordinates in the plane. What makes this state of matter unconventional is that despite not breaking any symmetry and

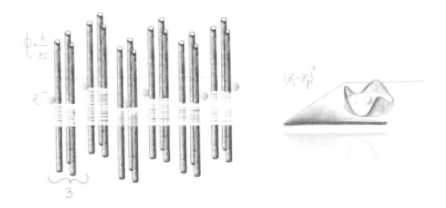

Fig. 1.   Schematic for a fractional quantum Hall liquid at $\nu = 1/3$. The system behaves as made of composites of one electron and three flux quanta. Any pair of electrons is obliged to be in a state of relative angular momentum 3.

therefore looking completely disordered, it has, however, a hidden pattern. This pattern is associated with the fact that every pair of electrons in the state (1) is obliged to have a relative angular momentum 3. Using Wen's language[2] we can think of state (1) as a correlated dancing among the electrons where every electron has to perform three steps (the three nodes of the relative wave function) when going around any other electron, no matter what the distance between them. Organizing such a dancing with local means would be an impossible task. We would need to have a global vision of the system to make sure that every electron is obeying the dancing rules. It is this hidden pattern with global character what we called topological order.

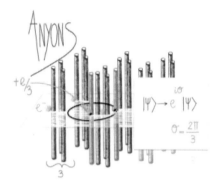

Fig. 2.   Schematic for anyons in a Laughlin liquid with $\nu = 1/3$. A quasihole with charge $e/3$ is created when adding a single flux quantum. The composite objects flux plus fractional charge are anyons with fractional statistics.

How could we detect this type of order? How can we learn about this number 3 characterizing the dancing pattern? There are two manifestations of topological order. The first one is its defects, which are fractional quasiparticles with fractional statistics. If we imagine piercing the sample with a single quantum of flux the system will feel it as if 1/3 of an electron was missing (see Fig. 2). This fractional quasihole will pick up a fractional phase when surrounding the flux quantum of another quasiparticle. The other manifestation, which is indeed a consequence of the existence of anyons, is more subtle. When the system is put onto a surface with nontrivial topology like, for example, a plane with a hole, or a torus, the ground state is degenerate with a degeneracy equal to 3 times the genus (number of handles) of the surface. This can be understood intuitively in the following way. Starting with a state like (1) in a plane with a hole (see Fig. 3), let us pierce the hole with a quantum of flux. The energy of the system does not change, however, the state is different. The dancing pattern of the new state has an additional rule: the system as a whole has to make one step when surrounding the hole. The center of mass has a unit of angular momentum. By adding a second flux to the hole we obtain a third distinct state. Adding a third one, however, puts as back to the initial pattern: the system will feel that a new dancer has entered into the hole, but the dancing rules will be the same as in the initial state. These three states look locally identical, but they are however topologically different. In order to distinguish them we would need to make a complete turn around the hole, a highly non-local operation.

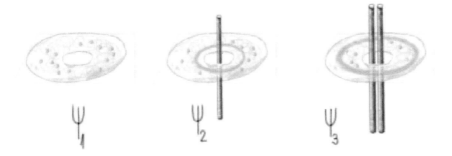

Fig. 3. Schematic of the three degenerate ground states of a fractional quantum Hall system at $\nu = 1/3$ in a plane with a hole. They correspond to different excitations of the center of mass and cannot be distinguished locally. To go from one state to the other one has to pierce a quantum of flux through the hole, an operation which is a highly non-local.

3. Topological order in a plaquette

In the following we will show that a plaquette, that is, a lattice with four sites and four spins is a minimum system in which we can find the manifestations of topological order that we have described above, namely, the existence of degenerate ground states that cannot be locally distinguishable and anyons. The fundamental reason behind the fact that we can find the same properties as in a macroscopic system is topological: a plaquette has the same topology as a plane with a hole (Fig. 4).

Considering a plaquette is motivated by recent experiments with optical

Fig. 4.   Topological equivalence between a plane with a hole and a plaquette.

superlattices in which an array of independent double wells can be created with full control over the parameters of the double well. In a similar way, we can envision a situation in which two superlattices are created in the two perpendicular directions so that an array of independent plaquettes is created (see Fig. 5).

Fig. 5.   Schematic of optical lattice setup. Using an optical superlattice configuration along two orthogonal lattice directions, an array of decoupled plaquettes can be created. By controlling the two optical superlattices independently, different potential biases, Δ_x and Δ_y, can be introduced along the x and y direction, leading to different site energy offsets μ_i as well as different vibrational level splittings at the lattice sites.

3.1. *Topological degeneracy and protected qubit in a plaquette*

A plaquette is the minimum system in which we can write down a Hamiltonian with two degenerate ground states which are locally indistinguishable. To show this let us consider the following Hamiltonian:

$$H = \sum_{ij} X_{ij} = (\vec{S}_1 + \vec{S}_2 + \vec{S}_3 + \vec{S}_4)^2, \tag{2}$$

consisting of the sum of all possible exchange operators

$$X_{ij} = \frac{1}{2} + 2\vec{S}_i \cdot \vec{S}_j \tag{3}$$

between two spins in the plaquette, which is identical to the square of the total spin operator of the system. We will be only interested in this Hamiltonian for theoretical purposes, to analyze the properties of its ground state. It is clear that the ground state subspace corresponds to the subspace of total singlet states, that is, states with total spin equal to zero. This subspace is doubly degenerate and is generated by the states:

$$|\mathrm{B}_x\rangle = s_{1,2}^\dagger s_{4,3}^\dagger |0\rangle \tag{4}$$

$$|\mathrm{B}_y\rangle = s_{2,3}^\dagger s_{1,4}^\dagger |0\rangle. \tag{5}$$

Here, the operator $s_{i,j}^\dagger$ creates a singlet between the spins on sites i and j, so that states (4) and (5) consist of two singlets along the x and y directions, respectively. The states (4) and (5) are not orthogonal. Let us consider the orthogonal basis:

$$|\Psi_-\rangle = |\mathrm{B}_x\rangle - |\mathrm{B}_y\rangle \tag{6}$$

$$|\Psi_+\rangle = \frac{1}{\sqrt{3}}\left(|\mathrm{B}_x\rangle + |\mathrm{B}_y\rangle\right). \tag{7}$$

These states are the ones we are looking for (Fig. 6). They are disordered states that do not break any symmetries: as total singlets, they are both spin rotationally invariant, and additionally, they are also invariant under spatial rotations of the plaquette. They are locally indistinguishable, since any local measurement, which will be generally described by an operator of the form $\sum_{i,\alpha} b_i^\alpha S_i^\alpha$ (with b_i^α arbitrary coefficients), will give the same output for both states:

$$\langle\Psi_-|S_i^\alpha|\Psi_-\rangle = \langle\Psi_+|S_i^\alpha|\Psi_+\rangle = 0. \tag{8}$$

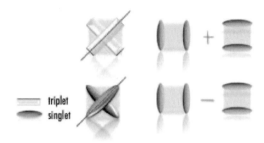

Fig. 6.   Schematic representation of the two resonating valence bond states in a plaque-tte. They have a hidden order along the diagonal bonds.

However, the states (6) and (7) are distinct. They behave differently when the plaquette is rotated by $\pi/2$: $|\Psi_+\rangle$ is symmetric (has s-wave symmetry) whereas $|\Psi_-\rangle$ is antisymmetric (has d-wave symmetry). It is interesting to note that this rotation operation, which can be written as $R_{\pi/2} = X_{12}X_{14}X_{34}$ is equivalent within the total singlet subspace we are considering to the spin exchange operator along the diagonals (see Fig. 7)

$$R_{\pi/2} \equiv X_{13} \equiv X_{24}. \tag{9}$$

This reveals a different symmetry of the two states along the diagonals: $|\Psi_-\rangle$ has singlets whereas $|\Psi_+\rangle$ has triplets. Thus, even though these states

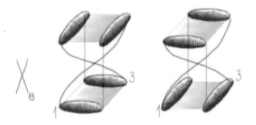

Fig. 7.   Equivalence within the total singlet subspace between exchange of spins along the diagonals X_{13} and rotation of the plaquette by an angle $\pi/2$.

do not have local order, they do exhibit a hidden order, which is as much hidden as it can be in a plaquette: it will be only revealed by a *two*-point measurement, where the *two* is for a plaquette the size of the system.

We could use this degenerate subspace to encode a qubit

$$|\Psi_-\rangle \equiv |\Downarrow\rangle \tag{10}$$

$$|\Psi_+\rangle \equiv |\Uparrow\rangle, \tag{11}$$

that will be immune to local decoherence. It is straightforward to see that the corresponding Pauli matrices are encoded in two-body operators of the form:

$$\sigma_z = X_{13} = X_{24} \tag{12}$$

$$\sigma_x = \frac{1}{\sqrt{3}}\left(X_{12} - X_{13}\right). \tag{13}$$

Therefore local errors will not affect (in first order perturbation) the state of the qubit.

We will show below how to realize this mini-quantum memory with current experimental techniques. To emphasize more the connection to a large topological system we will show first that these two states are minimum versions of the Laughlin states we discussed in the introduction.

3.2. Laughlin states in an plaquette

The topological character of the previous states becomes more explicit when we realize that they are indeed minimum versions of two indistinguishable Laughlin states. To see this, let us write the spin states (6) and (7) as states of two spin up particles in a background of spin down particles:

$$|\Psi_\pm\rangle = \sum_{x_1,x_2} \psi_\pm(x_1,x_2)S_{x_1}^+ S_{x_2}^+|\downarrow\downarrow\downarrow\downarrow\rangle. \tag{14}$$

Here S_x^+ is the spin raising operator on site $x = 1,\ldots,4$, and $|\downarrow\downarrow\downarrow\downarrow\rangle = a_{1\downarrow}^\dagger a_{2\downarrow}^\dagger a_{3\downarrow}^\dagger a_{4\downarrow}^\dagger|0\rangle$.

If we remove the background of spin down particles, that is, if we apply the operator $\sum_{i\neq j} a_{i\downarrow}a_{j\downarrow}$ to the states (14), we are left with a system of two polarized hard-core bosons with wave functions $\psi_\pm(x_1,x_2)$. These wave functions have the form

$$\psi_-(x_1,x_2) = z_1 z_2 (z_1 - z_2)^2 \tag{15}$$

$$\psi_+(x_1,x_2) = (z_1 + z_2)^2 z_1 z_2 (z_1 - z_2)^2, \tag{16}$$

where $z_i = e^{i\frac{\pi}{2}x_i}$, $x_i = 1,\ldots,4$. The state (15) is the Laughlin state for two particles (up to the gauge transformation $z_1 z_2$), whereas the state (16) is the same Laughlin state but with its center of mass excited with two units of angular momentum. We have then the same situation as for the

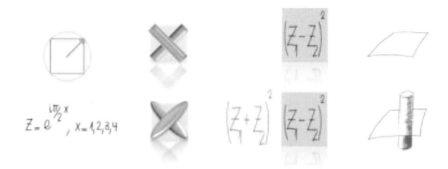

$$Z = e^{\frac{i\pi}{2}x}, \quad X = 1,2,3,4$$

Fig. 8.   Schematic of the correspondence between the two resonant valence bond states in a plaquette and two indistinguishable Laughlin states, which differ by a flux quantum piercing or not the hole.

macroscopic fractional quantum Hall system in a plane with a hole. Two indistinguishable states appear corresponding to whether or not a quantum of flux is pierced through the hole (Fig. 8).

4. Optical plaquette

We consider a system of atoms in two internal states $\sigma = \uparrow, \downarrow$, which for the case of e.g. ^{87}Rb atoms could correspond to the hyperfine states $|F = 1, m_F = +1\rangle$ and $|F = 1, m_F = -1\rangle$. The atoms are loaded into a two dimensional superlattice, which is produced by superimposing a long and a short period lattice[22] both in the x and $y$direction in such a way that an array of disconnected plaquettes is created (see Fig.5). The dynamics of atoms in a single plaquette is governed by the Hubbard Hamiltonian

$$H = -\sum_{\langle i,j\rangle,\sigma} t_{ij}(a_{i\sigma}^\dagger a_{j\sigma} + \text{H.c.}) + U\sum_{i,\sigma,\sigma'} n_{i\sigma}n_{i\sigma'} + \sum_{i,\sigma}\mu_{i\sigma}n_{i\sigma},$$

where $a_{i\sigma}$ and $n_{i\sigma}$ are, respectively, the bosonic annihilation and the particle number operator at site i and for spin σ. By controlling the superlattice structure, the tunneling amplitudes in the x and y direction, $t_x \equiv t_{12} = t_{34}$ and $t_y \equiv t_{23} = t_{14}$, can be tuned independently. Furthermore, the dependence of the offset energies $\mu_{i\sigma}$ on position and spin state can be designed using additional magnetic offsets or gradient fields. In the following we will make full use of the experimental ability to control these parameters, as already demonstrated in[22,23] for a single double well.

We are interested in a situation in which we have $N = 4$ atoms per plaquette and where the tunneling amplitudes t_x and t_y are very small in comparison to the on-site interaction energy U. Under these conditions, the particles are site localized and we are left with an effective system of four spins in a plaquette.

4.1. *Superexchange interaction*

In the optical plaquette described above the effective interaction between spins is the superexchange interaction. In the presence of a dominating interaction U, particles can only hop virtually to an already occupied site (see Fig. 9). This results in an exchange between the two spins, which is indeed nothing else but the Heisenberg interaction between them:

$$X_{ij} = \frac{1}{2} + 2\vec{S}_i \cdot \vec{S}_j \qquad (17)$$

The spin Hamiltonian that can be realized for spins in an optical plaquette

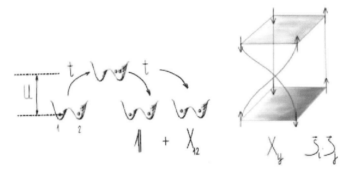

Fig. 9.   Superexchange interaction is realized by virtual tunneling to an occupied site.

is

$$H_S = J_x \left(X_{12} + X_{34} \right) + J_y \left(X_{23} + X_{14} \right), \qquad (18)$$

where the couplings $J_x = t_x^2/U$ and $J_y = t_y^2/U$ can be controlled independently both in sign and strength.

4.2. *Merging and splitting*

Additionally, the superlattice structure behind the optical plaquette allows us to merge and split double wells along the x or y direction (see Fig. 11). This will be a useful tool when creating and detecting the topological order in the plaquette.

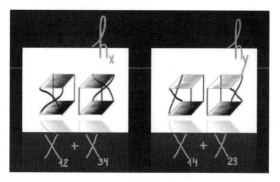

Fig. 10.   Schematic of basic Hamiltonians that can be realized in an optical plaquette by disconnecting the x or y bonds.

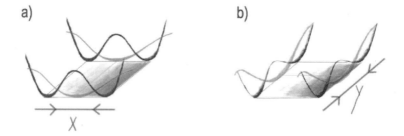

Fig. 11.   Splitting and merging double wells in the plaquette in the x and y direction.

5. Experimental realization of a mini topological quantum memory

We will show here how to realize the mini quantum memory that we introduced above using the optical plaquette described in the previous section. There are two basic Hamiltonians that can be naturally realized in the optical plaquette:

$$h_x = X_{12} + X_{34} \tag{19}$$
$$h_y = X_{23} + X_{14}, \tag{20}$$

which correspond, respectively, to a situation in which tunneling along the y or x bonds is suppressed. Combining gates based on these Hamiltonians together with merging of the wells of the plaquette we will see how writing and reading can be easily performed.

5.1. *Writing*

Let us show how to write in the plaquette a desired state of our encoded qubit. Let us assume that we have prepared the system in the state $|B_y\rangle$ in (4). As an illustrative example let us first prepare the state $|\Psi_-\rangle \equiv |\Downarrow\rangle$. It is clear that $|\Psi_-\rangle$, with singlets along the diagonals, is obtained from $|B_y\rangle$ by exchanging the two spins along one of the x bonds (see Fig. 12), so that we have:

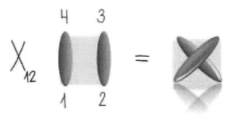

Fig. 12. By exchanging spins 1 and 2 or 3 and 4 the valence bond state is converted into the qubit down state.

$$| \Downarrow\rangle = e^{-ith_x}|B_y\rangle, \tag{21}$$

where the time scale is $t = \pi\hbar/4J_x$. In general the Pauli matrices in Eq. (13) can be generated by using the elementary Hamiltonians (19) and (20) in the following way. Within the total singlet subspace we have that $\sum_{ij} X_{ij} = 0$, since the total spin is zero, and that $X_{12}X_{34} = 1$ since we have specular symmetry. Therefore

$$X_{12} + X_{34} + X_{23} + X_{14} = X_{13} = X_{24}, \tag{22}$$

and

$$\sigma_z = h_x + h_y \tag{23}$$

$$\sigma_x = \frac{1}{\sqrt{3}}(h_x - h_y) \tag{24}$$

5.2. *Reading*

In order to distinguish the states $|\Uparrow\rangle$ and $|\Downarrow\rangle$ we have to probe their different symmetry (hidden) along the diagonals. However, the diagonal is not a natural direction for the square lattice. What we propose is to apply the operator X_{13} so that the symmetry is transferred to the vertical bonds. Then we can merge the wells along the vertical direction and monitor singlets using the band-mapping technique presented in Ref. 22.

6. Minimum lattice gauge theory in a plaquette

Lattice gauge theories[29] play an essential role in describing topological matter.[3] The building block Hamiltonian for a lattice gauge theory is the following four-body interaction among the spins of a plaquette:

$$H_\square = -J_\square \ S_1^x S_2^x S_3^x S_4^x. \tag{25}$$

We propose a way to realize this four-body interaction in a single plaquette via a two-body Hamiltonian on a time scale of the order of $J = t^2/U$. We start by considering the following two-body Hamiltonian:

$$H = J_z H_z + J_x H_x, \tag{26}$$

where

$$H_z = S_2^z S_3^z + S_3^z S_4^z + S_4^z S_1^z, \quad H_x = S_1^x S_4^x + S_2^x S_3^x. \tag{27}$$

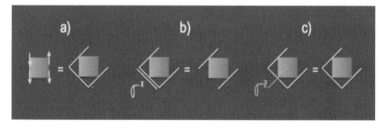

Fig. 13.   String representation of spin states and spin operators in a plaquette. A spin down is represented by a string at that site, whereas a spin up is represented by the absence of the string. The spin operator S_x is represented by a string, with the convention that two strings cancel each other. The spin operator S_z is represented by a string along the opposite diagonal, with the convention that two strings crossing each other give a minus sign.

The Hamiltonian H_z is engineered to give an appropriate energy distribution of the plaquette spin states.

As we can see in Fig. 14, every state is degenerate with its completely flipped one, but far away in energy from any state to which the perturbation

Hamiltonian H_x would couple it. If the coupling J_x is much smaller than J_z we have that H_x can only produce virtual transitions and end up in the completely flipped state. Effectively the Hamiltonian of the system is (25) with $J_\square = J_x^2/J_z$. By locally addressing the sites of the plaquette we can create anyonic excitations and test their fractional statistics as we have proposed in Ref. 31.

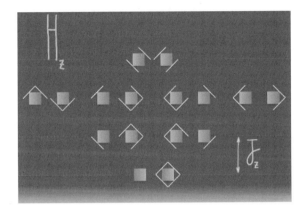

Fig. 14.   Spin states of a plaquette ordered by energy of the Hamiltonian H_z.

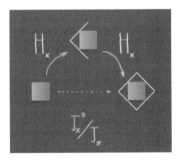

Fig. 15.   Effective plaquette interaction (closed string operator) via virtual processes.

References

1. F. Alet, A. M. Walczak and M. P. A. Fisher, *Physica A* **369**, 122 (2006).
2. X.-G. Wen, *Quantum Field Theory of Many-Body Systems* (Oxford University Press, Oxford, 2004).

3. S. Das Sarma, M. Freedman, C. Nayak, S. H. Simon and A. Stern, arXiv:0707.1889 (2007).

4. F. Wilczek, *Phys. Rev. Lett.* **49**, 957 (1982).

5. See, for example, S. Das Sarma and A. Pinczuk (eds.), *Perspectives in Quantum Hall Effect* (Wiley, New York, 1996).

6. S. Sachdev, *Rev. Mod. Phys.* **75**, 913 (2003).

7. R. B. Laughlin, *Phys. Rev. Lett.* **50**, 1395 (1983).

8. P. W. Anderson, *Science* **235**, 1196 (1987).

9. S. A. Kivelson, D. S. Rokhsar and J. P. Sethna, *Phys. Rev. B* **35**, 8865 (1987).

10. A. Kitaev, *Ann. Phys. (NY)* **303**, 2 (2003).

11. M. Levin and X.-G. Wen, *Rev. Mod. Phys.* **77**, 871 (2005), *Phys. Rev. B* **71**, 045110 (2005).

12. I. Bloch, J. Dalibard and W. Zwerger, *Rev. Mod. Phys.* (in press).

13. A. Micheli, G. K. Brennen and P. Zoller, *Nat. Phys.* **2**, 341 (2006).

14. S. Trebst, U. Schollwöck, M. Troyer and P. Zoller, *Phys. Rev. Lett.* **96**, 250402 (2006).

15. C. Zhang, V. W. Scarola, S. Tewari and S. Das Sarma, *Proc. Natl. Acad. Sci. USA* **104**, 18415 (2007).

16. L.-M. Duan, E. Demler and M. D. Lukin, *Phys. Rev. Lett.* **91**, 090402 (2003).

17. L. Santos, M. A. Baranov, J. I. Cirac, H.-U. Everts, H. Fehrmann and M. Lewenstein, *Phys. Rev. Lett.* **93**, 030601 (2004).

18. A. S. Sorensen, E. Demler and M. D. Lukin, *Phys. Rev. Lett.* **94**, 086803 (2005).

19. H. P. Büchler, M. Hermele, S. D. Huber, M. P. Fisher and P. Zoller, *Phys. Rev. Lett.* **95**, 040402 (2005).

20. J. Sebby-Strably, M. Anderlini, P. S. Jessen and J. V. Porto, *Phys. Rev. A* **73**, 033605 (2006).

21. M. Anderlini, P.J. Lee, B. L. Brown, J. Sebby-Strabley, W. D. Phillips and J. V. Porto, *Nature* **448**, 452 (2007).

22. S. Fölling, S. Trotzky, P. Cheinet, M. Feld, R. Saers, A, Widera, T. Müller and I. Bloch, *Nature* **448**, 1029 (2007).

23. S. Trotzky *et al.*, *Science* **319**, 295 (2008).

24. J. Sebby-Strabley, B. L. Brown, M. Anderlini, P. J. Lee, W. D. Phillips and J. V. Porto, *Phys. Rev. Lett.* **98**, 200405 (2007).

25. A. M. Rey, V. Gritsev, I. Bloch, E. Demler and M. Lukin, *Phys. Rev. Lett.* **99**, 140601 (2007).

26. V. Kalmeyer and R. B. Laughlin, *Phys. Rev. B* **39** 11879 (1989); R. B. Laughlin, *Science* **242**, 525 (1988).

27. E. Altman and A. Auerbach, *Phys. Rev. B* **65**, 104508 (2002).

28. H. Yao, W.-F. Tsai and S. A. Kivelson, *Phys. Rev. B* **76**, R161104 (2007).

29. J. B. Kogut, *Rev. Mod. Phys.* **51**, 659 (1979).

30. J. Pachos, W. Wieczorek, C. Schmid, N. Kiesel, R. Pohlner and H. Weinfurter, arXiv:0710.0895.

31. B. Paredes and I. Bloch, *Phys. Rev. A* **77**, 023603(2008).

STRONGLY CORRELATED BOSONS AND FERMIONS IN OPTICAL LATTICES

I. BLOCH

Institut fur Physik, University of Mainz, 55099 Mainz, Germany
E-mail: bloch@uni-mainz.de

Interacting bosons, fermions and Bose-Fermi mixtures in optical lattices form novel model systems for the investigation of fundamental quantum many body effects. This article summarizes some of our recent work on interacting bosonic and fermionic quantum gases in optical lattices. We show how the compressibility of a fermionic quantum gas mixture can be evaluated by measuring its size vs trap confinement. The results are compared to ab-initio Dynamical Mean Field Theory (DMFT) calculations, for which we find very good agreement with the experiment. Furthermore, quantum phase diffusion is introduced as a powerful method for the measurement of the renormalized Hubbard parameters underlying most lattice models.

Keywords: Optical lattices; Quantum Gases; Strong correlations.

1. Interacting Fermions with Repulsive Interactions in Optical Lattices

Interacting fermions in periodic potentials lie at the heart of modern condensed matter physics, presenting some of the most challenging problems to quantum many-body theory. A prominent example is high-T_c superconductivity in cuprate compounds.[1] In order to capture the essential physics of such systems, the fermionic Hubbard Hamiltonian[2] has been introduced as a fundamental model describing interacting electrons in a periodic potential.[1,3] In a real solid, however, the effects of interest are typically complicated by, e.g., multiple bands and orbital degrees of freedom, impurities, and the long-range nature of Coulomb interactions, which becomes especially relevant close to a metal to insulator transition. It is therefore crucial to probe this fundamental model Hamiltonian in a controllable and clean experimental setting. Ultracold atoms in optical lattices provide such a defect-free system,[4,5] in which the relevant parameters can be independently controlled, allowing quantitative comparisons of the experiment with modern quantum many-body theories. For the case of bosonic particles,[6,7] a series of experiments carried out in the regime of the superfluid to Mott insulator transition[8–10] have demonstrated the versatility of ultracold quantum gases in this respect. Recently, experiments with two-component fermionic quantum gases have extended this to the regime of the fermionic Hubbard model with attractive and repulsive interactions.[11–13] For both bosonic and fermionic systems, the entrance into a Mott insulating state is signaled by a vanishing compressibility, which can in principle be probed experimentally by testing the response of the system to a change in external confinement. This is probably the most straightforward way to identify the interaction-induced Mott insulator and to distinguish it, e.g., from a disorder induced Anderson insulator.[14–16]

Here we report on our work on non-interacting and repulsively interacting spin mixtures of fermionic atoms deep in the degenerate regime in a three-dimensional optical lattice. In the experiment, we are able to independently vary the interaction strength between the fermions using a Feshbach resonance, as well as the lattice depth and the external harmonic confinement of the quantum gas. By monitoring the in-trap density distribution of the fermionic atoms for increasing harmonic confinements, we directly probe the compressibility of the many-body system. This measurement allows us to clearly distinguish compressible metallic phases from globally incompressible states and reveals the strong influence of interactions on the density distribution. For non-interacting clouds the system changes continuously from a purely metallic state into a globally incompressible band-insulating state with increasing confinement. For repulsive interactions, we find the cloud size to be significantly larger than in the non-interacting case, indicating the resistance of the system to compression. For strong repulsion, the system evolves from a metallic state into a Mott insulating state and eventually a band insulator as the compression increases. In previous experiments, a suppression of the number of doubly occupied sites was demonstrated for increasing interaction strength for bosons[17] and fermions[12] at fixed harmonic confinement and used as an indicator to show that the system had entered a strongly interacting quantum phase.

The experimentally observed density distributions are compared to numerical calculations using Dynamical Mean Field Theory (DMFT).[18–21] DMFT is a central method of solid state theory and is widely used to obtain ab-initio descriptions of strongly correlated materials.[19]

1.1. *Hubbard Hamiltonian and trapping effects*

Restricting our discussion to the lowest energy band of a simple cubic 3D optical lattice, the fermionic quantum gas mixture can be modeled via the Hubbard-Hamiltonian[2] together with an additional term describing the potential energy due to the underlying harmonic potential:

$$
\hat{H} = -J \sum_{\langle i,j \rangle, \sigma} \hat{c}_{i,\sigma}^\dagger \hat{c}_{j,\sigma} + U \sum_i \hat{n}_{i,\downarrow} \hat{n}_{i,\uparrow}
$$
$$
+ V_t \sum_i (i_x^2 + i_y^2 + \gamma^2 i_z^2)(\hat{n}_{i,\downarrow} + \hat{n}_{i,\uparrow}). \tag{1}
$$

Here the indices i, j denote different lattice sites in the three-dimensional system ($i = (i_x, i_y, i_z)$), $\langle i,j \rangle$ neighboring lattice sites, $\sigma \in \{\downarrow, \uparrow\}$ the two different spin states, J the tunneling matrix element and U the effective on-site interaction energy. The operators $\hat{c}_{i,\sigma}(\hat{c}_{i,\sigma}^\dagger)$ correspond to the annihilation (creation) operators of a fermion in spin state σ on the ith lattice site and $\hat{n}_{i,\sigma}$ counts the number of corresponding atoms on the ith lattice site. The strength of the harmonic confinement is parameterized by the energy offset between two adjacent lattice sites at the trap center $V_t = \frac{1}{2} m \omega_\perp^2 d^2$, with $\omega_\perp = \omega_x = \omega_y \neq \omega_z$ being the horizontal trap frequency and d the lattice constant. The constant aspect ratio of the trap is denoted by $\gamma = \omega_z / \omega_\perp$. Due to the Pauli principle every lattice site can be occupied by at most one atom per spin state.

The quantum phases of the Hubbard model with harmonic confinement are governed by the interplay between three energy scales: kinetic energy, whose scale is given by the lattice bandwidth $12J$ in three dimensions, interaction energy U, and the strength of the harmonic confinement, which can conveniently be expressed by the *characteristic trap energy* $E_t = V_t(\gamma N_\sigma/(4\pi/3))^{2/3}$, which denotes the Fermi energy of a non-interacting cloud in the zero-tunneling limit with N_σ being the number of atoms per spin state ($N_\downarrow = N_\uparrow$). The characteristic trap energy depends both on atom number and trap frequency via $E_t \propto \omega_\perp^2 N_\sigma^{2/3}$ and describes the effective compression of the quantum gas, which is controlled in the experiment by changing the trapping potential.

Depending on which term in the Hamiltonian dominates, different kinds of many-body ground states can occur in the trap center (see Fig. 1). For the case of weak interactions in a shallow trap $U \ll E_t \ll 12J$ the Fermi energy is smaller than the lattice bandwidth ($E_F < 12J$) and the atoms are delocalized in order to minimize their kinetic energy. This leads to compressible metallic states with central filling $n_{0,\sigma} < 1$ (Fig. 1(a)), where

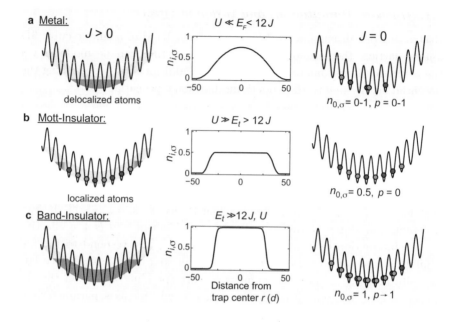

Fig. 1. Relevant phases of the Hubbard model with an inhomogeneous trapping potential for a spin mixture at $T = 0$. A schematic representation is shown in the left column. The center column displays the corresponding in-trap density profiles and the right column outlines the distribution of singly and doubly occupied lattice sites after a rapid projection into the zero tunneling limit $J = 0$, with p denoting the total fraction of atoms on doubly occupied lattice sites.

the local filling factor $n_{i,\sigma} = \langle \hat{n}_{i,\sigma} \rangle$ denotes the average occupation per spin state of a given lattice site. A dominating repulsive interaction $U \gg 12J$ and $U \gg E_t$ suppresses the double occupation of lattice sites and can lead to either Fermi-liquid ($n_{0,\sigma} < 1/2$) or Mott-insulating ($n_{0,\sigma} = 1/2$) states in the center of the trap (Fig.1(b)), depending on the ratio of kinetic to characteristic trap energy. Stronger compressions lead to higher filling factors, ultimately ($E_t \gg 12J$, $E_t \gg U$) resulting in an incompressible band insulator with unity central filling at $T = 0$ (Fig.1(c)).

Finite temperature reduces all filling factors and enlarges the cloud size, as the system needs to accommodate the corresponding entropy. Furthermore, in the trap, the filling always varies smoothly from a maximum at the trap center to zero at the edges of the cloud. For a dominating trap and a strong repulsive interaction at low temperature ($E_t > U > 12J$), the interplay between the different terms in the Hamiltonian gives rise to a wedding-cake like structure[13] consisting of a band-insulating core ($n_{0,\sigma} \approx 1$) surrounded by a metallic shell ($1/2 < n_{i,\sigma} < 1$), a Mott-insulating shell ($n_{i,\sigma} = 1/2$) and a further metallic shell ($n_{i,\sigma} < 1/2$).[20] The outermost shell remains always metallic, independent of interaction and confinement; only its thickness varies.

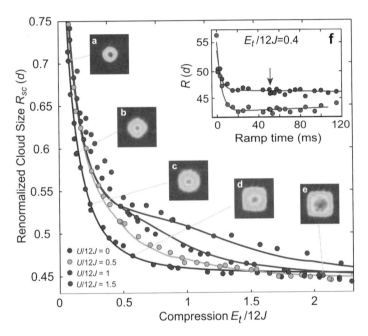

Fig. 2.   Cloud sizes of the interacting spin mixture versus compression. Measured cloud size R_{sc} in a $V_{lat} = 8\,E_r$ deep lattice as a function of the external trapping potential for various interactions $U/12J = 0$ (black), $U/12J = 0.5$ (green), $U/12J = 1$ (blue), $U/12J = 1.5$ (red)). Dots denote single experimental shots, lines the theoretical expectation from DMFT for $T/T_F = 0.15$ prior to loading. The insets (**a-e**) show the quasi-momentum distribution of the non-interacting clouds (averaged over several shots). (**f**) Resulting cloud size for different lattice ramp times at $E_t/12J = 0.4$ for a non-interacting and an interacting Fermi gas. The arrow marks the ramp time of 50 ms used in the experiment.

In the corresponding experiment, we first prepare a 50/50 spin mixture of ^{40}K in the spin states ($F = -9/2, m_F = -9/2$) and ($F = -9/2$, $m_F = -7/2$) at an initial temperature of $T/T_F \approx 0.15$ in a crossed optical dipole trap. Then a simple cubic three dimensional optical lattice potential is ramped up within 50 ms using three counterpropagating standing wave laser fields at an optical wavelength of $\lambda = 738$ nm. The red detuned crossed optical dipole trap allows for an independent control of the overall harmonic confinement of the trapped quantum gas. We use phase contrast imaging to determine the in-trap density distribution of the quantum gas mixture at different external confinements. The behaviour of the system can be quantified by plotting the cloud size R_{sc} (dots) in rescaled units as a function of the characteristic trap energy $E_t \propto \omega_\perp^2 N_\sigma^{2/3}$ (see Fig. 2). For the non-interacting case we find the cloud sizes (Fig. 2, black dots) to decrease continuously with compression until the characteristic trap energy roughly equals the lattice bandwidth ($E_t/12J \sim 1$). For stronger confinements the compressibility of the quantum gas, determined by $\partial R_{sc}/\partial E_t$, approaches zero, as almost all atoms are in the band insulating regime while the surrounding metallic shell becomes negligible. The corresponding quasi-momentum distribution (Fig. 2a-e) changes gradually from a partially filled first Brillouin zone, characteristic for a metal, to an almost evenly filled first Brillouin zone for increasing compressions, as expected for a band insulator. The measurements shown here directly demonstrate the global incompressibility of the fermionic band insulator, in excellent agreement with the theoretical expectation for a non-interacting Fermi gas (black line).

The green, blue and red dots in Fig. 2 represent the size of repulsively interacting clouds with $U/12J = 0.5, 1$ and 1.5 in comparison with the DMFT calculations (lines). For moderately repulsive interaction ($U/12J = 0.5, 1$) the cloud size is clearly bigger than in the non-interacting case but eventually reaches the size of the band insulator. For stronger repulsive interactions ($U/12J = 1.5$) we find the onset of a region ($0.5 < E_t/12J < 0.7$) where the cloud size decreases only slightly with increasing harmonic confinement, denoting a very small compressibility, whereas for stronger confinements the compressibility increases again. This is consistent with the formation of an incompressible Mott-insulating core with half filling in the center of the trap, surrounded by a compressible metallic shell.[13] For higher confinements an additional metallic core ($1/2 < n_{i,\sigma} < 1$) starts to form in the center of the trap. When the system is compressed even further, all cloud sizes approach that of a band insulating state and all compressibilities tend to zero.

Overall, we find the measured cloud sizes to be in very good quantitative agreement with the theoretical calculations up to $U/12J = 1.5$ ($B = 175$ G). Nevertheless, for repulsive interactions and medium compression ($E_t/12J \approx 0.5$) the cloud size is slightly bigger than the theoretical expectation. The discrepancies become more prominent for stronger interactions, i.e. when tuning the scattering length to even more positive values below the Feshbach resonance. This could be caused by non-equilibrium dynamics in the formation of a Mott-insulating state for strong interactions or may be an effect not covered by the simple single-band Hubbard model or the DMFT calculations and requires further investigation.

2. Quantum Phase Diffusion as a Probe for Strongly Interacting Atoms in Optical Lattices

When an initial coherent state of a Bose-Einstein condensate with N atoms is split adiabatically into two internal or external modes, the resulting atom number distribution in each of the modes becomes number squeezed as a result of the non-linear repulsive interactions between the particles. In the extreme limit of dominating repulsive interactions, this results in a Mott insulator like transition, where each mode is now populated by a Fock state with $N/2$ atoms. In case of a rapid non-adiabatic splitting process into two decoupled modes (see Fig. 3), the condensate is split into two coherent matter waves with a fixed relative phase, each condensate exhibiting a binomial atom number distribution. This binomial distribution resembles that of a coherent state and represents an out-of-equilibrium quantum state of the total two-mode many-body system. Under the influence of the interactions between the particles, each of the coherent states in the different modes will undergo a quantum phase diffusion due to the non-linear interactions between the particles.[22-28] Both number squeezed states and the quantum phase diffusion in external and internal modes have recently been revealed in several experiments.[29-35]

The dynamical evolution of the matter wave field based on quantum phase diffusion can be easily explained for the case of repulsively interacting particles and only two modes into which the condensate is split. After the rapid splitting process, each mode is populated by a superposition of different atom numbers. For simplicity, we assume each superposition to have the form of a Glauber coherent state. In order to determine how such superpositions of atom number states evolve over time, the interactions between the atoms have to be taken into account. Let us first assume that all atoms within a subsystem occupy the ground state of their external confining potential. If the interaction energy is small compared to the vibrational spacing in this potential well, the Hamiltonian governing the behavior of

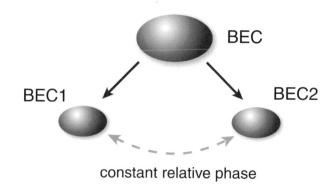

constant relative phase

Fig. 3.   A Bose-Einstein condensate is split into two parts with an initially constant phase between the two subsystems BEC1 and BEC2.

the atoms is given by:

$$H = \frac{1}{2}U\hat{n}(\hat{n}-1),$$ (2)

where U denotes the two-particle interaction energy in a single mode and $\hat{n}$ counts the number of particles in this mode.

Subsequently, under the action of this Hamiltonian, our initial coherent state $|\alpha\rangle$ will evolve according to:

$$|\alpha\rangle(t) = e^{-|\alpha|^2/2} \sum_n \frac{\alpha^n}{\sqrt{n!}} e^{-i\frac{1}{2}Un(n-1)t/\hbar}|n\rangle.$$ (3)

Here α is the amplitude and $|\alpha|^2$ corresponds to the average atom number of the coherent state (see e.g. Ref. 36).

The coherent matter wave field ψ in each of the subsystems can then be evaluated through $\psi = \langle\alpha(t)|\hat{a}|\alpha(t)\rangle$, which exhibits a series of collapses and revivals due to the quantum phase diffusion of Eq. 3. At first, the different phase evolutions of the atom number states lead to a collapse of ψ. However, at integer multiples in time of h/U, all phase factors in the above equation re-phase modulo 2π and thus lead to a revival of the initial coherent state (see also Fig. 4). Such a collapse and revival of the coherent matter wave field of a BEC is reminiscent of the collapse and revival of the Rabi oscillations in the interaction of a single atom with a single-mode electromagnetic field in cavity quantum electrodynamics.[37,38] There, the nonlinear atom-field interaction induces the collapse and revival of the Rabi oscillations whereas here the nonlinearity due to the interactions

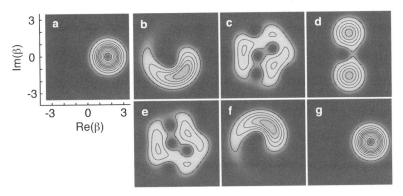

Fig. 4. Calculated quantum dynamics (represented via the Q-function) of an initial coherent state with an average number of three atoms. The dynamical evolution of the quantum state is caused by the coherent cold collisions between the atoms. The graphs show the overlap of the dynamically evolved input state with an arbitrary coherent state of amplitude β. Evolution times are **a** $0\,h/U$; **b** $0.1\,h/U$; **c** $0.4\,h/U$; **d** $0.5\,h/U$; **e** $0.6\,h/U$; **f** $0.9\,h/U$; and **g** h/U.

between the atoms themselves leads to the series of collapse and revivals of the matter wave field.

Up to four or five such collapses and revivals have been observed with ultracold bosonic atoms in an optical lattice.[30] Starting from a shallow lattice, where each lattice site is approximately occupied by a coherent state, the lattice potential was rapidly increased to a large lattice depth, for which the atomic wells were essentially decoupled and the interaction energy U dominated over the kinetic energy J. The resulting series of collapses and revivals, however, was damped due to inhomogeneities in the external confinement that precluded observing more periods of the quantum phase diffusion dynamics in the experiment. In our novel experimental setup used here, the combination of a blue detuned lattice potential together with an independent red detuned crossed optical dipole trap, has allowed us to control the harmonic confinement independent of the lattice depth. This laser combination has enabled us to realize conditions with a very weak external confinement, thus minimizing any inhomogeneities. Based on these experimental improvements, we are now able to observe up to 10 times longer time evolutions due to quantum phase diffusion. An example of such a trace can be seen in Fig. 5. In the inset of the same figure, the power spectrum of the time trace is shown, revealing three frequency components with different amplitudes that contribute to the collapse and revival dynamics. Such different frequencies can be expected if the onsite two-body interaction energy becomes dependent on the filling n on a lattice site. Within a single-band approximation, the two-particle interaction energy U is given by:

$$U = \frac{4\pi\hbar^2 a}{m} \int |w(\mathbf{x})|^4 d^3x, \qquad (4)$$

with a being the scattering length, m the mass of a single atom and $w(\mathbf{x})$ denoting the onsite Wannier function within the lowest energy band (see e.g. Ref. 5). For this case, the interaction energy U is independent of the local filling n. For interacting particles, higher energy bands can be admixed, yielding an effective broadening of the on-site wavefunction $w(\mathbf{x})$ due to the repulsive interaction between the particles. This becomes especially relevant when the interaction energy approaches the vibrational energy level splitting on a single lattice site. Such a wave function broadening will depend on the number of particles occupying a single lattice site n and essentially yields discrete values for the two-particle interaction energy: $U(2), U(3), U(4), \ldots U(n)$. These different interaction values result in the different collapse and revival frequencies observed in the experiment. Not only do they allow one to determine the effective renormalized Hubbard parameters $U(n)$, but their amplitudes also contains information on the population of the different atom number states that form the superposition state on a single lattice site. Just as in cavity QED experiments,[38,39] they can be used to reveal the atom number statistics of the atomic quantum states and are a direct proof of the quantum granularity present in the coherent matter wave field of a BEC.

Quantum phase diffusion experiments have allowed us to precisely determine the effect of interactions on the atomic wave functions in a lattice. The resulting effective renormalized Hubbard parameters can be compared to ab-initio calculations and yield valuable input for multi-band Hubbard physics which can become especially relevant for the case of Bose-Fermi mixtures in a lattice. There a strong shift of the bosonic superfluid to Mott insulator transition has been observed as fermionic atoms with attractive interactions are added to the system.[40–42] In a recent experiment we have found that the observed shift is in fact in very good agreement with a very strong renormalization of the Hubbard parameters.[42,43] Quantum phase diffusion enables one to quantitively measure these shifts in a direct way using fundamental quantum optics effects.

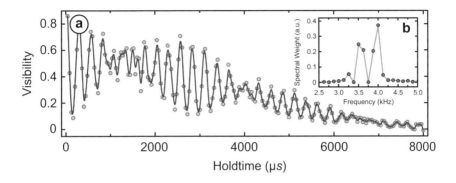

Fig. 5. Experimentally detected visibility vs hold time, exhibiting long-time collapse and revival dynamics (**a**). The different frequency components visibile in the time trace can be attributed to the discrete interaction strengths that occur for increasing bosonic fillings on a lattice site. The inset shows the power spectrum of the time trace, with three detected frequency components (**b**), caused by the different two-particle interaction energies for different on-site fillings.

Acknowledgments

I. Bloch acknowledges helpful discussions with U. Schneider, B. Paredes, A. Rosch, E. Demler and S. Trotzky and funding by the DFG, EU (SCALA, OLAQUI), AFOSR and DARPA within the OLE program.

References

1. P. Lee, N. Nagaosa and X.-G. Wen, *Rev. Mod. Phys.* **78**, 17 (2006).
2. J. Hubbard, *Proc. R. Soc. Lond. A* **276**, 238 (1963).
3. F. Gebhard, *The Mott Metal-Insulator Transition — Models and Methods* (Springer, New York, 1997).
4. D. Jaksch and P. Zoller, *Ann. Phys. (NY)* **315**, 52 (2005).
5. I. Bloch, J. Dalibard and W. Zwerger, *Rev. Mod. Phys.* **80**, 885 (2008).
6. M. P. A. Fisher, P. B. Weichman, G. Grinstein and D. S. Fisher, *Phys. Rev. B* **40**, 546 (1989).
7. D. Jaksch, C. Bruder, J. I. Cirac, C. W. Gardiner and P. Zoller, *Phys. Rev. Lett* **81**, 3108 (1998).
8. M. Greiner, M. O. Mandel, T. Esslinger, T. Hänsch and I. Bloch, *Nature* **415**, 39 (2002).
9. T. Stöferle, H. Moritz, C. Schori, M. Köhl and T. Esslinger, *Phys. Phys. Lett.* **92**, 130403 (2004).
10. I. B. Spielman, W. D. Phillips and J. V. Porto, *Phys. Rev. Lett.* **98**, 080404 (2007).
11. J. Chin, D. Miller, Y. Liu, C. Stan, W. Setiawan, C. Sanner, K. Xu and W. Ketterle, *Nature* **443**, 961 (2006).

12. R. Jördens, N. Strohmaier, K. Günter, H. Moritz and T. Esslinger, *Nature* **455**, 204 (2008).

13. U. Schneider, L. Hackermüller, S. Will, T. Best, I. Bloch, T. Costi, R. Helmes, D. Rasch and A. Rosch, arXiv:0809.1464.

14. P. W. Anderson, *Phys. Rev.* **109**, 1492 (1958).

15. J. Billy, V. Josse, Z. Zuo, A. Bernard, B. Hambrecht, P. Lugan, D. Clement, L. Sanchez-Palencia, P. Bouyer and A. Aspect, *Nature* **453**, 891 (2008).

16. G. Roati, C. D'Errico, L. Fallani, M. Fattori, C. Fort, M. Zaccanti, G. Modugno, M. Modugno and M. Inguscio, *Nature* **453**, 895 (2008).

17. F. Gerbier, S. Fölling, A. Widera, O. Mandel and I. Bloch, *Phys. Rev. Lett.* **96**, 090401 (2006).

18. A. Georges, G. Kotliar, W. Krauth and M. J. Rozenberg, *Rev. Mod. Phys.* **68**, 13 (1996).

19. G. Kotliar and D. Vollhardt, *Phys. Today* **57**, 53 (2004).

20. R. W. Helmes, T. A. Costi and A. Rosch, *Phys. Rev. Lett.* **100**, 056403 (2008).

21. L. DeLeo, C. Kollath, A. Georges, M. Ferrero and O. Parcollet, arXiv:0807.0790.

22. B. Yurke and D. Stoler, *Phys. Rev. Lett.* **57**, 13 (1986).

23. V. Buzek, H. Moya-Cessa, P. L. Knight and S. J. D. Phoenix, *Phys. Rev. A* **45**, 8190 (1992).

24. F. Sols, *Physica B* **194-196**, 1389 (1994).

25. E. M. Wright, D. F. Walls and J. C. Garrison, *Phys. Rev. Lett.* **77**, 2158 (1996).

26. A. Imamoglu, M. Lewenstein and L. You, *Phys. Rev. Lett.* **78**, 2511 (1997).

27. Y. Castin and J. Dalibard, *Phys. Rev. A* **55**, 4330 (1997).

28. J. A. Dunningham, M. J. Collett and D. F. Walls, *Phys. Lett. A* **245**, 49 (1998).

29. C. Orzel, A. K. Tuchmann, K. Fenselau, M. Yasuda and M. A. Kasevich, *Science* **291**, 2386 (2001).

30. M. Greiner, M. O. Mandel, T. Hänsch and I. Bloch, *Nature* **419**, 51 (2002).

31. G.-B. Jo, Y. Shin, S. Will, T. Pasquini, M. Saba, W. Ketterle, D. Pritchard, M. Vengalattore and M. Prentiss, *Phys. Rev. Lett.* **98**, 030407 (2007).

32. W. Li, A. K. Tuchman, H.-C. Chien and M. A. Kasevich, *Phys. Rev. Lett.* **98**, 040402 (2007).

33. P. Cheinet, S. Trotzky, M. Feld, U. Schnorrberger, M. Moreno-Cardoner, S. Fölling and I. Bloch, *Phys. Rev. Lett.* **101**, 090404 (2008).

34. J. Esteve, C. Gross, A. Weller, S. Giovanazzi and M. K. Oberthaler, *Nature* **455**, 1216 (2008).

35. A. Widera, S. Trotzky, P. Cheinet, S. Fölling, F. Gerbier, I. Bloch, V. Gritsev, M. D. Lukin and E. Demler, *Phys. Rev. Lett.* **100**, 140401 (2008).

36. D. F. Walls and G. J. Milburn, *Quantum Optics* (Springer, Berlin, 1994).

37. G. Rempe, H. Walther and N. Klein, *Phys. Rev. Lett.* **58**, 353 (1987).

38. M. Brune, F. Schmidt-Kaler, A. Maali, J. Dreyer, E. Hagley, J. M. Raimond and S. Haroche, *Phys. Rev. Lett.* **76**, 1800 (1996).

39. S. Haroche and J.-M. Raimond, *Exploring the Quantum* (Oxford University Press, 2006).

40. C. Ospelkaus, S. Ospelkaus, L. Humbert, P. Ernst, K. Sengstock and K. Bongs, *Phys. Rev. Lett.* **97**, 120402 (2006).
41. K. Günter, T. Stöferle, H. Moritz, M. Kohl and T. Esslinger, *Phys. Rev. Lett.* **96**, 180402 (2006).
42. T. Best, S. Will, U. Schneider, L. Hackermüller, D.-S. Lühmann, D. van Oosten and I. Bloch, arXiv:0807.4504.
43. D.-S. Lühmann, K. Bongs, K. Sengstock and D. Pfannkuche, *Phys. Rev. Lett.* **101**, 050402 (2008).

LASER COOLING OF MOLECULES

P. PILLET[1*], M. VITEAU[1], A. CHOTIA[1], D. SOFIKITIS[1], M. ALLEGRINI[1,2],

N. BOULOUFA[1], O. DULIEU[1] and D. COMPARAT[1]

[1]*Laboratoire Aimé Cotton, CNRS, Université Paris-Sud,
Building 505, 91405 Orsay cedex, Fance*
[2]*Dipartimento di Fisica "E.Fermi"- Università di Pisa
Largo Pontecorvo 3, 56127 PISA, Italy*
* *E-mail: pierre.pillet@lac.u-psud.fr*
www.lac.u-psud.fr

The cold-molecules field is very active trying to transfer laser cooling techniques from atoms to molecules. Photoassociation of cold atoms, followed by spontaneous emission of the electronically excited molecules, produces translationally cold molecules, but in several vibrational levels v of the ground state. We have recently shown that vibrational cooling can be obtained by optical pumping with a shaped broadband femtosecond laser. The broadband laser electronically excites the molecules, leading via a few absorption - spontaneous emission cycles to a redistribution of the vibrational population in the ground state. By removing the laser frequencies corresponding to the excitation of the $v = 0$ level, a dark state is produced by the so-shaped laser, yielding with successive laser pulses an accumulation of the molecules in the $v = 0$ level.

Keywords: Cold molecules, photoassociation of cold atoms, optical pumping, laser cooling, shaped femtosecond laser.

1. Introduction

Full control of the dynamics of a quantum system is crucial in both physics and chemistry.[1] For atoms, precise control of both internal and external degrees of freedom has been achieved thereby opening fascinating new fields.[2] Extension to molecules is not straightforward, but the impetus to prepare robust samples of trapped ultracold ground-state molecules with neither vibration nor rotation is strong. Indeed, significant advances are expected[3] in molecular spectroscopy, molecular clocks, fundamental tests of physics, super or controlled photo-chemistry, and also in quantum computation based on polar molecules. Slowing pre-existing molecules, for instance by buffer-gas cooling or supersonic beam deceleration, typically delivers molecules with translational temperatures down to a few millikelvins.[4] Temperatures in the sub-millikelvin range can only be achieved starting from cold atoms.[5,6] In quantum degenerate gases, a magneto-association step via Feshbach resonances followed by adiabatic population transfer was recently found to successfully form ultracold molecules in a single deeply bound level.[7,8] Photoassociation (PA) of ultracold atoms from a standard magneto-optical trap (MOT) is a well known efficient process to produce ultracold molecules with a rate as high as $10^6 - 10^7$ s^{-1}. Among the current methods used to prepare dense and ultracold samples of molecules,[9] here we report on photoassociation (PA) of laser cooled Cs atoms. In particular, we demonstrate efficient production of deeply-bound cesium dimers and vibrational cooling in the $v = 0$ level of the ground electronic single state. For cesium, the photoassociation process corresponds to the reaction

$$Cs(6s, F) + Cs(6s, F) + h\nu_L \longrightarrow Cs_2^*(\Omega\,(6s + 6p_j)\,; v, J) \qquad (1)$$

Two colliding atoms in a hyperfine level F of their $6s$ ground state absorb one laser photon at frequency $h\nu_L$ red-detuned from the atomic resonance frequency $(6s + 6p_j, j = 1/2$ or $3/2)$ and form a molecule in a well defined rovibrational level (v, J) of an excited molecular state Ω correlated to one of the asymptotes $(6s + 6p_{1/2})$ or $(6s + 6p_{3/2})$. The resolution of the PA process is limited by the width $(\sim k_B T)$ of the statistical distribution of the relative kinetic energies of the colliding atoms. For ultracold atoms, $(k_B T \sim h \times 2\,\mathrm{MHz}$ at $T \sim 100\,\mu\mathrm{K})$, it is smaller than any other relevant energy spacing of the system. Thus, cold atom PA is a powerful tool for high-resolution molecular spectroscopy. It has given access to previously unexplored domain of molecular dynamics at distances well beyond those of well-known chemical bonds. A pair of identical ground state atoms interacts at large interatomic distances R through their R^{-6} van der Waals interaction while for an excited atom pair the R^{-3} dipole-dipole interaction is dominant. Vibrational levels with a very large elongation (from a few tens up to a few hundred atomic units) are then efficiently populated by photoassociation. From a classical point of view, the ideal vibrational motion of the PA molecule should slow down in the intermediate distance range (say, around $15 - 20$ atomic units) to let spontaneous decay occur before going back towards long range. This was demonstrated[10] by using the 0_g^- and 1_u double-well potential curves correlated to the Cs$(6s)$+Cs$(6p_{3/2})$ limit, shown in Fig. 1. These peculiar states, known as pure long-range molecules, result from the competition between the spin-orbit interaction and the long-range dipole-dipole interaction. The slow $R-$variation of the left edge of the outer potential well induces a "speed bump" in the distance range appropriate for radiative decay towards stable vibrational levels.

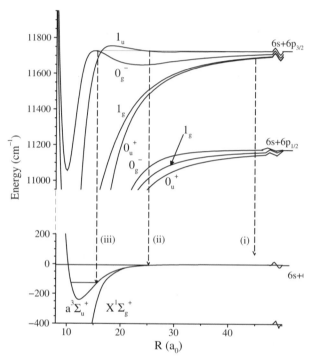

Fig. 1.   Long-range Cs_2 attractive molecular potential curves correlated to the $(6s+6s)$ and $(6s + 6p_{1/2,3/2})$ dissociation limits. The attractive 0_u^- and 2_u states, forbidden for dipole transitions, are not shown. Following spontaneous emission, the formation of cold molecules is pictured by dashed arrows (ii) and (iii), and dissociation by arrow (i).

2. Formation and detection of Cs_2 molecules

Within their short lifetime of a few tens of nanoseconds, the electronically excited molecules created by photoassociation, most often spontaneously decay back to a pair of "hot" atoms, i.e. with a large relative kinetic energy. In magneto-optical traps, the fluorescence intensity decrease due to the escape of the hot atoms from the trap provides a simple detection method. Fig. 2 displays a typical PA spectrum from a Cs MOT when the PA laser frequency (in our setup a cw Ti:Sa laser, intensity 300 W cm^{-2}, pumped by an argon-ion laser) is scanned below the $6s + 6p_{3/2}$ dissociation limit. The upper trace corresponds to the fluorescence spectrum due to trap loss, revealing rovibrational progressions for all attractive potentials which can be reached by photoassociation (cf. Fig. 1). Decays into either a pair of hot atoms or into a stable molecule induce a decrease of the trap fluorescence. The lower trace is the Cs_2^+ yield obtained by photoionization of the stable Cs_2 cold molecules created either in the ground state, $X^1\Sigma_g^+$, (through 1_u excitation) or in the lowest triplet state, $a^3\Sigma_u^+$, (through 0_g^- excitation). The electronically excited molecules formed by photoassociation have too short a lifetime to give a significant contribution to the photoionization signal. The $0_g^-(6s + 6p_{3/2})$ and $1_u(6s + 6p_{3/2})$ double-well potentials are the only states which contribute significantly to the molecular ion signal.[11,12] In the former case, the cold molecules are distributed over several rovibrational levels with binding energies in the middle of the lowest triplet state $a^3\Sigma_u^+$, while in the latter case, the populated levels are close to the dissociation limit, where the gerade (g) and ungerade (u) characters are no longer good symmetries. The Cs_2^+ ion spectrum of Fig. 2 exhibits 133 well resolved structures assigned to the vibrational progression of the $0_g^-(6s+6p_{3/2})$ outer well, starting at $v = 0$. The rotational structure, shown for $v = 10$ in the inset, is observed up to $J = 8$ for most of the vibrational levels below $v = 74$. The double-well route via the 0_g^- potential correlated to the $6s+6p_{3/2}$ limit was also demonstrated for rubidium. For other alkalies the 0_g^- outer well is still present, but located too far out to provide a Condon point in the desired distance range, and no significant formation of ultracold molecules can occur. The double-well route is an optimized compromise between an efficient photoassociation at long range and a quite reasonable branching ratio for spontaneous decay at intermediate distances.

The previous mechanism relies on the specific shape of the molecular potential curves. It is limited to Rb_2 and Cs_2 for the 0_g^- symmetry, and to Cs_2 for the 1_u symmetry. Another possibility is provided by the more general pattern of interactions in diatomic molecules, taking place when two (or more) molecular states interact together at a given interatomic distance. Such an example is given by the $0_u^+(6s+6p_{1/2})$ and $0_u^+(6s+6p_{3/2})$ coupled states in Cs_2. A detailed analysis of this mechanism[13] suggests that a resonant coupling between vibrational levels of the $0_u^+(6s + 6p_{1/2})$ and $0_u^+(6s+6p_{3/2})$ states enables formation of ultracold molecules. As with the double-well configuration, the PA excitation occurs at large distance while

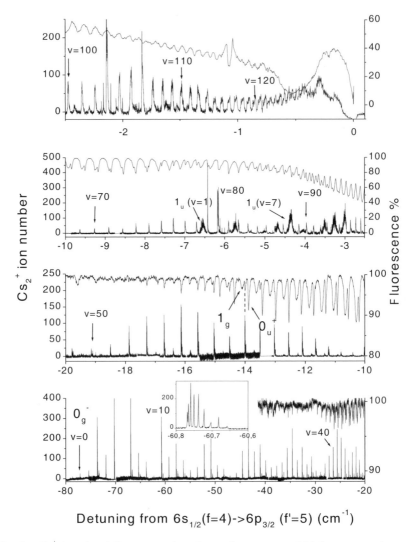

Fig. 2. Cs_2^+ ion signal (lower curve) and trap fluorescence yield (upper curve) versus detuning of the PA laser relative to the $6s + 6p_{3/2}$ dissociation limit. The rotational progression of the $v = 10$ level of the 0_g^- long-range potential well is shown in the inset. The dashed line indicates the correspondence of a vibrational level of the 0_g^- state on both spectra.

the coupling acts at short distances, favoring spontaneous emission towards stable bound levels of the $X^1\Sigma_g^+$ Cs_2 ground state.

The detection of cold molecules is made through REMPI (Resonant Enhanced MultiPhoton Ionization) detection,[11] whose sensitivity can reach one single ion.[14] The cold molecules are ionized into Cs_2^+ ions by a pulsed dye laser (7 ns duration, 1 mJ energy, focused spot $\sim 1\,mm^2$, 10 GHz resolution) pumped by the second harmonic of a Nd-YAG laser at 10 Hz repetition rate. The top of Fig. 3 shows the REMPI spectrum recorded by scanning the frequency of the pulsed REMPI laser in the range $13500 - 14500\,cm^{-1}$, when the molecules are formed via the photoassociation of the vibrational level $v = 79$, of the state $0_g^-(6s + 6p_{3/2})$. For frequencies above $13850\,cm^{-1}$, the REMPI process consists of a first resonant step corresponding to transitions between ro-vibrational levels of the lowest triplet state $a^3\Sigma_u^+$ and the $2^3\Pi_g$ state converging to the $6s + 5d$ dissociation limit. The second step corresponds to a one-photon ionization of the intermediate ro-vibrational level of the $2^3\Pi_g$ state. After the photoionization laser pulse, a pulsed high electric field (3 kV/cm, $0.5\mu s$) is applied at the trap position by a pair of grids spaced by 15 mm. Ions extracted from the photoassociation region cross a 6 cm free field zone, acting as a time-of-flight mass spectrometer to separate the Cs_2^+ ions from spurious Cs^+ ions, and are detected by a pair of microchannel plates. The global efficiency of the process is limited by the ion collection efficiency (80%) and the microchannel plate efficiency (35%). The REMPI efficiency is roughly estimated around 10% by comparing the Cs_2^+ ion signal with the trap loss fluorescence signal and by using calculated branching ratios between bound-bound and bound-free transitions for the photoassociated molecules. This relatively low rate for the REMPI process is related to its resonant character. It means that only a few initially populated ro-vibrational levels, for which the two-photon process is resonant, are efficiently ionized. For frequencies below 13850 cm^{-1}, the scan yields a well resolved spectrum (bottom of Fig. 3). The first resonant step of the REMPI process corresponds closely to transitions between ro-vibrational levels of the lowest triplet state $a^3\Sigma_u^+$ and the $2^3\Sigma_g^+$ state converging to the $6s + 5d$ dissociation limit. The analysis shows that the vibrational levels between $v = 12$ and 16 of the lowest triplet state, $a^3\Sigma_u^+$, are populated. Evidently, the detected cold molecules are vibrationally excited in the lowest triplet state. Molecules in the singlet ground state, such as those obtained via photoassociation into the $1_u(6s + 6p_{3/2})$ and $0_u^+(6s + 6p_{1/2})$ states, are also detected in highly excited vibrational levels. Thus, two questions arise: Can molecules be formed in low vibrational levels of the singlet ground state? How can such molecules be efficiently detected?

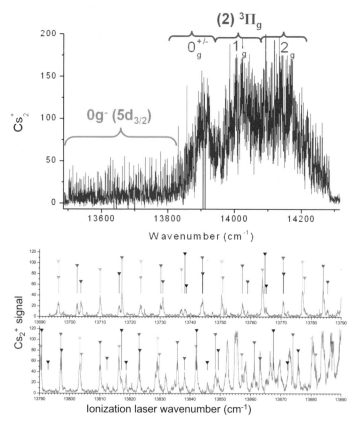

Fig. 3.   Photoionization spectrum of Cs_2 cold molecules formed after spontaneous emission from the vibrational level, $v = 79$, in $0_g^-(6s + 6p_{3/2})$ state. Each transition is indicated by a triangle.

3. Novel scheme for the formation of cold Cs_2 molecules

A recent novelty of our experiment is the modification of the detection of ground-state molecules after spontaneous decay of the photoassociated molecules. A broadband (FHWM ~ 25 cm^{-1}) dye (LDS751) laser, pumped by the second harmonic of a pulsed Nd:YAG laser, excites the molecules on the vibrational transitions $X^1\Sigma_g^+(v) \longrightarrow B^1\Pi_u(v')$. The second harmonic of the pulsed Nd:YAG laser ionizes these excited molecules and the Cs_2^+ ions are detected through a time-of-flight mass spectrometer. Such a broadband laser, tunable over a wide range (13500 - 14500 cm^{-1}), is able to detect cold molecules in a broad distribution of ro-vibrational levels. By scanning the cw Ti:Sa PA laser, we have found several new pathways for the formation

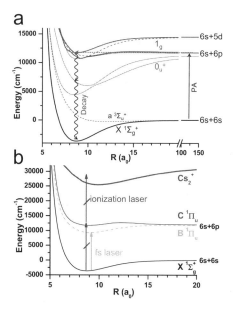

Fig. 4.   Relevant molecular states of the cesium dimer, converging towards the dissociation limits 6s+6s, 6s+6p, and 6s+5d. (a) Photoassociation of cold atoms and formation of cold molecules. For the potentials 1_g, long-range radial wavefunction is coupled to short range radial wavefunction by internal coupling of the potentials. The ground-state molecules, $X^1\Sigma_g^+$, are formed in a cascading spontaneous emission via the 0_u^+ potentials. (b) REMPI detection process via the $C^1\Pi_u$ state, and transition ($X^1\Sigma_g^+$ towards $B^1\Pi_u$) induced by the fs laser.

of cold molecules in the singlet ground state. Both the experimental and theoretical studies of these different mechanisms are in progress. Here, we focus our attention on one pathway involving the excitation of the 1_g state (Fig. 4a). The PA cw laser, $\sim 1\,\mathrm{cm}^{-1}$ to the red of the atomic transition $6s_{1/2} \longrightarrow 6p_{3/2}$, is tuned to a chosen vibrational level (v, J) of a state, spectroscopically labelled 1_g, converging towards the electronically excited limit $6s_{1/2} + 6p_{3/2}$. The process is

$$2Cs(6s) + h\nu_{PA} \longrightarrow Cs_2(1_g(6s + 6p); v, J).$$

The complete mechanism of molecule formation in the singlet ground state is complex and not described in detail here. It involves an internal coupling[13] with the $1g(6s+5d)$ state. Although these photoassociated molecules spontaneously decay mainly towards the triplet ground state, a reasonable branching ratio allows a two-photon spontaneous cascading towards the singlet ground state via the 0_u^+ states, i.e. the process

$$Cs_2(1_g) \xrightarrow{sp.em.} Cs_2(0_u^+) \xrightarrow{sp.em.} Cs_2(X^1\Sigma_g^+)$$

To analyze the vibrational distribution of these cold molecules in the singlet ground state, we use the REMPI process via the $C^1\Pi_u$ electronic excited state, depicted in Fig. 4(b). The transition frequencies and their Franck-Condon factors are well known. Fig. 5(a) shows the observed spectrum. Different vibrational levels, from $v = 1$ to $v = 10$, are populated, but no molecules are present in the vibrational level $v = 0$.

4. Efficient vibrational cooling by optical pumping with a broadband laser

Different schemes have been proposed to favor the formation of cold molecules in their lowest vibrational level. A few $v = 0$ (no vibration) ultracold ground-state potassium dimers have been observed[15] by two-photon PA, but several other vibrational levels were populated as well. By transferring a given vibrational level into the lowest vibrational one, cold ground state rubidium-cesium molecules have been prepared.[16] Raman PA for preparing ultracold molecules in a well defined level has been studied by different groups. Its efficiency is unfortunately limited, because the so-prepared molecules can be excited again, and spontaneously decay toward other vibrational levels.

Several theoretical approaches have also been proposed to favor the spontaneous emission towards the lowest ro-vibrational level. For instance the use of an external cavity has been envisaged.[17] The vibration of the molecule can also be manipulated through quantum interferences between the different transitions. Interplay of control laser fields and spontaneous emission has been investigated for rotational or vibrational cooling.[18–20] As in this theoretical proposal, our recent approach uses a shaped laser. However, in our case the coherence of the field does not play any role since it is a simple incoherent optical pumping process that uses femtosecond pulses spectrally broad enough to excite all relevant vibrational levels.

We have demonstrated the transfer of population from the vibrational levels of cold singlet-ground-state Cs_2 molecules prepared via photoassociation, towards the level, $v = 0$, with no vibration. The main idea is to use a broadband femtosecond laser tuned to the transitions $X^1\Sigma_g^+(v_X)$ towards $B^1\Pi_u(v_B)$ encompassing different vibrational levels of the ground state and the electronically excited one. With successive laser pulses, the absorption-spontaneous emission cycles lead through optical pumping to a

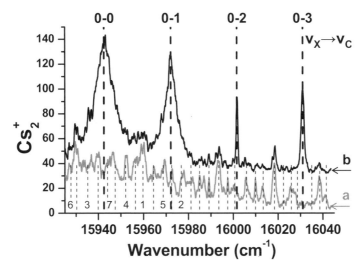

Fig. 5. Cs_2^+ ion spectra: (a) without the shaped laser, (b) with the shaped laser. The numbers indicate vibrational levels v_X. The dashed lines indicate the resonance lines for vibrational transitions (v_X towards v_C) between the $X^1\Sigma_g^+$ and the $C^1\Pi_u$ state.

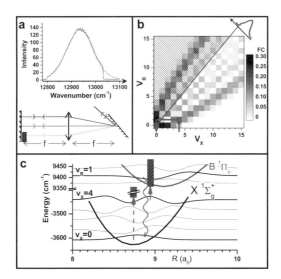

Fig. 6. (a) Shaping of the femtosecond laser (see text) (b) Franck-Condon parabola indicating the importance of the Franck-Condon Factor (level of grey). (c) Scheme of the optical pumping.

redistribution of the vibrational population in the ground state, as sketched in Fig. 6(c). By shaping the laser to remove frequencies corresponding to the excitation of the $v_X = 0$ level (Fig. 6(a)), this state become dark and molecules are accumulated in the $X^1\Sigma_g^+(v_0)$, meaning vibrational laser cooling (Fig. 5(b)). The fs laser is a Tsunami mode locked laser from Spectra Physics with repetition rate 80 MHz, average power 2 W, pulse duration 100 fs and σ-gaussian bandwidth 54 cm^{-1}. By tuning the laser wavelength, low vibrational levels ($v_X < 10$) are excited to vibrational levels v_B of the B state that through optical pumping redistribute the populations among the different vibrational levels of the ground state. Without shaping the fs laser, we observe a wavelength dependent modification of the molecular resonance lines, interpreted as a transfer of population between vibrational levels through optical pumping (Fig. 5(a)). Figure 6(b) shows the Condon parabola of the Franck-Condon factors for the $X^1\Sigma_g^+(v) \longrightarrow B^1\Pi_u(v')$ transitions. The importance of the Franck-Condon factors is indicated by the level of grey. Figure 6(b) shows the overlap between the radial wavefunctions, showing how the Franck-Condon factors depend on the square of the overlap of the wavefunctions in the X and B wells. The transitions from the $v_X = 0$ level towards v_B levels are at frequencies above 13030 cm^{-1}. The spectral shape of the laser does not contain any transitions from $v_X = 0$ (see Fig. 6(a)). Our home made shaper is a simple 4-f line using a grating (1800 lines per mm) for diffracting the laser beam. In this way, the vibrational level, $v_X = 0$, becomes a dark state for the shaped laser. If we consider for instance $v_X = 4$, it is essentially excited in $v_B = 1$, which decays with a probability of about 30% into the dark level, $v_X = 0$, and with a probability of 70% essentially into the levels $v_X = 3$, 4 or 5. More generally, after a few cycles of absorption of laser light then spontaneous emission, a large fraction of the molecules can be accumulated in the lowest vibrational level $v_X = 0$.

5. Vibrational cooling by a shaped femtosecond laser

Figure 5(b) shows the experimental results. The shaped laser strongly modifies the spectrum. The resonance lines corresponding to the transitions $v_X = 0 \longrightarrow v_B = 0 - 3$, mostly absent in the spectrum of Fig. 5(a), are here very strong. Their broadening corresponds to the saturation of the resonance in the REMPI process. The intensity of the lines indicates a very efficient transfer of the molecules into the lowest vibrational level, meaning a vibrational laser-cooling of the molecules. Figure 7(a) shows the time evolution of the population in the different vibrational levels. Population

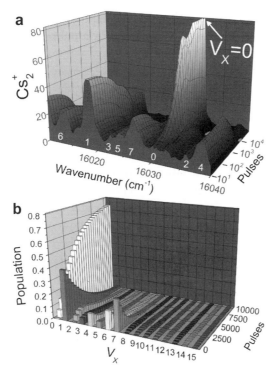

Fig. 7. (a) Temporal evolution of the population transfer. (b) Simulation of the vibrational laser cooling.

transfer into the $v_X = 0$ level is almost saturated after the application of 1000 pulses, which requires ten microseconds. The depumping of the $v_X = 1$ level is not very efficient due to the rough shaping of the laser pulse, and could be improved with further development of the shaper. Taking into account the efficiency of the detection ($<10\%$), the detected ion signal corresponds to about one thousand molecules in the $v_X = 0$ level, corresponding to a flux of $v_X = 0$ molecules of more than 10^5 per second.

We have modeled the optical pumping in a very simple way. Using the known $X^1\Sigma_g^+$ and $B^1\Pi_u$ potential curves and their rotational constants,[21,22] we have calculated the ro-vibrational energy levels. In the perturbative regime, we assume that the excitation probability is simply proportional to the laser spectral density at the transition frequency, to the Franck-Condon factor,[22] and to the Hönl-London factor.[23] We assume a laser spectrum shape very close to the experimental one: average intensity of 150 mW/cm^2; Gaussian shape centered at 12940 cm^{-1} with a Gaussian linewidth $\sigma = 54$ cm^{-1}; and all spectral components above 13030 cm^{-1} removed by the shaping. After being excited by a pulse and before the arrival of the next pulse, we assume a total decay of the excited state population with branching ratios given by the Franck-Condon factors and the Hönl-London factors. The perturbative regime and the $\sim$ 15 ns lifetime of the electronically excited state, close to the 12.5 ns period of the pulses, make reasonable the hypothesis of neglecting any accumulation of coherence due to the excitation by a train of ultrashort pulses.[24] This simple model shows that the molecules make a random walk, mostly in low vibrational levels, until reaching the $v_X = 0$ vibrational level. The accumulation of many molecules in the lowest vibrational level occurs with near unit transfer efficiency. Fig. 7b shows a simulation of the transfer of 70% of the population into the $v_X = 0$ level after 1000 pulses when the molecules are initially in a distribution of vibrational levels simulating the experimental one. The model agrees well with the data. It indicates that only about 5 cycles of absorption-spontaneous emission (the number of necessary laser pulses depends on the intensity) are enough to transfer into $v_X = 0$ all molecules initially in $v_X < 10$ vibrational levels. The limitation of this mechanism is the optical pumping towards high vibrational levels. A broader bandwidth laser could probably increase the population in $v_X = 0$. Nevertheless the Franck-Condon factors favor the accumulation of population in low vibrational levels.

6. Conclusion

We have demonstrated that optical pumping of molecules modifies the distribution of populations in the vibrational levels. By making $v = 0$ a dark state, we enable the accumulation of the population in this level, hence we have demonstrated laser cooling of the molecular vibration. Work is in progress to develop a specific high resolution detection for the analysis of rotational populations of vibration-free ground state molecules. The theoretical model indicates that the shaped lasers do not produce a significant heating of the molecular rotation. Rotational cooling can be performed in a similar way, provided the laser bandwidth and laser shaping match the rotational energy spread. The trapping of molecules can provide a way to prepare a sample of one million or more molecules. Similar results could also be reached for hetero-nuclear systems and the formation of polar molecules, opening exciting prospects in quantum information.

The result obtained is at the frontier of the ultracold and ultrafast fields. The use of femtosecond optical sources and pulse-shaping techniques[25] has enabled several advances in the manipulation of the internal degrees of freedom in both atoms[26] and molecules.[27,28] For example, performing photoassociation while controlling the whole dynamics of the reaction in a pump-dump experiment is a very exciting challenge. The broadband character of the femtosecond laser is however, the essential property of our experiment. Easy shaping is also an important argument for using it, but a continuous broadband laser such as fiber laser or diode laser might offer the same capability and should be tested in the near future.

Optical pumping of molecules has other important implications,[29] for instance for the combined control and cooling of vibration and rotation. Particularly interesting is the cooling of internal degrees of freedom of polar molecules loaded in an electrostatic trap, after velocity filtering of an effusive molecular beam.[30,31]

This work is supported by the "Institut Francilien de Recherche sur les Atomes Froids" (IFRAF).

References

1. H. Rabitz, R. de Vivie-Riedle, M. Motzkus and K. Kompa, *Science* **288**, 824 (2000).
2. I. Osborne and R. Coontz, *Science* **319**, 1201 (2008).
3. O. Dulieu, M. Raoult and E. Tiemann, *J. Phys. B* **39** (2006).
4. J. Doyle, B. Friedrich, R. V. Krems and F. Masnou-Seeuws, *Eur. Phys. J. D* **31**, 149 (2004).
5. R. V. Krems, *Int. Rev. Phys. Chem.* **24**, 99 (2005).
6. J. M. Hutson and P. Soldan, *Int. Rev. Phys. Chem.* **25**, p. 497 (2006).
7. J. G. Danzl, E. Haller, M. Gustavsson, M. J. Mark, R. Hart, N. Bouloufa, O. Dulieu, H. Ritsch and H.-C. Nägerl, *Science* **321**, 1062 (2008).
8. K.-K. Ni, S. Ospelkaus, M. H. G. de Miranda, A. Pe'er, B. Neyenhuis, J. J. Zirbel, S. Kotochigova, P. S. Julienne, D. S. Jin and J. Ye, *Science* **322**, 231 (2008).
9. I. W. M. Smith, *Low Temperatures and Cold Molecules* (Imperial College Press, 2008).
10. A. Fioretti, D. Comparat, A. Crubellier, O. Dulieu, F. Masnou-Seeuws and P. Pillet, *Phys. Rev. Lett.* **80**, 4402 (1998).
11. A. Fioretti, D. Comparat, C. Drag, C. Amiot, O. Dulieu, F. Masnou-Seeuws and P. Pillet, *Eur. Phys. J. D* **5**, 389 (1999).
12. D. Comparat, C. Drag, B. L. Tolra, A. Fioretti, P. Pillet, A. Crubellier, O. Dulieu and F. Masnou-Seeuws, *Eur. Phys. J. D* **11**, p. 59 (2000).
13. C. M. Dion, C. Drag, O. Dulieu, B. Laburthe Tolra, F. Masnou-Seeuws and P. Pillet, *Phys. Rev. Lett.* **86**, 2253 (2001).
14. C. M. Dion, O. Dulieu, D. Comparat, W. de Souza Melo, N. Vanhaecke, P. Pillet, R. Beuc, S. Milošević and G. Pichler, *Eur. Phys. J. D* **18**, 365 (2002).

15. A. N. Nikolov, E. E. Eyler, X. T. Wang, J. Li, H. Wang, W. C. Stwalley and P. L. Gould, *Phys. Rev. Lett.* **82**, 703 (1999).
16. J. M. Sage, S. Sainis, T. Bergeman and D. Demille, *Phys. Rev. Lett.* **94**, 203001 (2005).
17. G. Morigi, P. W. H. Pinkse, M. Kowalewski and R. de Vivie-Riedle, *Phys. Rev. Lett.* **99**, 073001 (2007).
18. D. Tannor and A. Bartana, *J. Phys. Chem. A* **103**, 10359 (1999).
19. A. Bartana, R. Kosloff and D. J. Tannor, *J. Chem. Phys.* **106**, 1435 (1997).
20. S. G. Schirmer, *Phys. Rev. A* **63**, 013407 (2000).
21. W. Weickenmeier, U. Diemer, M. Wahl, M. Raab, W. Demtröder and W. Müller, *J. Chem. Phys.* **82**, 5354 (1985).
22. U. Diemer, R. Duchowicz, M. Ertel, E. Mehdizadeh and W. Demtröder, *Chem. Phys. Lett.* **164**, 419 (1989).
23. G. Hertzberg, *Atomic Spectra and Atomic Structure* (Dover Publications, New York, 1944), p. 208.
24. D. Felinto, C. A. C. Bosco, L. H. Acioli and S. S. Vianna, *Opt. Commun.* **215**, 69 (2003).
25. A. M. Weiner, *Rev. Sci. Instrum.* **71**, 1929 (2000).
26. T. C. Weinacht, J. Ahn and P. H. Bucksbaum, *Nature (London)* **397**, 233 (1999).
27. C. Bardeen, *Chem. Phys. Lett.* **280**, 151 (1997).
28. A. Sharan and D. Goswam, *Curre. Sci.* **82**, p. 30 (2002).
29. J. T. Bahns, W. C. Stwalley and P. L. Gould, *J. Chem. Phys.* **104**, 9689 (1996).
30. S. A. Rangwala, T. Junglen, T. Rieger, P. W. Pinkse and G. Rempe, *Phys. Rev. A* **67**, 043406 (2003).
31. T. Rieger, T. Junglen, S. A. Rangwala, P. W. Pinkse and G. Rempe, *Phys. Rev. Lett.* **95**, 173002 (2005).

A DISSIPATIVE TONKS-GIRARDEAU
GAS OF MOLECULES

S. DÜRR,[1] N. SYASSEN,[1] D. M. BAUER,[1] M. LETTNER,[1] T. VOLZ,[1,*]

D. DIETZE,[1,†] J.-J. GARCÍA-RIPOLL,[1,2] J. I. CIRAC,[1] and G. REMPE[1]

[1] *Max-Planck-Institut für Quantenoptik, Hans-Kopfermann-Straße 1, 85748 Garching,*
Germany
[2] *Universidad Complutense, Facultad de Físicas, Ciudad Universitaria s/n, Madrid*
28040, Spain

Strongly correlated states in many-body systems are traditionally created using elastic interparticle interactions. Here we show that inelastic interactions
between particles can also drive a system into the strongly correlated regime.
This is shown by an experimental realization of a specific strongly correlated
system, namely a one-dimensional molecular Tonks-Girardeau gas.

Keywords: Strong correlations, optical lattices, one dimensional systems,
Tonks-Girardeau gas.

Strong correlations give rise to many fascinating quantum phenomena in
many-body systems, such as high-temperature superconductivity,[1] excitations with fractional statistics,[2] topological quantum computation,[3] and a
variety of exotic behaviors in magnetic systems.[4] Such strong correlations
are typically the result of a repulsive, elastic interparticle interaction. This
makes it energetically unfavorable for particles to be at the same position
and thus the wave function tends to vanish at those positions.

In this work we show that inelastic interactions offer an alternative route
into the strongly correlated regime. The inelastic collisions also lead to a
situation in which the wave function tends to vanish at positions where two
particles are at the same position.

---

*Present address: Institute of Quantum Electronics, ETH-Hönggerberg, 8093 Zürich,
Switzerland
†Present address: Institut für Photonik, Technische Universität Wien, Gußhausstr. 25-29,
1040 Wien, Austria

This behavior might seem counter-intuitive, but it can be understood in terms of an analogy in classical optics: consider an electromagnetic wave in a medium with refractive index n_1 impinging at normal incidence onto a surface with another medium with refractive index n_2. Fresnel's formula yields that a fraction

$$R = \left| \frac{n_1 - n_2}{n_1 + n_2} \right|^2 \tag{1}$$

of the light intensity is reflected. This formula also holds if absorption occurs in the media. In a nonmagnetic medium, dispersion and absorption are expressed by the real and imaginary parts, $\mathrm{Re}(\chi)$ and $\mathrm{Im}(\chi)$, of the electrical susceptibility χ, which determines the refractive index $n = \sqrt{1 + \chi}$.

In Fresnel's formula, the limit $|n_2| \to \infty$ yields $R \to 1$, which is called index mismatch. This result is independent of whether n_2 is real or complex. Even for strong pure absorption, $\mathrm{Im}(\chi_2) \to \infty$, we obtain $R \to 1$. In this limit, the light would be absorbed very quickly once inside the medium, but the index mismatch prevents it from getting there.

We add that the result $R \to 1$ in the limit of strong loss is quite different from the well-known result $R \approx 1$ for light impinging onto a metal. In a metal with negligible ohmic resistivity, the interaction is purely dispersive, so that χ_2 is purely real. At frequencies below the so-called plasma frequency, a metal has $\chi_2 < -1$ so that n_2 is purely imaginary, thus causing $R = 1$ for real n_1 and for any value of $\mathrm{Im}(n_2)$. This reflection is due to surface charges that build up. They cancel the field inside the metal, which results in a reflection of the wave from the surface. In the static limit, we obtain the well-known result that a static electric field cannot penetrate a conductor because of surface charges. For comparison, for reflection in the limit of strong loss, $\mathrm{Im}(\chi_2) \to \infty$, we obtain $n_2 \approx e^{i\pi/4}\sqrt{|\chi_2|}$ so that $\mathrm{Re}(n_2) \approx \mathrm{Im}(n_2)$. Furthermore, $R \approx 1$ is only obtained for large $\mathrm{Im}(\chi_2)$.

These reflection properties of light waves have a one-to-one analogy in the reflection of matter waves. This becomes obvious when defining the refractive index for matter waves obeying the Schrödinger equation as[5] $n(\mathbf{x}) = \sqrt{1 - V(\mathbf{x})/E}$, where $V(\mathbf{x})$ is the potential and $E > 0$ is the energy of the particles. One can show fairly easily that the fraction of particles that are reflected from a potential step at normal incidence is also given by Eq. (1). Loss of particles is expressed by $\mathrm{Im}(V) < 0$ and strong loss, $\mathrm{Im}(V_2) \to -\infty$, yields $R \to 1$, in full analogy to classical optics.

What we are really interested in is a system in which the loss is not caused by a static medium, but by inelastic interparticle interactions. The analogy to the case of reflection from the surface of a static medium is clear: once the wave functions of two particles overlapped, they would be lost very quickly. This corresponds to an index mismatch which prevents the wave functions from overlapping. The particles are thus reflected from one another, causing the wave function to vanish at the positions where they would overlap, just like in the case of strong elastic interactions.

An example of a strongly-correlated system that is well known in the field of ultracold gases is a Mott insulator of atoms in an optical lattice.[6] In this system, strong elastic, repulsive interactions between the particles make it energetically favorable that each site of an optical lattice contains the exact same number of particles. In a recent experiment,[7] we demonstrated that this interaction-induced property of an atomic gas can be maintained when converting atom pairs to molecules. To this end, an atomic Mott isolator was prepared such that it contained exactly two atoms at each site of the central region of the optical lattice. These atom pairs were then associated into molecules[8] using a Feshbach resonance.[9] In Ref. 7 the lattice was so deep that tunneling of molecules was negligible on the time scale of the experiment. The association thus mapped the strong correlations of the interaction-induced atomic Mott insulator to a corresponding quantum state of molecules. But this state was independent of any molecule-molecule interactions.

Studies of the excitation spectrum of this system showed that the molecule-molecule interactions are predominantly inelastic.[10] In the following, we thus neglect the elastic part of the molecule-molecule interactions. The inelastic character of the collisions arises from the fact that the molecules formed with the Feshbach resonance are in a highly excited ro-vibrational state. If two such molecules collide, it is possible that one of them falls down on the vibrational ladder, thus releasing binding energy into kinetic energy of the relative motion of the colliding particles. The released energy is typically much larger then the trap depth, so that all collision partners are quickly lost from the sample.

A natural question to ask is: What happens to this quantum state if the lattice depth is reduced so much that tunneling of molecules becomes significant? We addressed this question experimentally in Ref. 11 and we discuss the results of this experiment in the following. As hinted above, we find that for sufficiently strong inelastic interactions the system remains strongly correlated. It is well known that a reduction of the dimensionality of the system makes it easier to reach the strongly-correlated regime. We thus choose to lower the depth of the three-dimensional (3D) optical lattice only along one direction, as shown in Fig. 1. The strongly correlated

Fig. 1.   2D scheme of the 3D experiment in the optical lattice. The lattice depth along one dimension is lowered to zero so that the molecules are free to move along the resulting 1D tubes.

gas of repulsively interacting particles in the resulting 1D tubes is known as a Tonks-Girardeau gas[12,13] and has been observed in previous experiments with atoms.[14,15] Instead of elastic interactions, our experiment relies on inelastic interactions to reach the strongly correlated regime and thus represents a dissipative Tonks-Girardeau gas. Another novel aspect of our experiment is that it realizes the first Tonks-Girardeau gas of molecules.

A characteristic property of the Tonks-Girardeau gas is that the probability to find two particles at the same position is strongly suppressed.[16,17] A mathematical expression that captures this property is the pair-correlation function

$$g^{(2)}(\mathbf{x}_1, \mathbf{x}_2) = \frac{\langle \Psi^\dagger(\mathbf{x}_1) \Psi^\dagger(\mathbf{x}_2) \Psi(\mathbf{x}_1) \Psi(\mathbf{x}_2) \rangle}{\langle \Psi^\dagger(\mathbf{x}_1) \Psi(\mathbf{x}_1) \rangle \langle \Psi^\dagger(\mathbf{x}_2) \Psi(\mathbf{x}_2) \rangle}, \tag{2}$$

where $\Psi(\mathbf{x})$ is the bosonic field operator that annihilates a molecule at position $\mathbf{x}$. The probability to find two particles at the same position $\mathbf{x}$ is proportional to $g^{(2)}(\mathbf{x}, \mathbf{x})$. For a homogeneous system, this quantity is independent of $\mathbf{x}$ and we denote it simply as $g^{(2)}$. For an uncorrelated system $g^{(2)} = 1$.

Loss of particles due to inelastic two-body collisions occurs only if the particles come close together. The rate at which the loss occurs thus depends on $g^{(2)}$; more quantitatively[11]

$$\frac{dn}{dt} = -Kn^2 g^{(2)}, \tag{3}$$

where n is the 1D density of particles and K is a rate coefficient, which can be determined from independent measurements. A measurement of the

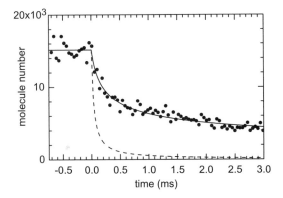

Fig. 2. Time-resolved loss of the number of molecules in 1D tubes. If the system were uncorrelated the loss would be expected to follow the dashed line, which is way off the experimental data (•). A fit to the data (solid line) reveals that the probability to find two particles at the same position is reduced by a factor of ~ 10 compared to an uncorrelated system, thus showing that the system is strongly correlated. Reproduced from Ref. 11.

loss rate can thus serve as a probe whether the strongly correlated regime is reached.

The following experimental procedure is used to study this effect: First, a state with exactly one molecule at each lattice site is prepared as in Ref. 7. Second, the lattice depth along one direction is lowered to zero in 0.5 ms. Third, the system is allowed to evolve for a variable hold time, during which the relevant loss occurs, and finally, the molecule number is measured.

Figure 2 shows experimental data (•) for the decay of the molecule number as a function of the hold time. No noticeable loss is observed during the lattice ramp down, which ends at $t = 0$. The subsequent loss differs significantly from the expectation for an uncorrelated system (dashed line), which is calculated from the independently determined parameters of the system, including a measurement of the 3D loss rate coefficient in Ref. 18. The solid line shows a fit to the data that reveals a value of $g^{(2)} = 0.11\pm0.01$, see Ref. 11 for details. The fact that $g^{(2)}$ differs from 1 by a large factor shows that the system is strongly correlated, thus realizing a dissipative Tonks-Girardeau gas.

An interesting variation of this experiment is obtained when considering the situation where the lattice depth $V_\parallel$ along the 1D tubes is lowered to a nonzero value. Of course, this is closely related to the above experiment, but there are three aspects that make this system interesting: first,

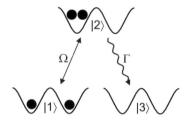

Fig. 3.   Understanding the loss in terms of the quantum Zeno effect. The initial state $|1\rangle$ contains exactly one particle at each site of a double-well potential. Tunneling with amplitude Ω coherently couples this state to state $|2\rangle$, where both particles occupy the same site. In this configuration, the particles can collide inelastically, resulting in loss of both particles, thus transferring the system into state $|3\rangle$. The rate coefficient for this incoherent loss is Γ. In the limit $\Omega \ll \Gamma$, loss from the initial state occurs at an effective rate $\Gamma_{\mathrm{eff}} = \Omega^2/\Gamma$.[19] If Γ is large, then Γ_{eff} becomes small. Fast dissipation thus freezes the system in its initial state, which can be interpreted as a manifestation of the continuous quantum Zeno effect.[20] Reproduced from Ref. 11.

the case $V_\parallel \neq 0$ offers new physical insight because the reduction of the loss can be interpreted in terms of the quantum Zeno effect as illustrated in Fig. 3; second, time-resolved calculations of the dynamics of the loss become numerically feasible; and third, a much larger suppression of $g^{(2)}$ is obtained.

In the following, we concentrate on the last two aspects. The pair-correlation function can again be determined from time-resolved measurements of the loss of molecule number. Results ($\bullet$) are shown in Fig. 4 as a function of $V_\parallel/E_r$, where E_r is the molecular recoil energy. The solid line shows an analytical model discussed in Ref. 11 that represents essentially the Zeno effect illustrated in Fig. 3. In addition, we performed time-resolved numerical calculations that make much fewer approximations than the analytical model. The numerical results are also shown in Fig. 4 and agree well with the analytical model and the experimental data. The lowest value of $g^{(2)}$ measured here is $\sim 1/2000$.

To summarize, inelastic collisions can be used to drive a many-body system into the strongly correlated regime, much like elastic collisions. This general concept is illustrated in an experimental realization of a dissipative Tonks-Girardeau gas. The suppression of the loss rate due to the correlations is used to measure the pair-correlation function, which quantifies the degree of correlation in the experiment. The physical origin of the suppression can be interpreted in terms of index mismatch in Fresnel's formula or in terms of the quantum Zeno effect.

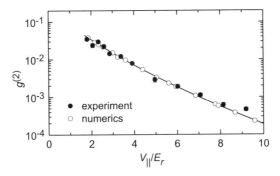

Fig. 4. Pair-correlation function as a function of the lattice depth applied along the one dimension. Experimental data (●), numerical results (○), and analytical model (solid line) agree well with each other. Reproduced from Ref. 11.

Acknowledgments

We acknowledge financial support of the German Excellence Initiative via the program Nanosystems Initiative Munich and of the Deutsche Forschungsgemeinschaft via SFB 631. JJGR acknowledges financial support from the Ramon y Cajal Program and the Spanish projects FIS2006-04885 and CAM-UCM/910758.

References

1. P. W. Anderson, *Science* **235**, 1196 (1987).
2. F. Wilczek, *Phys. Rev. Lett.* **48**, 1144 (1982).
3. A. Y. Kitaev, *Ann. Phys. (NY)* **303**, 2 (2003).
4. A. Auerbach, *Interacting Electrons and Quantum Magnetism* (Springer, Berlin, 1994).
5. C. S. Adams, M. Sigel and J. Mlynek, *Phys. Rep.* **240**, 145 (1994).
6. M. Greiner, O. Mandel, T. Esslinger, T. W. Hänsch and I. Bloch, *Nature* **415**, 39 (2002).
7. T. Volz, N. Syassen, D. M. Bauer, E. Hansis, S. Dürr and G. Rempe, *Nature Phys.* **2**, 692 (2006).
8. S. Dürr, T. Volz, A. Marte and G. Rempe, *Phys. Rev. Lett.* **92**, 020406 (2004).
9. A. Marte, T. Volz, J. Schuster, S. Dürr, G. Rempe, E. G. van Kempen and B. J. Verhaar, *Phys. Rev. Lett.* **89**, 283202 (2002).
10. S. Dürr, T. Volz, N. Syassen, D. M. Bauer, E. Hansis and G. Rempe, A Mott-like state of molecules, in *Atomic Physics 20: XXth International Conference on Atomic Physics*, eds. C. Roos, H. Häffner and R. Blatt, AIP Conf. Proc. No. 869 (AIP, New York, 2006), pp. 278–283.
11. N. Syassen, D. M. Bauer, M. Lettner, T. Volz, D. Dietze, J. J. García-Ripoll, J. I. Cirac, G. Rempe and S. Dürr, *Science* **320**, 1329 (2008).

12. L. Tonks, *Phys. Rev.* **50**, 955 (1936).
13. M. Girardeau, *J. Math. Phys.* **1**, 516 (1960).
14. B. Paredes, A. Widera, V. Murg, O. Mandel, S. Fölling, I. Cirac, G. V. Shlyapnikov, T. W. Hänsch and I. Bloch, *Nature* **429**, 277 (2004).
15. T. Kinoshita, T. Wenger and D. S. Weiss, *Science* **305**, 1125 (2004).
16. D. M. Gangardt and G. V. Shlyapnikov, *Phys. Rev. Lett.* **90**, 010401 (2003).
17. T. Kinoshita, T. Wenger and D. S. Weiss, *Phys. Rev. Lett.* **95**, 190406 (2005).
18. N. Syassen, T. Volz, S. Teichmann, S. Dürr and G. Rempe, *Phys. Rev. A* **74**, 062706 (2006).
19. C. Cohen-Tannoudji, J. Dupont-Roc and G. Grynberg, *Atom-Photon Interactions* (Wiley, New York, 1992), pp. 49–59.
20. B. Misra and E. C. G. Sudarshan, *J. Math. Phys.* **18**, 756 (1977).

SPECTROSCOPY OF ULTRACOLD KRB MOLECULES

WILLIAM C. STWALLEY

Department of Physics, University of Connecticut
Storrs, CT 06269-3046, U.S.A.

When ultracold (T < 1 mK) molecules are formed by photoassociation (PA) followed by spontaneous emission, they commonly are formed in high vibrational levels, typically within 30 cm^{-1} of the dissociation limit.[1,2] Such levels are difficult to produce in other ways, and, because their bands include only a few rotational quantum numbers (typically < 5 for KRb at 0.2 mK), their electronic spectra are readily assignable, especially when accurate *ab initio* calculations are also available. Research at the University of Connecticut has demonstrated how the spectroscopy of KRb ultracold molecules formed by PA[3,4] can be studied using multiple resonance spectroscopy.[1,5–7]

Keywords: Ultracold molecules; KRb; photoassociation; spectroscopy.

1. Introduction

Among the alkali metals, the energetically most similar are K and Rb. Thus the molecule KRb has a small dipole moment and is to some extent a "pseudo-homonuclear" molecule. The long range potential energy curves at the K(4s)+Rb(5p) asymptotes are all strongly attractive, while those at the K(4p)+Rb(5s) asymptotes are all strongly repulsive.[8] Moreover, these four asymptotes are closely spaced at 12578.950, 12816.545, 12985.186, and 13042.896 cm^{-1}, a range of only 464 cm^{-1}.

The photoassociation (PA) of ultracold K and Rb atoms to form electronically excited KRb molecules (and then X- and a-state molecules by spontaneous emission) has been studied extensively,[3,4] the latter reference giving an extensive discussion of previous KRb spectroscopy and the detailed assignments of the eight electronic states observed.[4] These states are labelled in Hund's case c notation as 2(0$^+$), 2(0$^-$), 2(1), 3(0$^+$), 3(0$^-$), 4(1), and 1(2), where the number in parentheses is the Ω value and the number before the parentheses indicates the energy order at long range.

In particular, we used a relatively low resolution tunable pulsed laser with significant amplified spontaneous emission for detection. This produced ion signal at virtually any wavelength within a few nanometers of 602.5 nm, which was the central wavelength used in most experiments.[3,4] This laser ionized $X^1\Sigma^+$ ground state and $a^3\Sigma^+$ metastable state molecules by two-photon REMPI (resonance-enhanced multiphoton ionization). The KRb^+ molecular ions produced were then detected by time-of-flight mass spectroscopy. In this way, other ions (e. g. Rb_2^+) produced by the simultaneous Rb_2 PA and REMPI did not interfere.

An example of a recently assigned PA spectrum is shown in Figure 1. Two previously assigned bands of the 2(1) and the 3(0$^+$) states correspond to the stronger lines, marked above the spectrum. Two newly assigned bands of the 4(1) and the 5(1) states correspond to the weaker lines, marked below the spectrum. The 4(1) and 5(1) states become the $1^1\Pi$ and the $2^1\Pi$ states at short distance. These two states have been previously studied by conventional spectroscopy.[12–14] Thus the vibrational assignments (v = 60 and 17) are unambiguous.

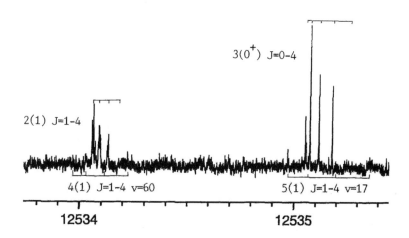

Fig. 1.   Fragment of the photoassociation spectrum of $^{39}K^{85}Rb$ in the 12533.8-12535.4 cm^{-1} range, showing two previously assigned strong bands of the 2(1) and 3(0$^+$) states[4] and two newly assigned weaker bands of the 4(1) ($1^1\Pi$) and 5(1) ($2^1\Pi$) states, corresponding to v = 60 and v = 17, respectively.

The 4(1) ($1^1\Pi$) v = 61, 62, and 63 levels were previously observed,[3,4] so observation of v = 60 is not unexpected. The 5(1) ($2^1\Pi$) state, however, would not normally be expected to be observed in PA since it has a potential barrier at long range (shown in Figure 2). Thus the wavefunction of the v = 17 level just below the K(4s)+Rb($5p_{1/2}$) asymptote would be expected to decay rapidly outside its outer turning point at 12 a_o, and overlap with the continuum wavefunction of the colliding K + Rb pair would be very small. However, the v = 60 level of the $1^1\Pi$ state is perturbed ("resonantly coupled") by the v = 17 level of the $2^1\Pi$ state,[13] so PA to the $2^1\Pi$ state becomes possible via the coupled $1^1\Pi$ (4(1)) state, which can be formed by PA at long range.

Figure 2 also illustrates a very promising expectation for the $2^1\Pi$ v = 17 level: namely, PA to this level should spontaneously emit to v = 0 of the X ground state. Moreover, stimulated Raman transfer via the resonantly coupled v = 60 and 17 levels to v = 0 of the X state should also be possible starting in high vibrational levels of the X and a states, these last levels being formed by PA (or by magnetoassociation (MA) via Feshbach resonances).

2. Vibrational-level-selective Detection

By switching to an improved pulsed laser with a bandwidth of 0.2 cm^{-1} and a low level of amplified spontaneous emission, we were able to obtain higher resolution spectra which allowed us to detect ultracold molecules with vibrational level selectivity.[1,5] The very small rotational spacings were not resolved, however. In particular, we were able to identify X-state levels v = 86-92 (in excellent agreement with the predicted levels[10] and the 3(0^+)-X Franck-Condon factors[1]), and the a-state levels v = 16-22 (in excellent agreement with the subsequent spectra of Ref. 11). [Note that the tentative a-state vibrational numbering reported in Ref. 1 has been corrected.] These high vibrational levels of the X and a states are all within 30 cm^{-1} of dissociation. It is also worth noting that the X-state levels still possess significant dipole moments[10] (0.033 and 0.008 atomic units for v = 86 and 92, respectively, compared to 0.257 for v = 0 and 3.2×10^{-5} for the last level, v = 98).

The REMPI spectra of the X- and a-state molecules allowed assignment of a large number of vibrational levels of the $4^1\Sigma^+$ and the $5^1\Sigma^+$ excited states, and the $4^3\Sigma^+$ and the $3^3\Pi$ excited states, respectively.[5] Eigenvalue calculations of vibrational levels and spacings of *ab initio* potential curves of KRb for these states[9] were in very good agreement with these observations.[5] Subsequently, we have observed a variety of lower energy excited states (which all happen to have a potential barrier to dissociation): $2^3\Pi$ ($\Omega = 0^+$, 0^-,1, and 2), $3^3\Sigma^+$, and the 5(0^+) (the outer well of the $2^3\Pi$ ($\Omega = 0^+$)) state.[7] The $3^3\Sigma^+$ state has also been observed in a spin-forbidden

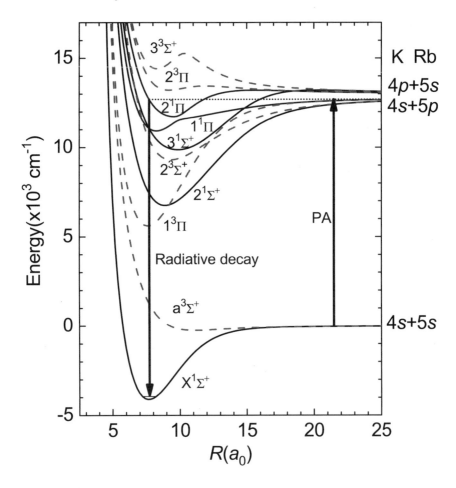

Fig. 2.   Potential energy curves (in 10^3 cm^{-1} vs R in a$_o$) of KRb, based on the high quality *ab initio* calculations of Rousseau et al.[9] The horizontal dotted line represents the K(4s)+Rb(5p$_{1/2}$) asymptote. The vertical line at 7.7 a$_o$ indicates the center of the region of overlap of the X-state v″ = 0 wavefunction, e.g. with the inner turning point region of the 2$^1\Pi$ v′ = 17 level.

transition from the X state,[15] in very good agreement with our results, as were eigenvalue calculations using the *ab initio* potential.[11] It is clear that such spectroscopy is a convenient complement to PA spectroscopy, potentially allowing observation of highly excited states and even autoionizing levels not readily reached with cw lasers.

Both ultracold molecule spectroscopy as just described and PA spectrscopy are particularly sensitive to potential curves at intermediate and long range, which are not readily studied in other ways. The ultracold molecule spectroscopy approach should also be applicable to more weakly bound "Feshbach molecules" formed by magnetoassociation (MA).

3. Rovibrational-level-selective Detection

In order to obtain higher resolution to resolve the rotational and hyperfine structure of vibrational levels near dissociation in the X and a states of KRb, we developed an ion depletion spectroscopic technique for ultracold molecules.[6] As before, a PA laser excites a pair of colliding K and Rb atoms to an excited state of KRb, which decays by spontaneous emission to the high vibrational levels of the X and a states of KRb within 30 cm^{-1} of dissociation. Molecules in a given X- or a-state vibrational level are then ionized by REMPI to form KRb$^+$, which is detected by time-of-flight mass spectroscopy. In our new depletion technique, we add a high resolution cw tunable laser which can deplete the population of a given high rovibrational level when it is tuned to a resonance, thereby decreasing the KRb$^+$ ion signal.

A simple application of this technique is the determination of the binding energy of a particular (v, J) level of the X ground state. Consider the case of a X-state high vibrational level (e.g. v = 87) formed from a particular level (e.g. v′, J′ = 1) of the excited 3(0$^+$) state formed by PA. One can simply tune the depletion laser back to the 3(0$^+$) (v′, 1) level. The difference between the depletion laser frequency (E$_3$(v′,1) – E$_X$(87,0)) cm^{-1} and the PA laser frequency (E$_3$(v′,1) – 2×10^{-5}) cm^{-1} is simply the binding energy of the (87,0) level with respect to dissociation (23.397 ± 0.002 cm^{-1}).[6] This may be compared to the corresponding binding energy of 23.492 cm^{-1} for the fit given in Ref. 11,[16] where the uncertainty is not given, but is probably ~0.01 cm^{-1}, the uncertainty in the dissociation energy in the fit.[11] Thus the two values do not agree and further experiments are needed to resolve the 0.1 cm^{-1} difference. Both binding energies can be combined with the separation of the v = 87 and v = 0 levels to obtain the dissociation energy, D$_o$, of KRb. The best value of (E$_X$(87,0) – E$_X$(0,0)) is 4156.475 cm^{-1},[16] which yields D$_o$ values of 4179.872 and 4179.967 cm^{-1}, respectively. A preliminary value of D$_o$ = 4179.92 ± 0.05 cm^{-1} is recommended.

Another application is confirmation of the long range selection rules. For example, for a 3(0$^+$) state J′ = 1 level, spontaneous emission is observed to J″ = 0 and 2 only, as expected.

An important application is to investigate possible intermediate levels for stimulated Raman transfer from the high vibrational levels of the X and a states to the $v'' = 0$ level of the X state (or other low-lying levels). In the X state case, the $3^1\Sigma^+$ state is promising and levels near $v' = 40$ have large Franck-Condon factors for both the PUMP and DUMP transitions involved in stimulated Raman transfer.[6] We have identified such levels by depletion spectroscopy, so the frequencies of selected PUMP and DUMP transitions are now accurately established.[6] For the a state case, the intermediate state is preferably a perturbed state of mixed singlet-triplet character,[17] as in the pioneering RbCs experiments of Ref. 18.

Thus we believe this depletion technique provides a powerful complement to the vibrational-level-selective technique discussed in the prior section, particularly where rotational and hyperfine structure plays a significant role.

Acknowledgements

The research reported here was carried out in collaboration with current members (Phil Gould, Ed Eyler, Hyewon Kim Pechkis, Michael Bellos, Jayita Ray Majumder, Ryan Carollo, Michael Mastroianni, Matthew Recore, and Warren Zemke) and former members (Dajun Wang, Jin-Tae Kim, Jianbing Qi, Mary Stone, Olga Nikolayeva, Court Ashbaugh, Brian Hattaway, Steve Gensemer, and He Wang) of the Ultracold Molecule Group at the University of Connecticut. It was supported by the National Science Foundation, the NATO Science for Peace Program, and the University of Connecticut Research Foundation. We also thank Eberhard Tiemann for further information and discussions concerning the results of Ref. 11.

References

1. D. Wang, E. E. Eyler, P. L. Gould and W. C. Stwalley, *Phys. Rev. A* **72**, 032052 (2005).
2. W. C. Stwalley, P. L. Gould and E. E. Eyler, "Ultracold molecule formation by photoassociation", to appear in *Cold Molecules: Theory, Experiment, Applications*, ed. R. Krems, W. C. Stwalley and B. Friedrich (Taylor and Francis, NY, 2009).
3. D. Wang, J. Qi, M. F. Stone, O. Nikolayeva, H. Wang, B. Hattaway, S. D. Gensemer, P. L. Gould, E. E. Eyler and W. C. Stwalley, *Phys. Rev. Lett.* **93**, 243005 (2004).
4. D. Wang, J. Qi, M. F. Stone. O. Nikolayeva, B. Hattaway, S. D. Gensemer, H. Wang, W. T. Zemke, P. L. Gould, E. E. Eyler and W. C. Stwalley, *Eur. Phys. J. D* **31**, 165 (2004).
5. D. Wang, E. E. Eyler, P. L. Gould and W. C. Stwalley, *J. Phys. B: At. Mol. Opt. Phys.* **39**, S849 (2006).

6. D. Wang, J. T. Kim, C. Ashbaugh, E. E. Eyler, P. L. Gould and W. C. Stwalley, *Phys. Rev. A* **75**, 032511 (2007).

7. J. T. Kim, D. Wang, E. E. Eyler, P. L. Gould and W. C. Stwalley, "Spectroscopy of KRb triplet excited states using ultracold $a^3\Sigma^+$ molecules formed by photoassociation", submitted to *New J. Phys.*

8. H. Wang and W. C. Stwalley, *J. Chem. Phys.* **108**, 5767 (1998).

9. S. Rousseau, A. R. Allouche and M. Aubert-Frecon, *J. Mol. Spectrosc.* **203**, 235 (2000).

10. W. T. Zemke and W. C. Stwalley, *J. Chem. Phys.* **120**, 88 (2004).

11. A. Pashov, O. Docenko, M. Tamanis, R. Ferber, H. Knöckel and E. Tiemann, *Phys. Rev. A* **76**, 022511 (2007).

12. N. Okada, S. Kasahara, T. Ebi, M. Baba and H. Kato, *J. Chem. Phys.* **105**, 3458 (1996).

13. S. Kasahara, C. Fujiwara, N. Okada and H. Kato, *J. Chem. Phys.* **111**, 8857 (1999).

14. C. Amiot, J. Verges, C. Effantin and J. d'Incan, *Chem. Phys. Lett.* **321**, 21 (2000).

15. Y. Lee, Y. Yoon, B. Kim, L. Li and S. Lee, *J. Chem. Phys.* **120**, 6551 (2004).

16. E. Tiemann, private communication (2008).

17. W. C. Stwalley, *Eur. Phys. J. D* **31**, 221 (2004).

18. J. M. Sage, S. Sainis, T. Bergeman and D. DeMille, *Phys. Rev. Lett.* **94**, 203001 (2005).

COLD MOLECULAR IONS: SINGLE MOLECULE STUDIES

M. DREWSEN

QUANTOP – Danish National Research Foundation for Quantum Optics,
Department of Physics and Astronomy, University of Aarhus,
Ny Munkegade, Build. 1520, DK-8000 Aarhus C, Denmark

Single molecular ions can be sympathetically cooled to a temperature in the mK-range and become spatially localized within a few μm^3 through the Coulomb interaction with laser-cooled atomic ions, and hence be an excellent starting point for a variety of single molecule studies. By applying a rather simple, non-destructive technique for the identification of the individual molecular ions relying on an *in situ* mass measurement of the molecules, studies of the photofragmentation of singly-charged aniline ions ($C_6H_7N^+$) as well as investigations of isotope effects in reactions of Mg^+ ions with HD molecules have been carried out.

Keywords: Single molecular ions; laser-cooling; photofragmentation; isotope effects.

1. Introduction

For the past decade, research involving cold molecules has gone through an extremely rapid developing phase. For neutral molecules the advances have, in particular, been relying on developments within the following approaches: Photo association of laser-cooled atoms,[1-5] buffer-gas cooling of molecules held in magnetic traps,[6-8] deceleration,[9] filtering[10] and trapping of molecules by electrostatic fields,[11-14] Feshbach resonance generated molecules in degenerated quantum gasses,[15,16] and deceleration of molecules by intense laser pulses.[17] Cooling techniques for molecular ions have been developed in parallel, so that e.g. it has become a standard to work with molecular ions which are sympathetically cooled into Coulomb crystals through the Coulomb interaction with laser-cooled atomic ions.[18-25] In the past years, the technique of He buffer gas ion cooling[26] has furthermore been extended to molecular anions.[27] Most recently, even the combination of cold neutral and ionic molecular techniques has made the first progress.[25]

After a presentation of our non-destructive single molecular ion mass measurement technique,[20] the focus in this report will be on the latest results from our laboratory regarding experiments with single molecular ions.[23,24]

2. Non-destructive single-molecular ion-mass measurement

In order to experiment with single molecular ions, it has been necessary to develop a technique to identify the ion under investigation with high efficiency without destroying it. A schematic of such a non-destructive identification technique used in our single molecular ion experiments is presented in Figure 1. More detailed information on the technique can be found in.[20,28] The technique relies on the measurement of the resonant excitation frequency of one of the two axial oscillation modes of a trapped and crystallized linear two-ion system consisting of one laser-cooled atomic ion of known mass m1 and an *a priori* unknown molecular ion, whose mass m_2 has to be determined for ion identification. From this measured frequency, the mass m_2 of the unknown ion can be deduced from a simple relation between the frequency and the relative mass of the two ions:

$$\omega_{\pm}^2 = \left(1 + \frac{1}{\mu} + \sqrt{1 - \frac{1}{\mu} + \frac{1}{\mu^2}}\right)\omega_1^2, \tag{1}$$

where $\mu = m_2/m_1$, and ω_1 is the single ion oscillation frequency of the known ion. The solutions ω_+ and ω_- correspond to the mode with eigenvectors where the ions move in phase (COM mode) and out of phase (BR mode), respectively, with mass-dependent amplitudes.[29]

The crystallization of the two-ion system results from the sympathetic cooling of the molecular ion through the Coulomb interaction with the laser-cooled ion. This crystallization can be observed by imaging the fluorescence light emitted by the laser-cooled ion onto a CCD camera chip. Here, a well-localized spot appears with the atomic ion displaced a specific distance away from the trap center when it is trapped together with a non-fluorescing unknown ion. In the linear rf trap used in our experiments[19] the two-ion system is aligned along the traps main axis (the z-axis in Fig. 1). The resonant excitation can be promoted either by applying a sinusoidal electric field along this axis (through sinusoidal voltages applied to the end-electrodes of the trap), which will exert a force on both ions, or by periodically modulating the laser intensity of one of the cooling laser beams propagating along the main axis, which leads to a periodically varying scattering force on the laser-cooled ion. The resonance frequencies are determined by monitoring the fluorescence light from the laser-cooled ion

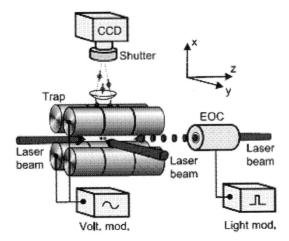

Fig. 1.   Sketch of the experimental setup. Shown on the figure are the linear rf trap electrodes, the cooling laser beams and the CCD camera used to monitor the fluorescence from the laser-cooled atomic ions. An image-intensifier based shutter that can be gated phase-locked to a periodic driving force is installed in front of the camera. A driving force is applied either in the form of a sinusoidal voltage applied to the end-electrodes of the trap or through modulation of the scattering force on the atomic ion by using an electro-optic chopper (EOC) for scattering force modulation.

by the CCD camera while scanning the period of the applied driving force. When the period is equal to the period of one of the two oscillation modes (the center-of-mass (COM) mode and the breathing (BR) mode), the motion of the ions is most highly excited (neglecting damping exerted by the cooling lasers. Details on that see Refs. 20, 28. For CCD camera exposure times larger than the oscillation period of the ions, an enlarged axial extension of the fluorescence spot is observed close to these mode resonance frequencies as seen in Fig. 2. This detection method easily leads to a relative mass resolution $\Delta m/m$ below 10^{-2}, and can for optimized conditions even lead to a resolution at the 10^{-4} level.[28]

For long measuring times, more precise mass measurements are expected when the phase of the motion of the laser-cooled ion is monitored. This can be done by gating the CCD camera such that only light emitted at a certain phase with respect to the phase of the driving force is detected. Examples of such measurements can be seen in Fig. 3. Due to systematic errors, such measurements have so far also been limited to mass resolutions $\Delta m/m$ of a few times 10^{-4}.[20,28]

In the following sections, a few recent single molecular ion experiments,

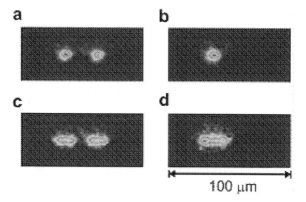

Fig. 2. (a) Fluorescence image of two laser-cooled Ca$^+$ ions. (b) An image of a laser-cooled Ca$^+$ ion trapped together with a sympathetically cooled *a priori* unknown singly charged ion. (c) and (d) are the images when the frequency of the electrical driving force is close to the COM mode frequency of the respective two-ion systems. For all images the camera integration time was 100 ms, which is much longer than the oscillation period of the ions of typical ~10 ms.

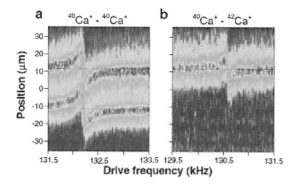

Fig. 3. The position resolved fluorescence along the trap axis (z axis) as a function of the drive frequency an intensity modulated cooling laser beam. (a) Two ^{40}Ca$^+$ ions. (b) One ^{40}Ca$^+$ ion and one ^{42}Ca$^+$ ion, where only the ^{40}Ca$^+$ ion is fluorescing. Each gray-scale (false-colored) contour plot is composed of axial projections of the fluorescence intensities in gated images recorded during the frequency scans. Dashed lines indicate equilibrium positions of the ions in the absence of modulation. The dark gray (red) areas near the dashed lines correspond to high intensity, while the dark (blue) areas are low intensity regions.

where the non-destructive mass measurement method is applied, will be discussed.

3. Consecutive photofragmentation of an aniline ion

The motivation for studying consecutive photofragmentation of aniline ions has been manifold. First, we wanted to prove that the non-destructive identification method described above allows detailed studies of the time evolution of light-induced consecutive fragmentation at long time scales (milliseconds up to several hours). Secondly, we wanted to prove that non-destructive detection of photofragments is a viable way for probabilistic preparation of a wealth of single-molecular ions which can be used as targets for other experiments (e.g., astrophysical studies). Thirdly, more generally, we wanted to prove that studies of spatially localized and very cold single ions can be extended from diatomic systems[20] to complex molecular systems. In all, this opens up new opportunities in molecular science, including molecular rotational dynamics and chemical reaction dynamics on long time scales.

In the experiments, a single aniline ion ($C_6H_5NH_2^+$) is irradiated by the combination of cw light at 397 nm (originating from the laser beams used to cool the calcium ions which provide the sympathetic cooling), and nanosecond pulses of light at 294 nm (used to produce the aniline ions in the first place through a 1+1 REMPI process).[23] In Fig. 4(a), the frequency, at which a specific molecular ion mass has been detected during a series of 77 experiments, is presented. As clearly seen, a series of molecular ions, and not only $C_6H_5NH_2^+$ ions (mass 93 amu), are produced. By repeatable mass scans we can follow the photofragmentation of the original aniline ion in time as the few examples in Fig. 4(b) show. In Fig. 5, a simplified picture is presented of how the photofragmentation is progressing. In a single experiment we were indeed able to monitor all the indicated molecular ions, but the initially produced aniline ion, as seen in Fig. 6.[23] While $C_6H_5NH_2^+$, $C_5H_6^+$, and $C_5H_5^+$ ions were only sometimes found to be stable against further photofragmentation (corresponding to a probabilistic preparation of those ions), the $C_3H_3^+$ ion was always stable.

How sequential breakage of larger molecules into lighter fragments takes place can potentially be monitored non-destructively on time scales ranging from less than a second to several hours by the applied technique.

4. Isotope effects in reaction of Mg^+ ions with HD molecules

Studies of isotope effects in chemical reactions can often help to obtain a better understanding of the underlying reaction dynamics. Resonance

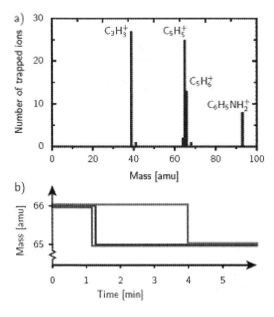

Fig. 4. (a) Molecular ion mass spectrum with aniline ions ($C_6H_5NH_2^+$ ions) produced through a 1+1 REMPI process as the starting point. The height of the bars indicates how often a specific ion mass was detected in a series of experiments. (b) Three recorded time sequences of the photodissociation of $C_5H_6^+$. The statistical nature of the dissociation process is clearly visible by the three single molecule experiments.

effects observed in the F + H_2 reaction and isotopic analogs have e.g. resulted in a much improved understanding of this benchmark reaction.[30–32] In a series of experiments, we have studied reactions between Mg^+ in the $3p^2P_{3/2}$ excited state (excitation energy of 4.4 eV) with isotopologues of molecular hydrogen at thermal energies. Due to the simple internal structure of the reaction partners, these reactions represent a particularly simple test case for reaction dynamics involving an electronically excited atomic collision partner, and hence can serve as good benchmark reactions for reaction dynamics simulations. In brief, we here consider only reactions of the types represented by the Eqs. (2) and (3). Reactions with other isotopologues of molecular hydrogen can be found in Ref. 24. From the results presented in Fig. 7, corresponding to a total of only about 300 single ion reactions, the branching ratio between the reactions

$$Mg^+ \left(3p^2P_{3/2}\right) + HD \rightarrow MgD^+ + H \tag{2}$$

$$Mg^+ \left(3p^2P_{3/2}\right) + HD \rightarrow MgH^+ + D \tag{3}$$

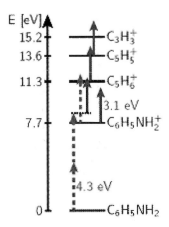

Fig. 5.   Energy level scheme with indication of the ground state energies of several relevant molecular ions relative to the ground state of neutral aniline. The dashed (solid) arrows indicate some of the photodissociation paths observed in the experiments due to the presence of light at 294 nm (397 nm).

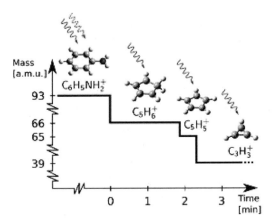

Fig. 6.   A consecutive photodissociation sequence observed in an experiment. The dashed (solid) wriggly arrows represent 294 nm (397 nm) photons responsible for the fragmentations.

has been found to be larger than 5. This strong isotope effect cannot be explained by a simple statistical model based on an assumption of an equal probability for populating energetically accessible states of MgH^+

and MgD$^+$, but must be attributed to a dynamical mechanism. In the ion beam experiments of Ref. 33 a similar isotope effect was observed in reactions between ground state Mg$^+$ ions and HD molecules at center-of-mass energies up to 11 eV. This was rationalized in terms of an impulsive interaction with a thermodynamic threshold.

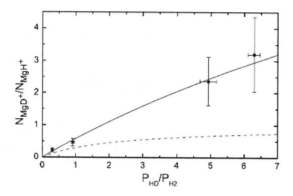

Fig. 7.   The ratio between the number of formed ^{26}MgD$^+$ and ^{26}MgH$^+$ ions as a function of the relative pressure of HD and H$_2$. From left to right, the data points correspond to the following numbers of times measuring MgD$^+$ / MgH$^+$: 17/75, 21/45, 33/14 and 32/10. The error bars represent statistical uncertainties. The blue (solid) and the red (dashed) curves represent results from a simple theoretical model based on Langevin capture (See Eq. (3) in Ref. 24). While the dashed curve is the expected result for equal probability of forming ^{26}MgD$^+$ and ^{26}MgH$^+$ after Langevin capture of HD, the solid curve shows the best fit to the experimental data. The error bars of the measurements taken into account, the fit suggests a ratio of the formation rate of ^{26}MgD$^+$ and ^{26}MgH$^+$ ions in the range of $\sim$ 6 to ∞ for the Mg$^+$(3p^{2}P$_{3/2}$) + HD reaction.

A schematic view of the potential surfaces involved in the reaction is shown in Fig. 8. Investigations of photofragmentation of MgD$_2^+$ indicate that the Mg$^+$ + HD reaction discussed here proceeds via the 1^{2}B$_2$ surface through a bond-stretch mechanism that eventually favors the formation of MgD$^+$.[34,35] To fully understand the transition from an MgHD$^+$ complex to a potential surface favoring the MgD$^+$ + H asymptote rather than the MgH$^+$ + D asymptote, a detailed theoretical study is required. It might be necessary to consider the details of the conical intersection which arises from the crossing of the 1^{2}A$_1$ and 1^{2}B$_2$ potential surfaces. Non-adiabatic couplings at the conical intersection could give rise to a preference of the MgD$^+$ channel over the MgH$^+$ channel. The same mechanism could as well be responsible for the isotope effect observed in reactions with ground state

Mg^+ ions.[33]

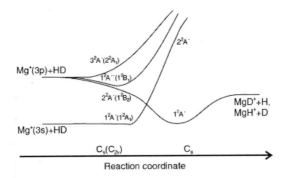

Fig. 8.   Sketch of relevant potential surfaces in C_s symmetry for the $Mg^+ + HD$ reaction proceeding by insertion of Mg^+ into the HD bond on the $2\,{}^2A'$ potential surface, followed by Mg^+-D or Mg^+-H bond formation. On the left-hand side the C_{2v} symmetry labels in parentheses are valid for the analogous reaction with H_2 or D_2.[35]

The above discussed reaction experiments demonstrate the prospects for similar single molecular ion studies using state prepared molecular ions,[36,37] complex molecular ions[22,23] or molecules of astrophysical interest.[38,39] The high detection efficiency can furthermore be useful for studies of reactions involving ions of rare species, e.g. super-heavy elements.[40]

Acknowledgements

The presented work has been support by the Danish National Research Foundation trough QUANTOP, the Danish Natural Science Research Council, and the European Science Foundation.

References

1. T. Takekoshi, B. M. Patterson and R. J. Knize, *Phys. Rev. Lett.* **81**, 5105 (1998).
2. A. Fioretti, D. Comparat, A. Crubellier, O. Dulieu, F. Masnou-Seeuws and P. Pillet, *Phys. Rev. Lett.* **80**, 4402 (1998).
3. A. N. Nikilov, E. E. Eyler, X. T. Wang, J. Li, H. Wang, W. C. Stwalley and P. L. Gould, *Phys. Rev. Lett.* **82**, 703 (1999).
4. M. J. Wright, J. A. Pechkis, J. L. Carini, S. Kallush, R. Kosloff and P. L. Gould, *Phys. Rev. A* **75**, 051401 (2007)
5. M. Viteau, A. Chotia, M. Allegrini, N. Bouloufa, O. Dulieu, D. Comparat and P. Pillet, *Science* **321**, 232 (2008).

6. J. D. Weinstein, R. deCarvalho, T. Guillet, B. Friedrich and J. M. Doyle, *Nature* **395**, 148 (1998).
7. D. Egorov, J. D. Weinstein1, D. Patterson1, B. Friedrich1 and J. M. Doyle, *Phys. Rev. A* **63**, 030501(R) (2001).
8. W. C. Campbell, G. C. Groenenboom, H.-I. Lu, E. Tsikata and J. M. Doyle, *Phys. Rev. Lett.* **100**, 083003 (2008).
9. H. L. Bethlem, G. Berden and G. Meijer, *Phys. Rev. Lett.* **83**, 1558 (1999).
10. S. A. Rangwala, T. Junglen, T. Rieger, P. W. Pinkse and G. Rempe, *Phys. Rev. A* **67**, 043406 (2003).
11. H. L. Bethlem, G. Berden, F. M. H. Crompvoets, R. T. Jongma, A. J. A. van Roij and G. Meijer, *Nature* **406**, 491 (2000).
12. F. M. H. Crompvoets, H. L. Bethlem, R. T. Jongma and G. Meijer, *Nature* **411**, 174 (2001).
13. T. Rieger, T. Junglen, S. A. Rangwala, P. W. Pinkse and G. Rempe, *Phys. Rev. Lett.* **95**, 173002 (2005).
14. S. Hoekstra, J. J. Gilijamse, B. Sartakov, N. Vanhaecke, L. Scharfenberg, S. Y. van de Meerakker and G. Meijer, *Phys. Rev. Lett.* **98**, 133001 (2007).
15. M. Greiner, C. A. Regal and D. S. Jin, *Nature* **426**, 537 (2003).
16. S. Jochim, M. Bartenstein, A. Altmeyer, G. Hendl, S. Riedl, C. Chin, J. H. Denschlag and R. Grimm, *Science* **302**, 2101 (2003).
17. R. Fulton, A. I. Bishop and P. F. Barker, *Phys. Rev. Lett.* **93**, 243004 (2004).
18. K. Mlhave and M. Drewsen, *Phys. Rev. A* **62**, 011401(R) (2000).
19. M. Drewsen, I. Jensen, J. Lindballe, N. Nissen, R. Martinussen, A. Mortensen, P. Staanum and D. Voigt, *Int. J. Mass. Spect.* **229**, 83 (2003).
20. M. Drewsen, A. Mortensen, R. Martinussen, P. Staanum and J. L. Srensen, *Phys. Rev. Lett.* **93**, 243201 (2004).
21. P. Blythe, B. Roth, U. Frhlich, H. Wenz and S. Schiller, *Phys. Rev. Lett.* **95**, 183002 (2005).
22. A. Ostendorf, C. B. Zhang, M. A. Wilson, D. Offenberg, B. Roth and S. Schiller, *Phys. Rev. Lett.* **97**, 243005 (2006).
23. K. Hjbjerre, D. Offenberg, C. Z. Bisgaard, H. Stapelfeldt, P. F. Staanum, A. Mortensen and M. Drewsen, *Phys. Rev. A* **77**, 030702(R) (2008).
24. P. F. Staanum, K. Hjbjerre, R. Wester and M. Drewsen, *Phys. Rev. Lett.* **100**, 243003 (2008).
25. S. Willitsch, M. T. Bell, A. D. Gingell, S. R. Procter and T. P. Softley, *Phys. Rev. Lett.* **100**, 043203 (2008).
26. See, for example, D. Gerlich, *Phys. Scr.* **59**, 256 (1995), and references therein.
27. S. Trippel, J. Mikosch, R. Berhane, R. Otto, M. Weidemller and R. Wester, *Phys. Rev. Lett.* **97**, 193003 (2006).
28. P. F. Staanum, K. Hjbjerre and M. Drewsen, to appear in *Practical Aspects of Trapped ion Mass Spectrometry IV*, ed. R. March and J. Todd (Taylor & Francis, 2008).
29. D. Kielpinski, B. E. King, C. J. Myatt, C. A. Sackett, Q. A. Turchette, W. M. Itano, C. Monroe, D. J. Wineland and W. H. Zurek, *Phys. Rev. A* **61**, 032310 (2000).

30. W. Hu and G. C. Schatz, *J. Chem. Phys* **125**, 132301 (2006), and references therein.
31. D. M. Neumark, A. M. Wodtke, G. N. Robinson, C. C. Hayden and Y. T. Lee, *Phys. Rev. Lett.* **53**, 226 (1984).
32. M. Qui, Z. Ren, L. Che, D. Dai, S. Harich, X. Wang, X. Yang, C. Xu, D. Xie, M. Gustafsson, R. T. Skodje, Z. Sun and D. H. Zhang, *Science* **311**, 1440 (2006).
33. N. Dalleska, K. Crellin and P. Armentrout, *J. Phys. Chem.* **97**, 3123 (1993).
34. L. N. Ding, M. A. Young, P. D. Kleiber and W. C. Stwalley, *J. Phys. Chem.* **97**, 2181 (1993).
35. P. D. Kleiber and J. Chen, *Int. Rev. Phys. Chem.* **17**, 1 (1998).
36. I. S. Vogelius, L. B. Madsen and M. Drewsen, *Phys. Rev. Lett.* **89**, 173003 (2002).
37. I. S. Vogelius, L. B. Madsen and M. Drewsen, *Phys. Rev. A* **70**, 053412 (2004).
38. D. Gerlich, E. Herbst and E. Roueff, *Planet. Space Sci.* **50**, 1275 (2002).
39. S. Trippel *et al.*, *Phys. Rev. Lett.* **97**, 193003 (2006).
40. M. Drewsen, *Eur. Phys. J. D* **45**, 125 (2007).

THE FRONTIERS OF ATTOSECOND PHYSICS

G. DOUMY, J. WHEELER, C. BLAGA, F. CATOIRE, R. CHIRLA, P. COLOSIMO,

A. M. MARCH, P. AGOSTINI, L. F. DIMAURO

Department of Physics, The Ohio State University, Columbus, OH 43210, USA
E-mail: dimauro@mps.ohio-state.edu, www.osu.edu

The genesis of light pulses with attosecond (10^{-18} seconds) durations signifies a new frontier in time-resolved physics. The scientific importance is obvious: the time-scale necessary for probing the motion of an electron(s) in the
ground state is attoseconds (atomic unit of time = 24 as). The availability
of attosecond pulses would allow, for the first time, the study of the time-
dependent dynamics of correlated electron systems by freezing the motion,
in essence exploring the structure with ultra-fast snapshots, then following
the subsequent evolution using pump-probe techniques. This paper examines
the fundamental principles of attosecond formation by Fourier synthesis of
a high harmonic comb and phase measurements using two-color techniques.
Quantum control of the spectral phase, critical to attosecond formation, has
its origin in the fundamental response of an atom to an intense electromagnetic field. We will interpret the laser-atom interaction using a semi-classical
trajectory model.

Keywords: Attophysics, spectral phase, harmonic generation, strong field
physics.

Introduction

The interaction of radiation with atoms or molecules, one of the basic tools
of quantum mechanics, is traditionally applied to uncover the structure
of matter. As sources of radiation, lasers have provided new spectroscopic
tools with fantastic resolution, given rise to nonlinear optics and to the control of an atom's external degrees of freedom. Strong field atomic physics
which includes Above-Threshold Ionization (ATI), High Harmonic Generation (HHG) and, recently, attophysics, belongs to a class of effects which
become observable only when the laser interaction energy is comparable or
larger than the atomic potential. One physical effect, the quiver motion of
a free electron in the electromagnetic field, provides a metric for the interaction: in atomic units the cycle-averaged kinetic energy, also dubbed

ponderomotive energy, of this motion is $U_P = I/4\omega^2$ where I is the intensity (equivalent to one atomic unit of field $= 3.5 \times 10^{16}$ W/cm^2) and ω the angular frequency (atomic unit$= 4.14 \times 10^{16}$ s^{-1}). The ratio $z = U_P/\omega$ is a dimensionless parameter which, when ≈ 1, indicates the limit of the strong field domain.[1] z is $\propto \omega^{-3}$ and reaches easily values > 1 at long wavelength. Another parameter, the ratio η of the interaction hamiltonian to the atomic hamiltonian is related to the so-called adiabaticity parameter $\gamma = \sqrt{\frac{E_B}{2U_P}}$ (E_B is the ionization energy) by $\eta = 1/\gamma^2$. The γ parameter was introduced by Keldysh[2] to measure the transition from multiphoton ionization ($\gamma > 1$) to tunneling or strong field ionization ($\gamma < 1$). Since $\gamma \propto \omega$, at constant intensity, the longer the wavelength (the smaller ω), the smaller γ and the more in the strong field regime.

Recent advances in laser technology have resulted in intense, short pulse, mid-infrared (i.e. with wavelength ranging between 1.5 and 4 microns with the current technology) lasers and open a new route to reach the strong field physics domain at lower intensities. Besides, and this is the main topic of this article, mid-infrared has a number of advantages in the production of photons and attosecond pulses: (i) As well known, the high harmonics cutoff energy increases as U_P, and mid-infrared long wavelength drivers help produce hard harmonic photons since $U_P \propto \omega^{-2}$. (ii) In the domain of attophysics, the duration of the periodic attosecond bursts of light emitted by the high harmonics is limited by the dispersion (linked to the quantum mechanical nature of the process) of the spectral group delay and the corresponding limitation of the effective spectral bandwidth. Long wavelengths allow the reduction of this group delay dispersion proportionally to the wavelength and therefore produce shorter attosecond pulses by permitting larger bandwidths. It results that such drivers are favorable to the generation of ultrashort soft/hard X-ray pulses, (which currently are at best around 100 as for a photon energy of 100 eV) and offer a promising route to produce pulses reaching the atomic unit of time (24 as).

In the following, Section 1 deals with the semiclassical description of strong field interaction, harmonic generation and attosecond pulse generation by Fourier synthesis. Section 2 briefly reviews the technology of mid-infrared femtosecond pulses as implemented at OSU. Section 3 discusses the scaling of high harmonics cutoff and yield with the fundamental wavelength, and finally that of the group delay dispersion or "attochirp".

1. High Harmonic generation of attosecond pulses

1.1. *Basic principles*

Following Keldysh,[2,3] ionization of, say, an hydrogen atom by a strong electric field $E \cos \omega t$ is described ($\hbar = m = e = 1$) as a transition from the ground state to a Volkov state which includes the quiver motion of the free electron in the field and has an asymptotic momentum $\vec{p}$. The transition rate is that of a tunneling process through the quasi-static potential barrier formed by the Coulomb potential $-1/r$ and the dipole interaction $-\vec{E} \bullet \vec{r}$. The free electron wavepacket oscillating in the field acquires a kinetic energy $\propto U_P$ which may be transformed into a hard harmonic photon if it recombines with the nucleus after a fraction of optical cycle. This is the essence of the non-perturbative theory of high harmonics.[4] Since this process reproduces every half-cycle it gives rise to a train of light bursts separated by π/ω which, in the frequency domain, corresponds to a series of odd harmonics separated by 2ω (the "plateau"). Theory and experiment agree that the spectrum extends up to a so-called cutoff frequency which is determined by the maximum kinetic energy of $3.17U_P$ acquired by the electron in the field.[5] Filtering a group of N consecutive harmonics yields a series of bursts of duration $\Delta t = \pi/2N\omega$ if the harmonic spectral phase is constant or linear.[6] Moreover, the theory predicts that among all possible electron quantum paths starting and ending in the vicinity of the nucleus, for each harmonic energy, the main contribution arises from two trajectories only, for which the quasi-classical action is stationary (those are dubbed the short and the long trajectories.[6] This forms the basis of attosecond pulse generation via Fourier synthesis.

1.2. *Attochirp*

The synthesis of the first attosecond pulse train by Paul *et al.*[7] showed that the spectral phase was approximately constant over a few harmonic orders in argon. To achieve agreement with the calculation[6] it was necessary to admit that the experiment somewhat filtered out the contribution of the long trajectories. This was clarified and systematically investigated a few years later.[8] Although the spectral phase appeared to be quasi linear, under careful examination Paul's data revealed a small quadratic term.[9] Since $d\varphi/d\omega$ is associated with a group delay, Δt_G, this term corresponds to a group delay dispersion $d\Delta t_G/d\omega = d^2\varphi/d\omega^2$ or attochirp.

What is the origin of the attochirp? Simply the different lengths of the electron trajectories that give rise to consecutive harmonics. As such, it can be calculated from classical mechanics.[9] A simple argument allows the determination of the scaling as a function of the driving wavelength and intensity (Fig. 1). A plot of the trajectory duration, i.e. the difference between the return time and the emission time, versus the harmonic energy is comprised in a rectangle of height $\propto \lambda/c$ and width $\propto I\lambda^2$ (c the speed of light and I the intensity). The attochirp is the slope of the diagonal and is therefore $\propto 1/I\lambda$ (see Ref. 10).

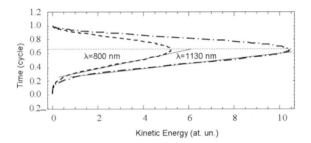

Fig. 1.   Group Delay, and Group Delay Dispersion and scaling with laser intensity (I) and wavelength (λ).

The attochirp has obvious consequences on the duration of the short-est burst of light emitted in the attosecond pulse train. To decrease this duration it is not sufficient to increase the bandwidth. Taking into ac-count the group delay dispersion (GDD) shows that there is an optimum bandwidth.[11] One way to compensate for the GDD is by propagating the pulses in a medium with an opposite group velocity dispersion (GVD): for the short trajectories the intrinsic attochirp is positive and must be com-pensated by a negative GVD found in metals or fully ionized plasmas.[8,10] Another way is to reduce as much as possible the intrinsic GDD by taking advantage of the wavelength/intensity scaling discussed above. The mid-infrared range is in principle well suited for that purpose with its high energy harmonic cutoffs as well as high energy photoelectrons.[12] As will be shown here, experiment confirms this expectation.[13]

Note that the attochirp is a concept valid in the harmonic plateau. For harmonics generated in the cutoff region the short and long trajectories coalesce and the spectral phase becomes constant.[14] This is one option to produce isolated attosecond pulses by combining sub-10 fs pump pulses and spectral filtering[15] although the drop of the spectral amplitude in that re-gion is a clear handicap. The production of single attosecond pulses is also possible in the plateau using the polarization gating technique[17] which reduces the number of effective optical cycles to one. Long wavelength drivers have another advantage in this case since the number of cycles for a given duration decreases with the optical period which is $\propto \lambda$.

2. Mid-Infrared technology

2.1. *The 2μm optical parametric amplifier*

Generation of the 2 μm radiation is based upon a commercial optical parametric amplifier (OPA) (Light Conversion HE-TOPAS-5/800). The OPA is pumped by 5 mJ, 50 fs Ti:S pulses derived from a homemade Ti:S system, using a small amount of pump light to generate a broadband spectrum via superfluorescence. A narrow spectral portion is then used as a seed with the remaining 0.8 μm light. After 6 passes in two BBO (beta-barium borate) nonlinear crystals, up to 600 μJ of 2 μm radiation is available. In the basic process of a parametric amplifier, one 0.8 μm photon is split into two lower energy ones: conservation of energy implies that close to 1 mJ of 1.3 μm idler radiation is also generated. The beam quality is fair and it can be focused down to a spot about twice the diffraction limit.

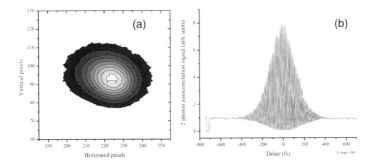

Fig. 2. (a) Focal spot image of the 3.6 μm source. (b) Interferometric autocorrelation trace

2.2. *The Difference Frequency Generation 3.6μm source*

In this case, two laser pulses are combined in a mixing non-linear crystal. The 3.6 μm radiation is generated through Difference Frequency Generation (DFG): the nonlinear crystal (KTA) converts the wavelengths of 0.815 μm ("pump") and that at 1.053μm "idler") into the "signal" at

3.6 μm. Typically, 2.5 mJ, 100 fs Ti:S pulses centered at 0.815 μm and 500 μJ, 16 ps Nd:YLF pulses centered at 1.053 μm are used in a 5 mm long KTA crystal to produce 160 μJ of MIR corresponding to 28% of the theoretical maximum possible MIR energy. At the mixing crystal, the Ti:S and Nd:YLF FWHM spot sizes are $\approx$ 2.5 mm. Mixing the amplified Ti:S and Nd:YLF beams requires locking the repetition rates of the two oscillators. To do so, the cavity length of one oscillator must be referenced to the other.[18] A typical spot image recorded with a thermal camera is displayed in Fig. 2(a). The pulse duration was measured by performing an interferometric autocorrelation, which yields a value of 115 fs FWHM (see Fig. 2(b)).

3. Harmonic cutoff with the 2μm laser

The harmonic spectrum is expected to extend to values between $IP + 3.2U_P$ (theoretical single atom response) and $\approx IP + 2U_P$ (when phase-matching is taken into account) where IP is the atom ionization potential.[5] The cutoff is in general not easy to measure because it is not abrupt, because the intensity is averaged over the interaction volume, because of the unknown transmission of the apparatus... The only previous measurement[19] at a wavelength other than 0.8 μm or 1 μm has been performed by Shan and Chang, and illustrated those difficulties. Our harmonic setup includes a soft X-ray spectrometer (Hettrick Scientific) with three different grazing incidence gratings. Differentially pumped with respect to the harmonic source chamber with the gas cell, it is placed at the spectrometer's entrance plane: The dispersed spectrum is detected on a back-illuminated soft-X-ray CCD (charge-coupled device) camera (Andor). Different metal filters (Al or Zr) are used to suppress the fundamental and low-order harmonics.

Excitation with 2 μm pulses produces a dense (consequence of the small photon energy) harmonic comb extending to the Al L-edge at $\approx$ 70 eV energy. Using a Zr filter instead, the argon harmonic comb is found to extend over the entire Zr filter transmission window (60-200 eV) (Fig. 3).[12] Measurements using the second Al transmission window establish that the cutoff is $\approx$ 220 eV, which is consistent with previous calculations and extends the argon cutoff well beyond the previous measurement at 1.5 μm.[19]

4. Harmonic yield

The above spectra qualitatively agree with the cutoff law and highlight the effectiveness of the U_P-scaling. The scaling of the yield on the laser

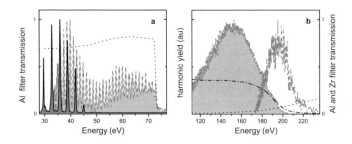

Fig. 3. Harmonic spectrum from the 2000 nm laser in the cutoff region obtained at an intensity of 1.8×10^{14} W/cm^2. Left: Spectrum with aluminum filter, compared with the spectrum at 0.8μm. Right: Spectrum through a Zirconium filter and through the Aluminum filter second transmission window.

wavelength however is not as clearly defined theoretically. One must specify what quantities are kept constant and what is meant by "yield" for a single harmonic/ over which bandwidth etc. For instance the usual understanding from the Lewenstein model[4] was that at a constant U_P the modulus square of the dipole scales as λ^{-3}. Actually this is misleading since it ignores the effect of the change of intensity required to maintain U_P constant while changing λ and it can be shown that a more dramatic decrease is to be expected.[20] At constant intensity the scaling found[11] from numerical solutions of the time-dependent Schrödinger equation (TDSE) is $\propto \lambda^{-5.5\pm0.5}$, ignoring the phase-matching effects. Experimentally, in argon the yield at constant intensity, atomic density and focusing, over the bandwidth $35 - 50$ eV drops by a factor 1000 between 0.8 and 2 μm, close to the TDSE prediction. This confirms the model but naturally does not prevent an independent optimization of the yield at 2 μm. Preliminary results suggest that the brightness of a 2 μm harmonic source can be comparable to that of an 0.8 μm one in the extreme-UV region($\omega \approx 50$ eV) and even higher at higher energies ($\omega \in (50 - 200)$ eV). Consistent with our findings, a recent theoretical study[21] shows that favorable phase matching conditions can be realized at mid-infrared wavelengths.

5. Attochirp

The attosecond metrology is usually based on photoionization of a target atom (with an ionization energy IP) by a superposition of the weak intensity harmonics and a relatively intense infrared beam.[7,8,15] This method called "RABBITT" (Reconstruction of Attosecond Beating By Interference of Two-photon Transitions) has been well validated and documented.[10,16] It requires, however, sophisticated equipment which is under construction at OSU but not yet available. Dudovich and coworkers[22] have demonstrated an all-optical method which is simpler to implement. A small amount of the driving field is converted to its second harmonic (2ω) and propagates with the fundamental beam to the atomic jet where the harmonics are generated. The superposition creates even harmonics (through transitions involving, for instance, an even number of ω-photons and one 2ω-photon). The amplitude of the even harmonics as a function of the relative phase ϕ between the ω and 2ω fields is modulated at a frequency 4ω and the phase $\phi_{max}(2q)$ which maximizes the even harmonic $2q$ depends on q (see Fig. 4). The method relies then on theory associating $\phi_{max}(2q)$ to the spectral GDD, with an arbitrary shift of the phase origin (fixed at zero for the harmonic generation cutoff in Ref. 22). In the original paper,[22] the calculation is based on the assumption that the second harmonic field only slightly perturbs the fundamental and that the even harmonics phases interpolate that of the "normal" odd-harmonics. This is the case as the intensity of the second harmonic is limited to less than 1% of the fundamental. Other calculations[23,24] drop that approximation and allow higher 2ω fields. It is the Dudovich method which has been implemented and used to measure the mid-infrared attochirp.

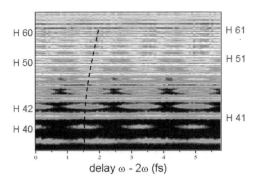

Fig. 4.   Typical 2D plot of the harmonics amplitudes vs the relative $\omega - 2\omega$ delay . The modulation of the even harmonics is clearly visible as well as the shift in the position of the maxima

The experimental setup is similar to the one of Dudovich et al. The OPA described above yields $550\mu J$ of 2 μm light with a pulse duration of 50 fs. A 300μm BBO crystal tuned out of phase-matching generates a small amount of second harmonic polarized perpendicularly to the fundamental.

The group delay accumulated between the two color pulses through propagation in air and glass, is compensated by calcite plates. A pair of fused silica wedges is used as well for compensation and to control the subcycle delay between the electric fields of the two pulses. A zero order half waveplate resets the two polarizations along the same direction. The two beams are focused by a silver-coated mirror into the harmonic generation atomic jet (argon or xenon). Phase matching conditions and spatial filtering of the on-axis radiation due to the very small acceptance angle of the Hettrick spectrometer combined to select the contribution of the short trajectory quantum path and eliminate the more diverging one due to the long trajectory. Following Dudovich et al., by varying the delay between the fundamental and the second harmonic fields the $\phi_{max}(q)$ is extracted to an arbitrary origin and then compared to a simple theoretical modeling in which the quantum paths are limited to the classical trajectories. This is obviously a great simplification and does not yield any information on the classically forbidden cutoff region. It is actually unclear whether the method works in that region.

We first validated the method at 0.8 μm with a result in agreement with both the RABBITT measurement[10] and the $\omega - 2\omega$ one.[22] The attochirp at 2 μm[13] is as expected from the λ-scaling, about 2.5 times smaller. Energy losses in the setup limited greatly the generation cutoff that we could obtain in Argon with the 2 μm driver, and our spectrometer resolution prevented us from observing all the harmonic orders generated. Those limitations combined in a full measurement over a limited total bandwidth (35-60 eV). In Fig. 5, we present the reconstructed attosecond pulse train corresponding to our measurements. Dudovich et al stress that their method, in contrast to RABBITT, which measures the harmonic phases on a target which might be located quite far from the source, yields the attochirp *in-situ*, i.e. where the harmonics are generated. This is both an advantage (to compare with theory) and a drawback (to evaluate the duration of attosecond pulses on target).

6. Conclusion: the attophysics frontier

Currently the record for attosecond pulses is of 130 *as* at $q\omega \approx 30$ eV[17] and 100 *as* at $q\omega \approx 90$ eV.[25] The mid-infrared driving lasers promise attosecond sources beyond the 800 nm harmonic cutoff that are both shorter and brighter. So far it has not been possible to generate harmonics with driving wavelengths longer than 2 μm. However ionization of argon or helium atoms with the 3.6 μm source has been observed[12] and therefore harmonic

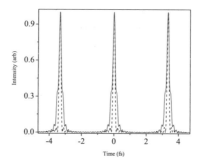

Fig. 5. An attosecond pulse train from the 2000 nm harmonics. The dotted line represents the transform limited pulses (135 as FWHM), while the solid line takes the measured attochirp into account (170 as FWHM).

generation will most likely be achieved in the near future. With the tremendous values of U_P created at that wavelength, the generation of hard harmonics (currently limited to about 100 eV with the Ti:Sapphire laser) becomes possible. The cutoff has already been pushed above 200 eV using a 2 μm driver. One atomic unit (24 as) of several hundreds eV of central frequency can be envisioned. The current mid-infrared sources are hindered by the low intensity of the commercial systems. The 2 μm OPCPA under construction at OSU should remove this limitation and finally open the road to intense X-ray attosecond pulses and thus provide a real chance for the spectacular applications that attophysics promises.

Acknowledgments

This work was performed with support from USDOE/BES under contracts DE-FG02- 04ER15614 and DE-FG02-06ER15833. L. F. Dimauro acknowledges support from the Hagenlocker chair. We are grateful to Dr. Linda Young (Argonne National Laboratory) for the loan of the Hettrick spectrometer.

References

1. H. Reiss, Effect of an intense electromagnetic field on a weakly bound system, *Phys. Rev. A* **22**, 1786 (1980).
2. L. V. Keldysh, Ionization in the field of a strong electromagnetic wave, *Sov. Phys. JETP* **20**, 1307–1314 (1965).
3. M. Fedorov, *Atomic and Free Electrons in a Strong Light Field* (World Scientific, Singapore 1991).
4. M. Lewenstein *et al.*, Theory of high-harmonic generation by low-frequency laser fields, *Phys. Rev. A* **49**, 2117–2132 (1994).

5. J. L. Krause, K. J. Schafer and K. C. Kulander, High-order harmonic generation from atoms and ions in the high intensity regime, *Phys. Rev. Lett.* **68**, 3535 (1992).

6. P. Antoine, A. L'Huillier and M. Lewenstein, Attosecond pulse trains using high-order harmonics, *Phys. Rev. Lett.* **77**, 1234–1237 (1996).

7. P. M. Paul *et al.*, Observation of a train of attosecond pulses from high harmonic generation, *Science* **292**, 1689–1692 (2001).

8. R. Lopez-Martens *et al.*, Amplitude and phase control of attosecond light pulses, *Phys. Rev. Lett.* **94**, 033001 (2005).

9. S. Kazamias and Ph. Balcou, Intrinsic chirp of attosecond pulses: Single-atom model versus experiment, *Phys. Rev. A* **69**, 063416 (2004).

10. Y. Mairesse *et al.*, Attosecond synchronization of high-harmonic soft X-rays, *Science* **302**, 1540–1543 (2003).

11. J. Tate *et al.*, Scaling of wave-packet dynamics in an intense midinfrared field, *Phys. Rev. Lett.* **98**, 013901–013904 (2007).

12. P. Colosimo *et al.*, Scaling strong-field interactions towards the classical limit, *Nat. Phys.* **4**, 387–389 (2008).

13. G. Doumy *et al.*, Attosecond synchronisation of harmonics from long wavelength drivers, submitted to *Science* (2008).

14. Y. Mairesse *et al.*, Optimization of attosecond pulse generation, *Phys. Rev. Lett.* **93**, 163901 (2004).

15. M. Hentschel *et al.*, Attosecond metrology, *Nature* **414**, 509–513 (2001).

16. H. G. Muller, Reconstruction of attosecond harmonic beating by interference of tao-photon transitions, *Appl. Phys. B* **74**, S17 (2002).

17. G. Sansone *et al.*, Isolated single-cycle attosecond pulses, *Science* **314**, 443–446 (2006).

18. P. Colosimo, A study of wavelength dependence of strong-field optical ionization, PhD Dissertation (Stony Brook, August 2007).

19. B. Shan and Z. Chang, Dramatic extension of the high-order harmonic cutoff by using a long-wavelength driving field, *Phys. Rev. A* **65** 011804–011807 (2001).

20. T. Auguste *et al.*, Harmonic yield scaling with driving laser wavelength, to be submitted to *Phys. Rev.* (2008).

21. V. S. Yakovlev, M. Ivanov and F. Krausz, Enhanced phase-matching for generation of soft X-ray harmonics and attosecond pulses in atomic gases, *Opt. Express* **15**, 15351–15364 (2007).

22. N. Dudovich *et al.*, Measuring and controlling the birth of attosecond XUV pulses, *Nat. Phys.* **2**, 781–786 (2006).

23. S. Long, W. Becker and J. K. McIver, Model calculations of polarization-dependent two-color high harmonic generation, *Phys. Rev. A* **52**, 2262–2278 (1995).

24. D. Milosevic and B. Piraux, High-order harmonic generation in a bichromatic elliptically polarized laser field, *Phys. Rev. A* **54**, 1522–1531 (1996).

25. E. Goulielmakis *et al.*, Single-cycle nonlinear optics, *Science* **320**, 1614 (2008).

STRONG-FIELD CONTROL OF X-RAY PROCESSES

L. YOUNG*,†, R. W. DUNFORD†, E. P. KANTER†, B. KRÄSSIG†,

R. SANTRA†,‡ and S. H. SOUTHWORTH†

†Argonne National Laboratory, Argonne, Illinois 60439, USA

‡Department of Physics, University of Chicago, Chicago, Illinois 60637, USA

*E-mail: young@anl.gov

Exploration of a new ultrafast-ultrasmall frontier in atomic and molecular physics has begun. Not only is it possible to control outer-shell electron dynamics with intense optical lasers, but now control of ultrafast inner-shell processes has become possible by combining strong optical laser fields with tunable sources of X-ray radiation. This marriage of strong-field laser and X-ray physics has led to the discovery of methods to control reversibly resonant X-ray absorption in atoms and molecules on ultrafast timescales. Here we describe three scenarios for control of resonant X-ray absorption: ultrafast field ionization, electromagnetically induced transparency in atoms and strong-field molecular alignment.

Keywords: Ultrafast X-rays, resonant X-ray absorption, strong-field laser interactions, electromagnetically induced transparency, molecular alignment.

1. Introduction

Control of X-ray processes using intense optical lasers represents an emerging scientific frontier — one which combines X-ray physics with strong-field laser control.[1] While the past decade has produced many examples where intense lasers at optical wavelengths are used to control molecular motions,[2-5] extension to the control of intraatomic inner-shell processes is quite new.[1,6-9] At first glance, it is an unusual concept to control X-ray processes using an optical or infrared radiation field since X-rays interact predominantly with inner-shell electrons, whereas longer wavelength radiation interacts with outer shell electrons. However, the inner and outer shells of atoms are coupled through resonant X-ray absorption, e.g. promotion of a K-shell electron to an empty outer shell orbital. Because outer shell electronic structure can be perturbed (dressed) by an optical radiation field, one can exert control over resonant X-ray absorption using optical lasers.

Reversible control is possible when the applied dressing field is gentle enough to significantly perturb outer-shell electronic structure, but is not intense enough to destroy (ionize) the atom.

Let us consider the optical field strength necessary to achieve this control, i.e. to induce outer-shell transitions at a rate comparable to inner-shell processes. If we take the simplest case, X-ray absorption by an atom ejects a K-shell electron to create a 1s hole. The resulting atom is unstable and decays via both radiative and non-radiative (Auger) channels.[10] These inner-shell decay rates increase with atomic number; for neon $Z = 10$ the lifetime of the $1s^{-1}$ hole state is 2.4 fs. In order to compete with the ultrafast inner-shell decay, transitions in the outer shell must be driven at a comparable rate. Transitions in a resonantly driven two-level system occur at the Rabi flopping frequency, $\Omega_{12} = \mu_{12}E/\hbar$ where $\mu_{12} = \langle 1|ez|2\rangle$ is the transition dipole matrix element between levels 1 and 2, and, E is the electric field amplitude. For the hydrogen $1s_{1/2} \rightarrow 2p_{1/2}$ transition, $\mu_{12} = 1.05ea_0$, and a driving field of amplitude $E = 1$ atomic unit, the Rabi flopping frequency greatly exceeds the Ne $1s^{-1}$ decay rate; i.e. $\Omega_{12} = 1.05/t_0 \sim (1/0.024)$ fs^{-1}, where t_0 is the atomic unit of time. For $E = 1$ atomic unit (51 V/Å), the equivalent laser intensity is 3.5×10^{16} W/cm^2, a value routinely achieved by focusing modern ultrafast Ti:sapphire laser systems. The versatile Ti:sapphire medium permits engineering of laser pulse amplitude and phase on the femtosecond to hundreds of picosecond timescale thus enabling exposure of atoms and molecules to arbitrary pulse shapes at high intensity.

After the strong optical laser field, the next ingredient needed to study control of ultrafast inner-shell processes is a tunable X-ray source. Synchrotrons provide a convenient source of pulsed, tunable, polarized radiation from 10 eV — 100,000 eV. This range covers inner-shell edges of all elements. In the left panel of Fig. 1 the three dominant photoprocesses, photoabsorption, elastic (Rayleigh) scattering and inelastic (Compton) scattering are shown for krypton. Photoabsorption cross sections greatly exceed scattering cross sections over energy ranges from below to far above the respective K edges of each atom. Zooming in on the region near the K-edge, resonances occur due to the presence of unoccupied outer shell orbitals (Rydberg states), as shown in Fig. 1 for krypton in various stages of ionization.

In order to probe resonant X-ray interactions in atoms and molecules in the presence of strong pulsed optical fields one must have precision overlap in a five-dimensional space, namely three spatial dimensions, time

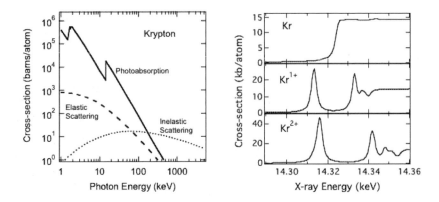

Fig. 1.   Left: X-ray photoprocesses in krypton. Right: Resonances near the K edge for krypton neutral, Kr^{1+} and Kr^{2+}

and X-ray energy. For this purpose, we have pioneered development of a synchrotron-based X-ray microprobe.[6] In the X-ray microprobe, laser and X-ray pulses are overlapped with precisions of $\sim$ 2 microns, $\sim$ 5 picoseconds and the x rays are tuned to an absorption resonance (bandwidth $\sim \Delta E/E = 10^{-4}$). For typical synchrotron X-ray pulses of 100 ps duration one can probe atoms and molecules at laser intensities of 10^{12} W/cm^2 with millijoule laser pulse energies focused to tens of microns. The powerful characteristics of synchrotron radiation, i.e. user-controlled continuous tunability of photon energy, polarization and bandwidth combined with exquisite energy and pulselength stability at high average flux (10^{14}/s at $\Delta E/E = 10^{-3}$) are ideal for these investigations. To probe targets subjected to higher laser intensities shorter X-ray pulses are required. X-ray pulse lengths, $\sim$ 100 fs, are currently available at synchrotron sources[12–14] using laser slicing techniques,[11] albeit at much reduced flux (10^6/s at $\Delta E/E = 10^{-3}$). Importantly, even using X-ray pulses with durations longer than the laser pulse new insights into ultrafast processes can be revealed.

In the next three sections we discuss examples of laser-controlled X-ray processes in free atoms and molecules for three regimes of laser intensity: Section 2 — ultrafast field ionization of atoms at $10^{14} - 10^{15}$ W/cm^2; Section 3 — laser-dressed atoms at $10^{12} - 10^{13}$ W/cm^2; Section 4 — laser-aligned molecules at $10^{11} - 10^{12}$ W/cm^2.

2. Orbital alignment in ultrafast field ionization

When atoms are exposed to strong optical fields ($\sim 10^{14}$–10^{15} W/cm^2), they rapidly ionize losing their least bound electrons through tunnel ionization.[15] Questions naturally arise. What is the nature of the initial electron distribution in the residual ion? Will the valence hole orbital that is created by tunnel ionization be aligned with the polarization axis of the laser? How does the hole orbital distribution evolve in time? These questions can be directly answered using an ultrafast resonant X-ray absorption microprobe.[6] Unlike laser-only experiments which detect ejected particles resulting from the high-field laser/atom interaction, i.e. ions, electrons and high-order harmonic photons, the X-ray microprobe method directly addresses the quantum state distribution in the residual ion using a well-understood process — resonant X-ray absorption.

We examine ultrafast field ionization in krypton.[6] For Kr, single ionization saturates at an intensity of 1.6×10^{14} W/cm^2. We saturate ionization in Kr gas using a 2 mJ, 45 fs Ti:sapphire laser pulse to produce a sample of $\sim 10^7$ Kr ions. Relative to Kr neutral, Kr^{1+} ions contain a new absorption feature, the $1s \rightarrow 4p$ resonance, due to the creation of a $4p$ hole as shown in the right panel of Fig. 1. We use $K\alpha$ fluorescence, which occurs within 0.2 fs, as a collision-free signature of X-ray absorption.

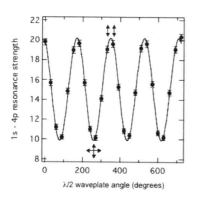
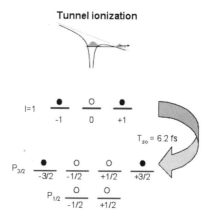

Fig. 2. Left: X-ray absorption as a function of the angle between the X-ray and laser polarization axes. Adapted from Ref. 6. Right: Initially populated $m_l = 0$ hole states redistribute over the spin-orbit timescale to $m_j = \pm 1/2$ hole states.

These experiments with linearly polarized light show a strong degree of atomic alignment persisting for times far beyond the spin-orbit time scale (6.2 fs) in Kr.[6] The nonrelativistic tunnel ionization model predicts a strong propensity for removing a $4p$, $m_l = 0$ electron,[16] where the quantization axis is taken along the linear polarization direction of the laser field, as shown pictorially in the right panel of Fig. 2. A dependence of the $1s \rightarrow 4p$ cross-section on the angle between the linear polarization vectors of the laser and X-rays is expected and observed, but the measured parallel-to-perpendicular cross-section ratio (≈ 2) is much smaller than predicted by the nonrelativistic models by nearly a factor of 20![6] Inclusion of the effects of spin-orbit coupling leads to calculated cross-section ratios in reasonable agreement with the measured ratio.[6,16] One may qualitatively understand the observed cross-section ratio; $p_{3/2}$ and $p_{1/2}$ states are probed simultaneously and the $p_{1/2}$ state cannot be aligned. More detailed analyses of the quantum state distribution of the residual ion may be found in Refs. 17 and 18. New methods to visualize the electron distribution in the residual ion using controlled electron recollisions are now under development.[19]

3. Electromagnetically induced transparency for X-rays

Electromagnetically induced transparency in the optical regime has been widely studied.[20,21] In a Λ-type medium characterized by atomic levels $|1\rangle, |2\rangle$, and $|3\rangle$ with energies $E_1 < E_2 < E_3$, resonant absorption on the $|1\rangle \rightarrow |3\rangle$ transition can be strongly suppressed by simultaneously irradiating the medium with an intense laser that couples the levels $|2\rangle$ and $|3\rangle$. This phenomena, shown in Fig. 3a, is known as electromagnetically induced transparency, EIT.

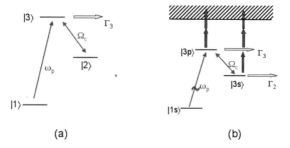

Fig. 3.   (a) Optical EIT. (b) X-ray EIT in neon.

In the X-ray regime, EIT is considerably more complex. In the optical regime, levels $|1\rangle$ and $|2\rangle$ are stable to electronic decay. However, in the

X-ray regime the core-excited states, $|2\rangle$ and $|3\rangle$ are metastable. For the quasi three-level system in neon, depicted in Fig. 3b, the lifetime for these core-excited states is 2.4 fs (0.27 eV). Thus, at an optical field strength sufficient to compete with inner-shell decay, multiphoton transitions to the continuum can also play a role. Inner-shell decay rates are denoted by Γ_2 and Γ_3 and multiphoton transitions to the continuum are denoted by red block arrows. Prior to calculation, it was not clear that the EIT effect would persist in the X-ray regime.

Neon is particularly advantageous because the $1s \rightarrow 3p$ transition is isolated from neighboring resonances. The calculated X-ray photoabsorption cross section for 800 nm laser dressing of the $1s \rightarrow 3p$ transition in neon at 10^{13} W/cm^2 with parallel and perpendicular laser/X-ray polarizations[8] is shown in Fig. 4. At this intensity the $1s \rightarrow 3p$ excitation at 867 eV is suppressed by a factor of 13 for the configuration in which the laser and X-ray polarizations are parallel. A fit to a three-level EIT model is shown to reproduce the general features of the calculated laser-dressed X-ray photoabsorption spectrum, Fig. 4. In the three-level model effective linewidths of $\Gamma_3 = 0.68$ eV and $\Gamma_2 = 0.54$ eV account for the laser-ionization broadening.

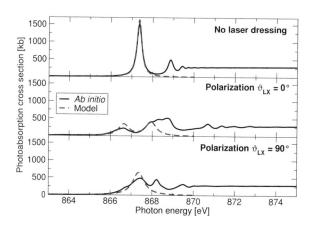

Fig. 4. X-ray photoabsorption cross-section of neon near the K edge. Results from an *ab initio* calculation and a three-level model are shown. From Ref. 8.

The ability to control X-ray absorption in Ne at the $1s \rightarrow 3p$ resonance allows one to imprint pulses shapes of the optical dressing laser onto long X-ray pulses.[8] This idea is illustrated in Fig. 5. With a 2-mm long gas cell containing 1 atmosphere of neon, the transmission of an X-ray pulse resonant with the $1s \rightarrow 3p$ transition will be only 0.07%. A typical X-ray pulse from a synchrotron source has a duration of 100 ps. Such an X-ray pulse may be overlapped in time and space with one or several, ultrashort intense laser pulses. Those portions of the X-ray pulse that overlap with the laser are transmitted through the gas cell. In the case shown in Fig. 5, where the two dressing laser pulses have a peak intensity of 10^{13} W/cm^2, the intensity of the two transmitted X-ray pulses is roughly 60% of the incoming pulse. The time delay between the two X-ray pulses can be controlled by changing the time delay between the two laser pulses, opening a route to ultrafast all X-ray pump-probe experiments. With an analogous strategy, controlled shaping of short-wavelength pulses might become a reality. Experimental efforts to demonstrate EIT for X-rays are ongoing at the Advanced Light Source's soft X-ray laser slicing beamline.[12]

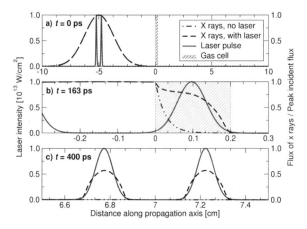

Fig. 5. Imprinting ultrashort laser pulses onto 100 ps X-ray pulses using EIT in a 2-mm gas cell containing 1 atm neon. From Ref. 8.

4. Controlled X-ray absorption by laser-aligned molecules

An entirely different mechanism provides strong field control of resonant X-ray absorption in molecules subjected to laser intensities 10^{11} – 10^{12} W/cm^2. A non-resonant, linearly polarized laser field will align a molecule by interaction with the molecule's anisotropic polarizability; the most polarizable axis within the molecule will align parallel to the laser polarization axis.[5] Since the laser polarization direction is under simple control with a waveplate, so is the direction of the molecule's most polarizable axis with respect to the X-ray polarization axis, which is fixed in the laboratory frame. It is well established that X-ray absorption resonances in the near-edge region, resulting from the promotion of a $1s$ electron to an empty σ^* or π^* orbital, are sensitive to the angle between the molecular axis and the X-ray polarization axis.[22] Thus, laser control of molecular alignment implies laser control of resonant X-ray absorption.

We have demonstrated this principle using the X-ray microprobe methodology,[9] as shown in Fig. 6. We aligned CF_3Br using 100 ps, 2 mJ pulses from a Ti:sapphire laser. The alignment was detected using polarized resonant X-ray absorption at the Br $1s \rightarrow \sigma^*$ resonance at 13.476 keV. The lowest unoccupied molecular orbital (LUMO) has σ^* symmetry and largely consists of the atomic Br $4p_z$ orbital. The Br $1s \rightarrow \sigma^*$ X-ray absorption resonance therefore has its transition dipole vector directed along the C–Br axis (the x and y components of this vector vanish), so the absorption cross section is sensitive to the angle between the X-ray polarization vector and the C–Br axis.

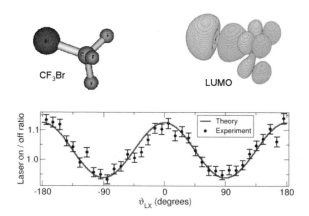

Fig. 6. Top: CF_3Br molecule and its lowest unoccupied molecular orbital. Bottom: X-ray absorption by CF_3Br as a function of ϑ_{LX}, defined as the angle between the polarization vectors of the laser and X-ray pulses. Adapted from Ref. 9.

Laser-controlled molecular alignment also enables control over X-ray diffraction; scattering from an ensemble of aligned molecules produces Bragg-like diffraction spots rather than the concentric rings observed in an isotropic gas. An important potential application is single biomolecule structure determination using coherent diffractive imaging with X-ray free-electron lasers.[23] The original concept[23] did not suggest aligned molecules, but rather proposed to scatter 10^{12} X-rays from a single biomolecule within 10 fs and collect a diffraction pattern with sufficient information for single-shot molecular orientation. Having pre-aligned molecules will vastly simplify the data collection and analysis — as pointed out by Spence and Doak in the context of electron scattering.[24] While these proposals[23,24] focus on scattering from a *single* large molecule, and multiple repetition to build up statistics, i.e. "serial crystallography", our work focuses on X-ray probing of an *ensemble* of 10^8 small molecules in the gas phase which have been aligned with laser techniques.[9] This strategy will allow one to acquire X-ray diffraction patterns of aligned, non-interacting molecules and thus obtain Ångstrom-level molecular images using existing synchrotron sources.

5. Outlook

The era in which characteristic X-ray processes can be considered invariant is at an end. We have demonstrated that placing atoms and molecules in strong optical fields can significantly affect resonant X-ray absorption. We can reversibly control X-ray absorption with application of a strong-optical field to a gaseous medium; the control mechanism is EIT in atoms and laser-constrained rotation in molecules. One may be able to create an X-ray amplitude pulse shaper using these tools. X-ray scattering from aligned molecules is not far off as we harness high repetition rate methods to utilize the full flux of x rays at synchrotron sources. In addition, we look forward using ultraintense X-ray radiation to alter characteristic X-ray processes, such hollow atom formation[25] and the competition between Auger decay and X-ray induced Rabi flopping,[26] when the world's first X-ray free electron laser, the Linac Coherent Light Source, becomes operational in 2009.

Acknowledgments

We would like to thank D. A. Arms, C. Buth, E. M. Dufresne, C. Höhr, D. L. Ederer, E. C. Landahl, E. R. Peterson, S. T. Pratt, N. Rohringer, J. Rudati, and D. A. Walko for fruitful collaboration. This work and the use of the Advanced Photon Source were supported by the U.S. Department of Energy, Office of Science, Office of Basic Energy Sciences, under Contract No. DE-AC02-06CH11357.

References

1. R. Santra, R. W. Dunford, E. P. Kanter, B. Krässig and L. Young, Strong field control of X-ray processes in *Advances in Atomic, Molecular and Optical Physics*, Vol. 56, Ch. 5, p. 218 (Elsevier, Amsterdam, Netherlands, 2008).

2. S. A. Rice and M. Zhao, *Optical Control of Molecular Dynamics* (Wiley, New York, 2000).

3. M. Shapiro and P. Brumer, *Principles of the Quantum Control of Molecular Processes* (Wiley-Interscience, New York, 2003).

4. H. Rabitz, R. de Vivie-Riedle, M. Motzkus and K. Kompa, *Science* **288**, 824 (2000).

5. H. Stapelfeldt and T. Seideman, *Rev. Mod. Phys.* **75**, 543 (2003).

6. L. Young, D. A. Arms, E. M. Dufresne, R. W. Dunford, D. L. Ederer, C. Höhr, E. P. Kanter, B. Krässig, E. C. Landahl, E. R. Peterson, J. Rudati, R. Santra and S. H. Southworth, *Phys. Rev. Lett.* **97**, 083601 (2006).

7. C. Buth and R. Santra, *Phys. Rev. A* **75**, 033412 (2007).

8. C. Buth, R. Santra and L. Young, *Phys. Rev. Lett.* **98**, 253001 (2007).

9. E. R. Peterson, C. Buth, D. A. Arms, R. W. Dunford, E. P. Kanter, B. Krässig, E. C. Landahl, S. T. Pratt, R. Santra, S. H. Southworth and L. Young, *Appl. Phys. Lett.* **92**, 094106 (2008).

10. M. O. Krause, *J. Phys. Chem. Ref. Data* **8**, 307 (1979).

11. A. A. Zholents and M. S. Zolotorev, *Phys. Rev. Lett.* **76**, 912 (1995).

12. R. W. Schoenlein, S. Chattopadhyay, H. H. W. Chong, T. E. Glover, P. A. Heimann, C. V. Shank, A. A. Zholents and M. S. Zolotorev, *Science* **287**, 2237 (2000).

13. S. Khan, K. Holldack, T. Kachel, R. Mitzner and T. Quast, *Phys. Rev. Lett.* **97**, 074801 (2006).

14. P. Beaud, S. L. Johnson, A. Streun, R. Abela, D. Abramsohn, D. Grolimund, F. Krasniqi, T. Schmidt, V. Schlott and G. Ingold, *Phys. Rev. Lett.* **99**, 174801 (2007).

15. M. V. Ammosov, N. B. Delone and V. P. Krainov, *Sov. Phys. JETP* **64**, 1191 (1986).

16. R. Santra, R. W. Dunford and L. Young, *Phys. Rev. A* **74**, 043403 (2006).

17. S. H. Southworth, D. A. Arms, E. M. Dufresne, R. W. Dunford, D. L. Ederer, C. Höhr, E. P. Kanter, B. Krässig, E. C. Landahl, E. R. Peterson, J. Rudati, R. Santra, D. A. Walko and L. Young, *Phys. Rev. A* **76**, 043421 (2007).

18. Z.-H. Loh, M. Khalil, R. E. Correa, R. Santra, C. Buth and S. R. Leone, *Phys. Rev. Lett.* **98** 143601 (2007).

19. See N. Dudovich *et al.*, Invited talk ICAP 2008; *Nat. Phys.* **2**, 781 (2006).

20. S. E. Harris, J. E. Field and A. Imamoglu, *Phys. Rev. Lett.* **64**, 1107 (1990).

21. K. J. Boller, A. Imamoglu and S. E. Harris, *Phys. Rev. Lett.* **66**, 2593 (1991).

22. J. Stöhr, *NEXAFS Spectroscopy* (Springer, New York, 1996).

23. R. Neutze, R. Wouts, D. van der Spoel, E. Weckert and J. Hajdu, *Nature* **406**, 752 (2000).

24. J. C. H. Spence and R. B. Doak, *Phys. Rev. Lett.* **92**, 198102 (2004).

25. N. Rohringer and R. Santra, *Phys. Rev. A* **76**, 033416 (2007).

26. N. Rohringer and R. Santra, *Phys. Rev. A* **77**, 053404 (2008).

AUTHOR INDEX